Disaster Management and Risk Reduction: Multidisciplinary Perspectives and Approaches in the Indian Context

Sudip Mitra · Kaustubh Dasgupta · Arindam Dey ·
Rajshree Bedamatta
Editors

Disaster Management and Risk Reduction: Multidisciplinary Perspectives and Approaches in the Indian Context

Proceedings of NERC 2022

Springer

Editors
Sudip Mitra
Centre for Disaster Management and Research
Indian Institute of Technology Guwahati
Guwahati, Assam, India

Kaustubh Dasgupta
Centre for Disaster Management and Research
Indian Institute of Technology Guwahati
Guwahati, Assam, India

Arindam Dey
Centre for Disaster Management and Research
Indian Institute of Technology Guwahati
Guwahati, Assam, India

Rajshree Bedamatta
Centre for Disaster Management and Research
Indian Institute of Technology Guwahati
Guwahati, Assam, India

ISBN 978-981-99-6394-2 ISBN 978-981-99-6395-9 (eBook)
https://doi.org/10.1007/978-981-99-6395-9

This Springer imprint is published by the registered company Springer Nature Singapore Pte Ltd.
The registered company address is: 152 Beach Road, #21-01/04 Gateway East, Singapore 189721, Singapore

Foreword

It is a matter of great satisfaction for me that the Indian Institute of Technology Guwahati successfully hosted the North-East Research Conclave (NERC) 2022 on May 20–22, 2022. The NERC 2022 was conducted on the theme "Sustainable Science and Technology". Concurrently, Assam Biotech Conclave (ABC) was also organized on May 21–22, 2022. Both events attracted huge participation from policymakers, researchers, industrialists, the army and students. Even the participation of schoolchildren was overwhelming.

NERC and ABC had many events including panel discussions, exhibitions, keynote lectures, competitions and paper presentations. Presentation of technical papers forms the core of any research conference. NERC attracted 879 research papers on various themes covering science, technology and humanities. Out of these, some select papers have been published by Springer Nature in the form of 15 volumes. These papers have been peer-reviewed and thoroughly edited by IIT Guwahati faculty members. I am sure that these volumes will prove to be excellent resource material for research. Most of the papers presented in these volumes highlight the special needs and aspirations of eight states of North-East India. I congratulate and thank authors, reviewers, editors and publisher for bringing out proceedings.

The motivation for organizing NERC came from none other than the Honorable Minister of Education, Government of India, Shri Dharmendra Pradhan Ji. It helped to bring policy-makers, researchers, industrialists, academicians, students and children into one forum. It is perhaps the rarest Conclave covering almost all possible research themes. For better readability, the Proceedings have been divided into 15 volumes, but each volume reflects diversity in terms of topics and researchers. The only common thread is the sustainable development of North-East India. Invariably, Sustainable North-East India is a prerequisite for sustainable India and the whole world.

In that sense, these 15 volumes will serve guiding and stimulating light for all the stakeholders of the development. I am pleased to dedicate these volumes to the nation as a part of Azadi ka Amrit Mahotsav.

T. G. Sitharam
Director
Indian Institute of Technology
Guwahati
Guwahati, Assam, India

Preface

The rapid pace of climate change is already imposing its deleterious effects on nature and natural systems. With the increasing human interventions in altering the natural geographical and geological systems, the waning outcome is now widespread. India has been already experiencing ever-growing disasters of natural and anthropogenic origin in the recent past. The widespread urbanization in all possible locations, ranging from the riverine alluvial banks to the difficult hilly terrains, has been responsible for the substantial altering of the prevalent natural systems to support the needs of human society and associated industrializations. However, with all these, the wrath of nature is also ever-growing and recurrent in terms of unwarranted rainfall and cloud bursts, unprecedented flooding, catastrophic landslides, dam breakages, glacial outbursts, snow avalanches, seismicity and its impacts, liquefaction, and wreaking environmental pollution leading to unimaginable toll on lives, property and economy. Such outcomes set up a question mark on the path of sustainable development. To address such issues and frame a refined path towards a sustainable future, a three-fold approach is immensely necessary—awareness, inferences and implementations. For this approach, it is ardently necessary to understand the core reasoning behind the disasters, their impact on the socio-economic contexts, and the ways to mitigate them. Thus, there is a need to devise a framework for disaster risk reduction and spreading the message of constructing a disaster-resilient society. This can only be achieved with proper dissemination of the knowledge behind the already occurred disasters and forecasting for the impending ones. This book presents a congregation of articles that would help to pave the way forward in this direction through a holistic and multidisciplinary approach.

Guwahati, India

Sudip Mitra
Lead Editor

Kaustubh Dasgupta
Editor

Arindam Dey
Editor

Rajshree Bedamatta
Editor

About IIT Guwahati

Indian Institute of Technology (IIT) Guwahati, established in 1994, completed 25 years of glorious existence in 2019. At present, the Institute has 11 departments, seven interdisciplinary academic centres and five academic schools covering all the major engineering, science, healthcare, management and humanities disciplines, offering B.Tech., B.Des., M.A., M.Des., M.Tech., M.Sc. and Ph.D. programmes. The institute presently offers a residential campus to 435 faculty members and more than 7500 students at present. Besides its laurels in teaching and research, IIT Guwahati has been able to fulfil the aspirations of people of the North-East region to a great extent since its inception in 1994. The picturesque campus is on a sprawling 285 hectares plot on the north bank of the Brahmaputra, around 20 km from the heart of the Guwahati city.

IIT Guwahati is the only academic institution in India that occupied a place among the top 100 world universities—under 50 years of age—ranked by the London-based Times Higher Education (THE) in the year 2014 and continues to maintain its superior position even today in various International Rankings. IIT Guwahati gained a rank of 32 globally in the 'Research Citations per Faculty' category and an overall 364th rank in the QS World University Rankings 2024 released recently. IIT Guwahati has retained the 7th position among the best engineering institutions in the country in the 'India Rankings 2021' declared by the National Institutional Ranking Framework (NIRF) of the Union Ministry of Education. IIT Guwahati has been also ranked 2nd in the 'Swachhata Ranking' conducted by the Government of India. Recently, IIT Guwahati has been ranked as the top-ranked University in 2019 for IT developers by HackerRank in the Asia-Pacific region.

Among other frontier areas of research and innovation, IIT Guwahati is working towards augmenting critical science research initiatives in Genomics, Developmental Biology, Health Care and Bioinformatics, Flexible Electronics, Advanced Functional Materials, Sustainable Polymers, Rural Technologies, Renewable Energy, Artificial Intelligence, Disaster Resilience and Risk Reduction and Water Resources and Management. In its silver jubilee year, IIT Guwahati is poised to scale newer heights through all-round growth and development.

Indian Institute of Technology Guwahati has dedicated itself to the cause of improving and empowering Northeast India through cutting-edge research, region relevant projects, innovations, individual and multilateral collaborations and special initiatives. Being the only IIT in the entire Northeastern region, IIT Guwahati has an immense amount of responsibility to develop the region and empower the people of the region.

While the entire country is celebrating the 'Azadi ka Amrit Mahotsav'—75 glorious years of Independence, and the great pride with which our nation of more than a billion people has been steadily growing today, IIT Guwahati is strongly committed to supporting that pace of growth for the entire NE so that we can keep pace along with the rest of the country. The specific areas of focus where IIT Guwahati has been contributing immensely to the region are:

(a) Infrastructure development across multiple sectors.
(b) Providing solutions for multiple natural disasters such as recurring floods, landslides, earthquakes, cyclones, hailstorms and other natural calamities.
(c) Improving the education sector and creating opportunities for employment.
(d) Internet, telecommunication and cultural integration.
(e) Technological intervention in interdisciplinary areas.
(f) Healthcare services and education.
(g) Renewable energy generation (solar, wind, biomass, hydro, geothermal).
(h) Overall industrialization, refining fossil fuels and setting up biorefineries.

Besides bringing in state-of-the-art technical know-how for most of the above sectors, the institute has been partnering with the local governments and enhancing the technological and educational interactions such that the next generation of youth is empowered with knowledge, skills and necessary entrepreneurial ability. These measures in Assam as well as all other northeast states will usher in a new era of growth and the opportunities it will provide for interaction with the ASEAN countries as part of the Act East Policy of the Government of India will bring prosperity to this region.

Prof. Parameswar K. Iyer
Dean, PRBR
Indian Institute of Technology
Guwahati
Guwahati, India

North East Research Conclave-2022: Toward Sustainable Science and Technology

It is extremely important and imperative to have knowledge-driven growth based on innovation in the case of academic higher education institutes of high repute. The North Eastern region endowed with rich biodiversity comprises eight states. However, the climatic conditions, limited connectivity, lack of research infrastructure/institutes, territorial conflicts and the mountainous terrain of these regions are major impediments to the research ecosystem in the North-East. Quality higher education focusing on industry-academia collaboration and translational research is extremely beneficial for society. It has also been rightly pointed out by the Hon'ble Prime Minister Shri Narendra Modi that, "*India cannot develop till Eastern India develops*".

With this idea and as India marks 75 years of Independence, Indian Institute of Technology Guwahati organized "The North-Eastern Research Conclave" from 20 to 22 May 2022. This grand event was jointly conducted with Science, Technology and Climate Change Department and the Department of Education, Government of Assam at IIT Guwahati Campus.

The mission behind the conclave was to showcase the best R&D activities from educational and research institutions across North-East India and to create an environment, conducive to development of local indigenous technologies and innovations, creating the scope and laying the foundation for entrepreneurship.

In order to attract people and spread awareness about the event, a roadshow was initiated from IIT Guwahati on 7th May 2022 in order to reach all the partnering academic institutes and make them an integral part of the mega event. The Director, IITG waved the NERC-2022 flag and sent off the road show vehicle from the institute. More than 400 students, staff and faculty participated actively in the roadshow.

A huge response was received by participants from throughout the country. The total number of Participating institutions in this conclave included 7 IITs, 10 NITs, 5 IIITs and other CFTIs, 23 Research Labs, 17 Central Funded Universities, 47 other Universities/Institutes along with about 100 schools. Eminent personalities from industries, start-ups, research councils and PSUs also joined in.

The presence of dignitaries from important Ministries was observed such as Shri Dharmendra Pradhan, Hon'ble Union Minister of Education and Minister of Skill Development and Entrepreneurship, GOI; Dr. Himanta Biswa Sarma, Hon'ble Chief Minister of Assam State; Dr. Ranoj Pegu, Hon'ble Minister of Education, Government of Assam; Dr. Rajkumar Ranjan Singh, Hon'ble Minister of State for Education, GOI; Dr. Subhas Sarkar, Hon'ble Minister of State for Education, GOI; Shri Keshab Mahanta, Hon'ble Minister of Science Technology and Climate Change, Government of Assam and many more.

The inauguration ceremony of the conclave was followed by the signing of an MoU between IIT Guwahati and the Government of Assam to establish 'The Assam Advanced Health Innovation Institute (AAHII)'. This MoU would prove to be a unique partnership between the Government of Assam and IIT Guwahati in order to set up a Research Institution to leverage advanced technologies to transform medical science. This joint venture company will be able to invite participation from intending parties including corporates/businesses/research institutions and philanthropic organizations.

The third edition of Assam Biotech Conclave 2022 was also held as part of NERC 2022. It brought together Biotech Entrepreneurs, industry leaders, researchers, academicians, Government Representatives, policymakers, innovators and investors together on one platform to explore the possibilities of Biotechnology in North-East India and to discuss the new opportunities in the transition.

Officers from the Indian Army also actively participated in the Conclave. A talk on “Atmanirbhar Bharat—Indian Army Initiatives towards Self Reliance” was delivered by Lt. Gen. D. S. Rana AVSM, YSM, SM General Officer Commanding, Gajraj Corps on 21st May 2022. The talk was aligned with the vision of the apex leadership of the Government of India and initiatives undertaken by the Indian Armed Forces with a focus on the integration of civil-military establishment in the field of self-reliance. He also elucidated that institutions such as IIT Guwahati which has many running research projects and elaborate student exchange and joint collaboration setup with a large number of Countries have the wherewithal to take up defence-related R&D and also facilitate delivery with Industry Partners. He also invited IIT Guwahati to participate in the EAST TECH Symposium planned at Kolkata in July 2022. This led to the signing of an MoU between the Indian Army Eastern Command and IIT Guwahati on 7th July 2022 during East Tech 2022. This would further impetus to Indigenisation and Raksha Atmanirbharta.

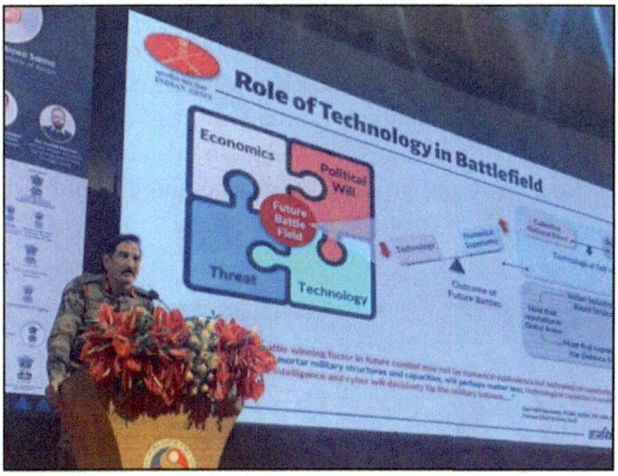

Royal Society of Chemistry, Global battery experiment was performed by more than 1300 students in three sessions starting from 20 May to 22 May at IIT Guwahati. Along with the Global Battery Experiment, Creating Skilful Educators (Teacher training programme) was also conducted in parallel sessions. Students had arrived from various schools across Assam and other North-Eastern states.

The Guwahati Declaration was launched at the valedictory ceremony of the conclave by Shri Lok Ranjan, Secretary, Ministry of Development of North Eastern Region (DoNER), in the presence of Shri Kailash Karthik, Deputy Commissioner, Kamrup. The Declaration is intended to create a set of guidelines, through which individual as well as a collective responsibility to promote and encourage innovation at the grass-root level and strive to stimulate and execute indigenization and entrepreneurship.

Science, education, research and innovation are the four pillars on which the development, as well as the work culture of a nation, rests. This was well articulated by the promising number of Exhibitors being seen participating from all across the NE states in the NERC 2022. All the NITs, CFTIs and CFIs were allocated two stalls each, where the delegates showcased the working models of their inventions. Distinctive pavilions were arranged for IIT, NIT, CFIs and CFTIs. Excellent response was obtained from the Start-Ups all across the NE states. The Federation of Industry Commerce of North Eastern Region (FINER) had partnered with NERC-2022 as an Industry Partner and they showcased 50 start-ups as a part of the Exhibition under the FINER Pavilion. Other significant organizations that came forward to showcase their allied R&D start-ups were the Oil and Natural Gas (Oil and Natural Gas Pavilion), Indian Army (Defense Pavilion) and NE-Railway (NE-Railway Pavilion).

Multifarious research work on topics of societal relevance was presented by researchers from different organizations/institutes. The presentations were conducted in oral and poster presentation modes. The thematic areas for these presentations were part of some of the Sustainable Development Goals (SDGs) such as SDG-3: Good health and wellbeing; SDG-7: Affordable and Clean Energy; SDG-9: Industry, Innovation and Infrastructure; SDG-11: Sustainable cities and communities and SDG-12: Responsible consumption and production. Some of the papers highlighted environmental sustainability, efficiency and management issues, which are important to be presented in the case of North East regions. Two awards were given under each technical category for these presentations. Overall, the technical sessions were a grand success due to the active cooperation from editors, chairpersons of all the sessions and student volunteers of IITG.

The government of India has taken various steps to encourage women in the field of science and technology. In this line, the IIT Guwahati Woman Researcher Award was approved to recognize the contribution of women Faculty members of IIT Guwahati fraternity. This prestigious award was conferred to Dr. Latha Rangan who is a Senior Professor in the Department of Biosciences and Bioengineering, Indian Institute of Technology Guwahati, India. Professor Rangan has played a key role in Plant Biotechnology and Sustainable development, especially in the areas of energy security, food security and medicinal crops.

The Conclave paved the way for creating mass awareness of Research and Innovation for developing a sustainable society. There was knowledge exchange and dissemination that led to the establishment of Centres of Excellence in Translational Collaborative Research and Innovation. This mega event led to the bridging of the gap between Industry–Academia and Creating Hand holding Pathways for setting up long-term collaboration for R&D innovations towards the goal of establishing sustainable NE India. The Conclave brought together over 8,000 participants including Hon'ble Ministers, Official Bureaucrats, Eminent Professors, Scientists, Renowned Industrialists, School Children/Teachers and Other delegates. This revolutionized the R&D road map of all the NE states through various dissemination of policies which will benefit the sustainable development of all NE states in the near future.

It is an honour and a moment of extreme pride to get the NERC proceedings published in the prestigious Springer volumes. We would like to thank and acknowledge the globally active publisher Springer for helping us being able to publish the articles in 15 broad areas. We would also like to thank all the authors for their contribution to the grand success of NERC 2022 and wish them great success in all of their future endeavours.

Prof. Vimal Katiyar
Dean, R&D
Department of Chemical Engineering
Centre for the Sustainable Polymer
Indian Institute of Technology Guwahati
Guwahati, India
e-mail: vkatiyar@iitg.ac.in

Prof. Subhendu Sekhar Bag
Associate Dean, R&D
Department of Chemistry
Centre for the Environment
Indian Institute of Technology Guwahati
Guwahati, India
e-mail: ssbag75@iitg.ac.in

From the Desk of Chairman of Technical Committee of NERC 2022

North-East Research Conclave 2022 was successfully organized on May 20–22, 2022 with the participation of thousands of delegates. A total of 879 oral and poster papers were presented at the conference on 16 different tracks. The theme of the Conclave was Sustainable Science and Technology, which is very pertinent in the modern era of globalization. Science and technology have to address economic, environmental and social problems of the world. Technology and sustainability are not incompatible. In fact, technology can achieve the goal of sustainability, which also includes preserving our rich cultural heritage. Concurrently with the North-East Research Conclave (NERC), the Assam Biotech Conclave 2022 was also organized on May 21–22, 2022. These mega-events were organized at the Indian Institute of Technology Guwahati (IITG) in physical mode after two years of the pandemic period. Along with IITG, the Science, Technology and Climate Change Department and Department of Education, Government of Assam, were also organizers of these events under the patronage of Shri Dharmendra Pradhan Ji, Honorable Minister of Education and Minister of Skill Development and Entrepreneurship in the Government of India, and Shri Himanta Biswa Sarma Ji, Honorable Chief Minister of Assam.

It is a matter of great pleasure that Springer Nature is publishing the select papers from the conclave in 15 volumes. These are Advanced Functional Materials, Low-Cost Manufacturing Technologies, Agro and Food Processing Technologies, Artificial Intelligence and Data Science-based R&D interventions, Conservation of Biodiversity in the North Eastern States of India, Disaster Management, Healthcare Research and Related Technologies, Innovative Design for Societal Needs, Policies for Research and Innovation, Research and Innovation for Sustainable Development Goals, Sustainable Environment, Sustainable Energy Generation and Storage, Sustainable Transportation and Urban Development, Teaching and Learning Technologies and Technologies for Rural Development. These volumes are useful archival and reference materials for policy-makers, researchers and students.

As the Chairman of the Technical Committee, I am thankful to all Editors of all volumes, reviewers and student volunteers who have put tireless efforts to review, select and edit the papers of respective divisions, overcoming the time constraints.

Support provided by Convener, Prof. Vimal Katiyar, Dean R&D, IITG, and Co-conveners Prof. Subhendu Sekhar Bag, Associate Dean R&D, IITG and Shri Kailash Karthik N., IAS is commendable. It is difficult to express words of gratitude for the Director, IITG, Prof. T. G. Sitharam who has been motivating and guiding all the teams of NERC 2022 and ABC 2022.

Uday S. Dixit
Professor
Department of Mechanical Engineering and
Head Center for Indian Knowledge Systems

Contents

About the Editors

Dr. Sudip Mitra is the founding head of the Centre for Disaster Management and Research (CDMR) at IIT Guwahati and a faculty of the School of Agro and Rural Technology (SART). In his 23 years of professional career, he has been actively involved in climate change, adaptation, and disaster risk reduction (DRR) research, policy, and outreach. His present research looks at climate-smart agriculture and the application of indigenous resources and technologies for the enhancement of rural livelihood. He is a Fulbright Fellow and visiting faculty at the University of California, Davis, USA and Keio University, Japan. He served as a task force member of MNREGA convergence of schemes under the Ministry of Rural Development and member of the Global Technology Watch Group, TIFAC, Department of Science and Technology (DST), Government of India. He is an elected member of the National Academy of Science, India (NASI) and is also the editorial board member of international journals of repute.

Dr. Kaustubh Dasgupta is a faculty member at the Department of Civil Engineering and an adjunct faculty member at the Centre for Disaster Management and Research in Indian Institute of Technology Guwahati and completed his undergraduate studies in civil engineering from Bengal Engineering College Shibpur and master's and Ph.D. programs in civil engineering from Indian Institute of Technology Kanpur. After his undergraduate studies, he worked in Development Consultants Limited, Kolkata, and Reliance Industries Limited, Jamnagar, for three years. His areas of research interest are seismic behavior of reinforced concrete structures, seismic soil-structure interaction, and seismic vulnerability assessment.

Dr. Arindam Dey is currently an associate professor in Civil Engineering Department at IIT Guwahati and an adjunct faculty at the Centre for Disaster Management and Research (CDMR), IIT Guwahati. He obtained his B.E. from Jadavpur University Kolkata, M.Tech. and Ph.D. from IIT Kanpur, and was a post-doctoral researcher at University of Molise, Italy. He was a recipient of the "Fellowship for Young Indian Researchers" from the Ministry of Italy. He works in varied fields of geotechnical engineering, including foundation engineering, geotechnical earthquake engineering, geotechnical and geophysical investigation, and computational geotechnics. He holds special expertise in the fields of disaster management, earthquake and landslide hazard mappings and mitigations, liquefaction mitigation through ground improvement, and disasters originating from dam and embankment failures. Apart from being members of IGS, IST, and FOSET, he held a position in the prestigious TC-208 Committee on "Slope Stability and Landslides" hosted by International Society of Soil Mechanics and Geotechnical Engineering (ISSMGE) from 2018 to 2023. He is also an integral part of the Codal Committee of Indian Roads Congress (IRC) SP-102. He has published more than 250 technical documentations of national and international repute. He has received many research awards from IGS and ISSMGE. He has delivered several invited lectures in India and abroad and is a reviewer for several national and international journals.

Dr. Rajshree Bedamatta is a professor of Economics at the Department of Humanities and Social Sciences and an adjunct faculty at the Centre for Disaster Management and Research in Indian Institute of Technology Guwahati. She is trained in the field of development economics, and her research interests cut across the applied areas of food and agriculture, health, labor markets, education, and governance. One of her emerging engagements is action-research in disruptions caused by disasters and its socio-economic impact on vulnerable groups of population.

Landslide Early Warning and Susceptibility Zonation of Guwahati City Through Local and Regional Scale Slope Stability Analyses

Arindam Dey, Chiranjib Prasad Sarma, and A. Murali Krishna

Abstract Rainfall-triggered landslides are a major geohazard in the Guwahati region which cannot be convincingly addressed by conventional slope stability approaches. This study attempts to address this issue with the help of advanced analysis considering the concepts of unsaturated soil mechanics involving transient/steady-state phenomena aided by the variation of pore water pressures for providing a predictive model for the potential instability under rainfall infiltration process. It is observed that the matric suction and its effect on the variation of permeability and degree of saturation, as well as the intensity and duration of rainfall, are critical factors in assessing the state of instability. Although landslides in the stated region involve relatively smaller volumes of soil (~100–400 m^3), the presence of proximal urbanized areas can possibly induce significant damage to infrastructure and life loss. This study presents the applicability of physically-based models (TRIGRS, SHALSTAB and SINMAP) for assessing the regional-scale landslide susceptibility and hazard in the Guwahati region. The advantages and limitations of the different models are discussed, and the significant results are compared on the basis of the observed landslides. It is concluded from the study that physically-based models can incorporate the quantitative hydrogeological factors (e.g., soil characteristics, rainfall and areal topography) for a successful regional-scale rainfall-induced landslide hazard assessment.

Keywords Rainfall-induced landslide · Early warning · Susceptibility zonation · Local scale analyses · Regional scale analyses · TRIGRS · SHALSTAB · SINMAP

A. Dey (✉)
Indian Institute of Technology Guwahati, Assam, India
e-mail: arindam.dey@iitg.ac.in

C. P. Sarma
Jorhat Engineering College, Assam, India

A. M. Krishna
Indian Institute of Technology Tirupati, Andhra Pradesh, India

S. Mitra et al. (eds.), *Disaster Management and Risk Reduction: Multidisciplinary Perspectives and Approaches in the Indian Context*,
https://doi.org/10.1007/978-981-99-6395-9_1

1 Introduction

Historical records of landslide events suggest that the phenomena in the Guwahati region are rain-triggered. All landslides were reported to occur when the monsoon is at its peak or nearing completion, i.e., by the month of July–August to the end of September to mid-October [1, 2]. A total cumulative rainfall of 750 mm over three days, between 5 and 8th October 2004, triggered many landslides around Guwahati hillslope areas, causing the death of 17 persons and destruction of property worth millions of rupees. The same storm event triggered as many as 100 landslides of varying dimensions along NH 44, Guwahati–Shillong road within a reach of 80 km [3]. In the month of June 2012, several landslides occurred, triggered by intense rainfall events throughout the month. Though landslides have caused significant destruction to life and property in this region, still there is a lack of research directed towards this problem. The area still lacks a well-developed landslide inventory, which is the foremost requirement for undertaking any analysis for the prediction of landslide occurrences, not to mention, a historical record of the relation between rainfall and shallow landslides. A few studies were conducted to evaluate the landslide hazard of this area based on statistical or semi-heuristic methods [4, 5].

Speedy and unplanned urbanization within the city of Guwahati due to the phenomenal influx of population over the past four decades has resulted in continuous conversion of land from non-urban use to urban use, leading to rampant encroachment and earth-cutting on the hillslopes. Shrinkage of forest cover has brought about a change in the surface and subsurface hydrology. Economically incapacitated societies are increasingly reluctant to invest in structural measures that can reduce this natural risk. Hence, the new issue is to implement reliable decision support based on early warning systems aimed at minimizing the loss of lives and property without investing in long-term, costly projects of ground stabilization. The precursor to the development of such a system is to quantify the correlation of landslide occurrences to the destabilizing factors.

Several literatures highlight the influence of rainwater infiltration in triggering landslides [6–9]. An increase in water content due to rainwater infiltration leads to a decrement in matric suction, a rise in the bulk unit weight and a reduction in shear strength of hillslope soil. Hill slopes in Guwahati city mostly comprise unsaturated residual soils [10, 11]. An effective modelling of the unsaturated behaviour of slopes comprising residual soils is required for assessing the potential rainfall-induced landslide susceptibility. The study applies the principles of unsaturated soil mechanics for explaining the rainfall-induced landslides in this region with whatever limited literature is available.

In spite of the capacity of geotechnical slope stability models to predict potential triggering instability as a consequence of environmental and climatic changes [12–15], they are hardly used to develop landslide early warning systems. These models even exhibit the prediction capability where an extensive landslide inventory and rainfall record are unavailable [16]. The application becomes compounded when regional-scale landslide susceptibility and hazard assessment are taken into account. In such cases, physically-based models developed on a GIS-based platform aid in developing the scenario mappings for the region while incorporating the important geotechnical perspectives. Among various such available models, a few of the commonly used ones are Shallow Landslide Stability Model (SHALSTAB) [17], Stability Index Mapping (SINMAP) [18] and Transient Rainfall Infiltration and Grid-based Regional Slope-stability (TRIGRS) [19, 20]. The objective of the present study is, therefore, to apply the stated physically-based landslide hazard assessment models for the regional-scale landslide zonation of Guwahati city, while accounting for the hydrogeological aspects of the region.

2 Study Area

2.1 Location and Topography

Guwahati falls is approximately encompassed by the latitudes 91°33′E–91°52′6″E and longitudes 26°4′45″N–26°14′N, with an approximate area of 328 sq. km spread across both banks of the river Brahmaputra [21]. Hills composed of residual soils, dotting around the alluvial plains and the marshy wetlands, mainly comprise three prominent geomorphological features that can be easily identified. The hills, ranging in altitude from 100–300 m above mean sea level (MSL) and inclinations of 10°–70° [22] are interspersed among elongated low-lying alluvial plains. The region is interspersed with 9 major hill series including Nabagraha/Sunsali, Japorigog, Sonaighuli/Jutikuchi, Narakashur, Nilachal, Fatasil, Jalukbari, Khanapara and Agyathuri hills (Fig. 1).

2.2 Regional Hydro-geomorphology

In the hills of the region, gneiss and granite bodies are affected by joints and additional intrusions from quartz–feldspathic veins, aplite dykes and pegmatite. Thin bands or lenses of quartzite, amphibolites and biotite schists, are found parallel to the foliation [23, 24]. The regional geomorphology comprises well-drained regional zones with thick soil formation (up to 30 m) [10, 11] as well as moderate, imperfect or poorly drained regional zones with exposed rock layers. Physical and chemical weathering processes on the parent rock produce the residual soils, which are evident in the geological stratification of the regional hillslopes. A detailed field investigation exhibited two types of prominent soil layering in the hillslopes of the region. A laterite formation of reddish residual silty clay (thickness varying from a few centimetres to a few metres) overlies a saprolite formation of pale yellowish residual soil (poorly graded silty sand). A typical formation of the soil type is shown in Fig. 2 as obtained from a hillslope excavation at IIT Guwahati.

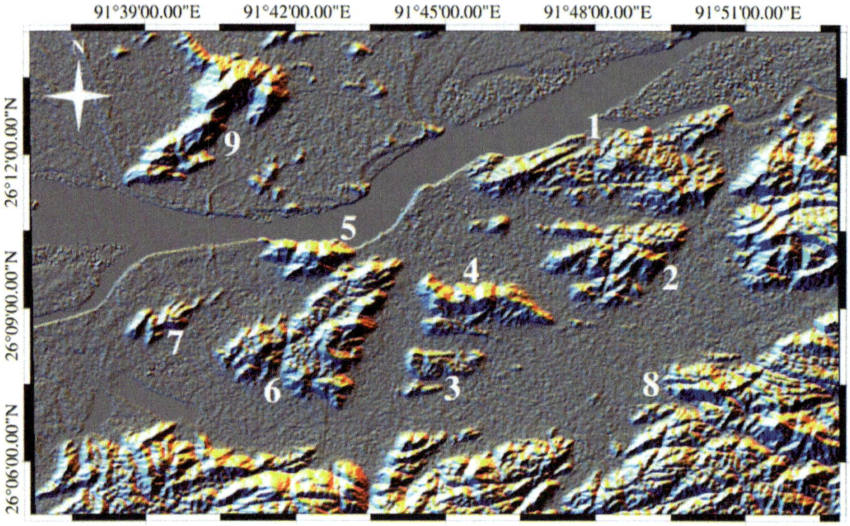

Fig. 1 A digital elevation model of Guwahati city depicting the hill series

Fig. 2 A typical cut slope at IIT Guwahati Campus showing lateritic and saprolitic soil profiles

The North Eastern Region of India experiences climatically harsh seasonal wetting and drying cycles. The wetting cycle spans between April and mid-October, while the drying cycle spans between November and March. Significant evaporation rate during the drying cycle generates substantial high matric suction in the residual soil slopes of the Guwahati region. However, in the following monsoon season, rainwater infiltration reduces the matric suction which mostly leads to evident slope failures [10, 11]. There is a lack of attention to this issue from an analytical standpoint. The major difficulty is the assessment of the soil behaviour in the unsaturated state. Understanding unsaturated soil behaviour and more importantly assessing the unsaturated soil parameters is imperative in formulating a correct analytical solution to the problem. In order to determine the decrease of the suction pressures, the relationship between the negative pore–water pressure of the soil and the water content, referred to as soil–water characteristic curve (SWCC) is required and can be obtained through laboratory tests. However, most of these methods are costly and time-intensive and would require skilled personnel for their measurement. More difficult than that is to experimentally obtain the unsaturated hydraulic conductivity curve (UHCC), which gives the relationship of soil permeability with matric suction. The shear strength parameters of unsaturated soils require suction-controlled direct shear or triaxial test to be conducted, which is another impasse towards the solution to this problem. Exhaustive treatises on unsaturated soil mechanics are found in literature but only a handful of them are concerned with the soils found in this region.

2.3 *Input Parameters for Local and Regional Scale Analyses*

Literature [10, 11] presents the properties of unsaturated soil properties of typical residual soils (RS—Red Silty Clay, PYS—Pale Yellow Silty Sand) commonly found in the regional hillslopes in the premise of Guwahati city; the same has been adopted in this study. For the local-scale analyses, Table 1 gives the index properties of the soils, while the relevant geotechnical parameters adopted for this study are presented in Table 2. The reported soil–water characteristic curves (SWCC) for both types of soils are presented in Fig. 3a. The unsaturated hydraulic conductivity curve (UHCC) is empirically derived by integrating along the entire SWCC curve [25] and programmed within the GeoStudio Software Suite [27]. Figure 3b presents the unsaturated hydraulic conductivity curve derived for both types of soils.

For the regional-scale analyses, Shuttle Radar Topography Mission (SRTM) Digital Elevation Model (DEM) of 1 arc-second resolution is used as the base map. Topographic elements, viz., the slope map (Fig. 4) and the upslope catchment area map (Fig. 5), required as input for the analysis are thus extracted from the DEM. Several researchers have noted that slope values decrease with coarser DEM resolutions [27, 28] and that grid size influences the critical uniform recharge rate [29]. However, there are no specific guidelines for choosing a particular resolution, and no

Table 1 Index properties of the soils used in this study [8, 9]

Property	RS	PYS
Colour	Reddish	Light yellowish
Specific gravity	2.44	2.64
Field bulk density (g/cc)	1.65	1.79
Field dry density (g/cc)	1.49	1.63
Plastic limit	27%	Non-plastic
Porosity	0.34	0.38
Classification	Silty clay	Poorly graded silty sand

Table 2 Geotechnical parameters of the soils adopted for local-scale analyses

Property	RS	PYS
Effective friction angle, φ'	31°	38.5°
Effective cohesion, c'	10 kPa	0 kPa
Friction angle related to matric suction, φ^b	16.7°	7.5°
Permeability, k_s (m/s)	1.86×10^{-6}	1.21×10^{-5}

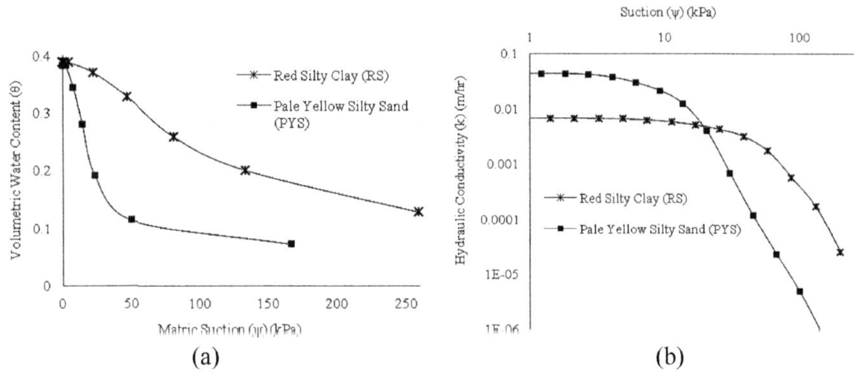

Fig. 3 **a** Soil water characteristic curves and **b** Unsaturated hydraulic conductivity curves for the two types of soil

resolution can represent the dimensions of all the possible slope failures [28]. The resolution adopted for the present study is, therefore, solely based on the condition of availability of the particular data set.

The range of important soil properties in the context of landslide studies within the Guwahati city region is summarized in Table 3 from the available literature [22]. Although the depth of soil cover has been found to vary from 0–30 m, field observation of landslides has revealed the shallow nature of such occurrences, which is mostly confined to depths of 1–4 m. For the TRIGRS model, other input parameters,

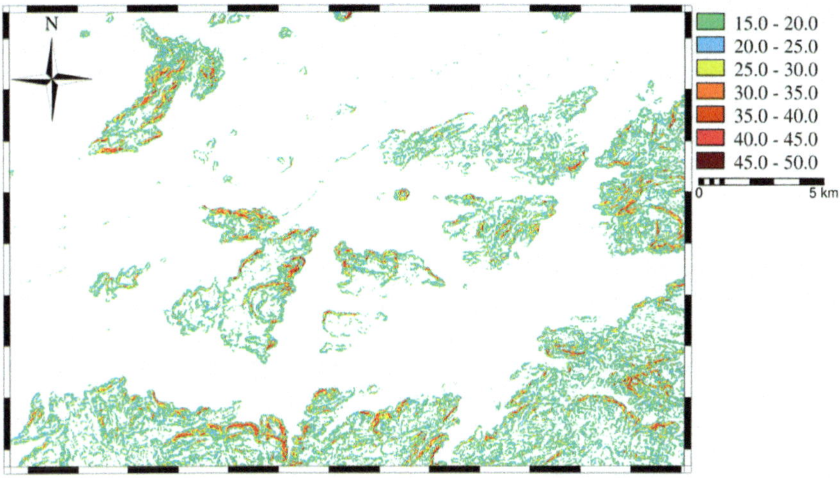

Fig. 4 Slope map (in degrees) of the study area

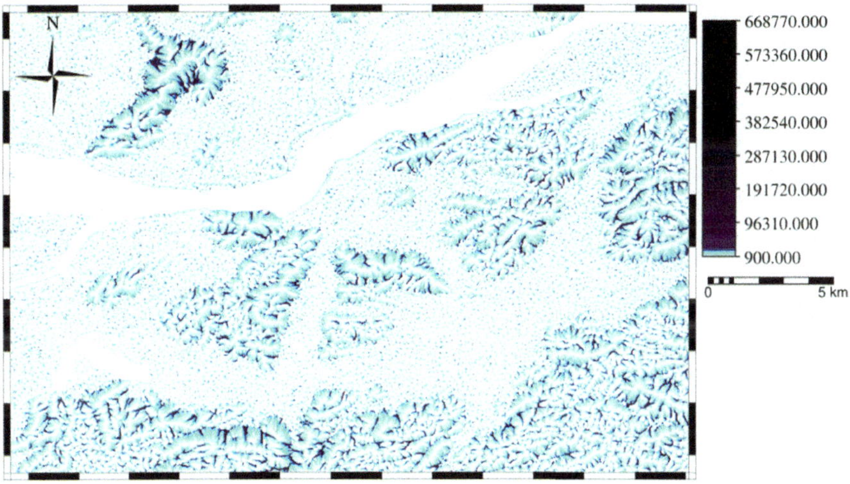

Fig. 5 Upslope catchment area map (in m^2) of the study area

viz., the soil–water characteristics curve (SWCC) and saturated hydraulic diffusivity are required to describe the transient infiltration and flow through the unsaturated surface soil and finally quantify the unsaturated shear strength. Experimental data on the unsaturated soil properties for this study is adopted from Das and Saikia [10, 11]. The reported SWCC used for this study is presented in Fig. 5. The data is fitted to the Gardner [30] model for soil–water characteristics and the fitting parameter α is determined. The saturated hydraulic diffusivity can be estimated from the ratio of the saturated hydraulic conductivity to the specific storage, which is the slope

Table 3 Range of geotechnical parameters of the soils comprising the hillslopes within the study area

Property	Minimum	Maximum
Angle of Internal Friction, ϕ (°)	30	40
Cohesion, c (kPa)	10	50
Density, ρ (kg/m^3)	1400	1900
Saturated Permeability, k_s (m/s)	10^{-6}	10^{-4}

of SWCC. For the present study, the saturated hydraulic diffusivity is taken as 5×10^{-5} m^2/s. The parameter α, the volumetric water content at saturation (θ_s), and the residual state (θ_r) are 1.5, 0.4 and 0.14, respectively.

The rainfall data used as input into the physically-based models is adopted from the Tropical Rainfall Measuring Mission (TRMM) 3-h rainfall estimate (3B42 V7) data set. The area averaged-time series rainfall data for a duration of 84 h (3 days and 12 h) is obtained from Goddard Earth Sciences Data and Information Services Center (GES DISC) maintained web portal Giovanni (http://disc.sci.gsfc.nasa.gov/giovanni).

The groundwater in the hills of the region is generally found to vary along the basal boundary, i.e., the bedrock is observed to be mostly in a saturated state, while the soil covers to be in an unsaturated state. Seepage is observed in some cases where the bedrock outcrops. Therefore, one can safely assume that groundwater level varies from surface level to a depth of 30 m. However, data on the spatial variation of the depth of the water table is not available for the hilly regions of the study area, and therefore, for the present study, the groundwater level is assumed to be at a uniform depth of 5 m. These assumptions are taken based on the observed depth of failure planes of landslide occurrences.

3 Methodology

3.1 Numerical Adaptation for Local-Scale Analysis

The numerical study for local-scale analyses is sequentially carried out in two steps. Initially, a transient seepage analysis to investigate the effect of infiltration rates on the infiltration mechanism and reduction in matric suction. The infiltration is modelled using SEEP/W of GeoStudio package, a module particularly used for seepage modelling [26]. The coefficient of permeability is a non-linear function of the matric suction within the soil under unsaturated conditions. SEEP/W applies the finite element method to solve the governing differential equation describing the flow through soil, which in this case is a modified form of Richard's equation. Subsequently, based on a limit equilibrium Morgenstern-Price method, a slope stability analysis is conducted to assess the temporal degradation of the stability using the Slope/W module of Geostudio package. The pore water pressures estimated in the

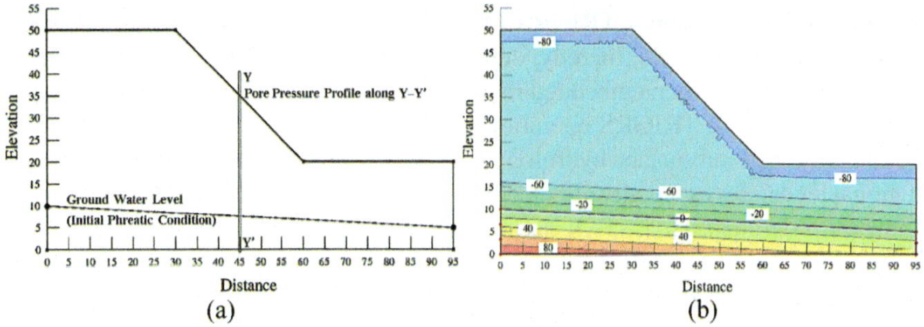

Fig. 6 Slope geometry used for local-scale analyses

previous step (i.e., seepage analysis) are fed to the stability analysis. The unsaturated shear strength conditions are used in the Slope/W module by modifying the basic Mohr–Coulomb material model [31–33].

The typical slope section considered for the study has a height of 30 m with an inclination angle of 45° (Fig. 6a). Two homogenous sections are investigated, each comprising isotropic residual soil having red silty clay (RS) and pale yellow silty sand (PYS) properties, respectively. This part of the study constitutes local-scale analyses and is primarily carried out to study the influence of complex hydrogeological conditions on the seepage analysis subjected to different rates of applied infiltration. The initial groundwater level was defined by a phreatic line as shown in Fig. 6a, while the maximum suction (i.e., negative pore water pressure) was limited to 80 kPa (Fig. 6b). Five different rates of infiltration (50, 100, 150, 200 and 250 mm/day) that are consistent to storm events were applied on the open boundaries of the slope. Further details of the finite element model can be found in Sarma et al. [34]. Finally, based on the Slope/W analyses, the temporal degradation of FoS is determined and represented at 6 h interval.

3.2 *Numerical Adaptation for Regional Scale Analysis*

Infinite Slope Stability Model: The adopted physically-based model(s) operate on an assumed infinite slope stability analysis [35] where the failure is defined through a comparative of the restoring strength and destabilizing stresses. The triggering of rainfall-induced landslides is represented in terms of the reduction in FoS due to the rise of the water table that is governed by the intensity and duration of rainfall [6]. In the present study, the physically-based model(s) accounts for the settings for spatial distribution on a grid-based geographic information system (GIS) framework for assessing the regional stability (in terms of a *FoS*–map) by accumulating the stability of each grid cell.

Physically-based Models: TRIGRS evaluates the transient pore pressure response to rainfall infiltration and, thereby, depicts the regional distribution of the rainfall-induced landslides when embedded in a GIS platform. Elaboration on the principles and mechanisms of TRIGRS operation can be found in standard literature [19, 20, 36]. Steady-state subsurface hydrology has a profound influence on the triggering of regional landslides especially affected by rainfall, and the same is used in the stability analyses programs SHALSTAB and SINMAP. SHALSTAB is used to classify the region in accordance with the amount of recharge required to initiate instability, wherein a lower value of the steady-state recharge indicates a greater inclination to instability. Further details regarding the correct use of SHALSTAB can be found in [17, 18]. SINMAP is a probabilistic model that provides the probability of failure on a regional scale in terms of Stability Index (SI) while considering the input parameters to follow a 'uniform' distribution. SINMAP also applies a steady-state hydrological model within a probabilistic framework. Further details about SINMAP can be found in Pack et al. [18].

4 Results and Discussions

Outcomes from Local Scale Analysis: Based on local-scale analyses, Fig. 7a, b depicts the temporal degradation of the factor of safety of the RS and PYS slope, respectively, based on various infiltration rates. It is clearly noticeable ignoring soil suction (as done in conventional soil mechanics) completely undermines the evolving mechanism of rainfall-induced landslides and exhibits an unrealistic estimate of slope stability. A comparison of the figures reveals that matric suction and soil cohesion is instrumental in imparting a much greater stability to the RS slope. Even after intense infiltration of 250 mm/day applied for the duration of 5 days, the RS slope is still stable, though very marginally. On the other hand, such infiltration caused the PYS slope to undergo failure within the duration of 1 day and 6 h. Moreover, it is to be noted that RS has a much higher suction value at a similar volumetric water content range that renders much greater stability to the RS slope.

Figure 8a, b depict the FoS for the corresponding slopes after it is subjected to 200 mm/day rainfall for 3 days. In Fig. 8a, the influence of cohesion is evident on the stability of RS slope. It depicts the development of a circular and moderately deep critical surface, while yet maintaining a FoS sufficient enough to prevent slope failure. Under the same scenario, the PYS slope shows FoS < 1, thereby indicating a shallow translational slide that is also evident from the shape and depth of the critical slip surface. For the PYS slope, the pore pressures developed at the instance of failure are plotted in Fig. 9a. Further, the total applied cumulative infiltration required to initiate failure was assessed and the same was plotted against the rate of applied infiltration in Fig. 9b, and a linear relation between the same was exhibited. This phenomenon illustrates the fact that with decreasing rate of applied infiltration, more amount rainwater can infiltrate into the slope, thereby minimizing the runoff and leading to subsequent instability.

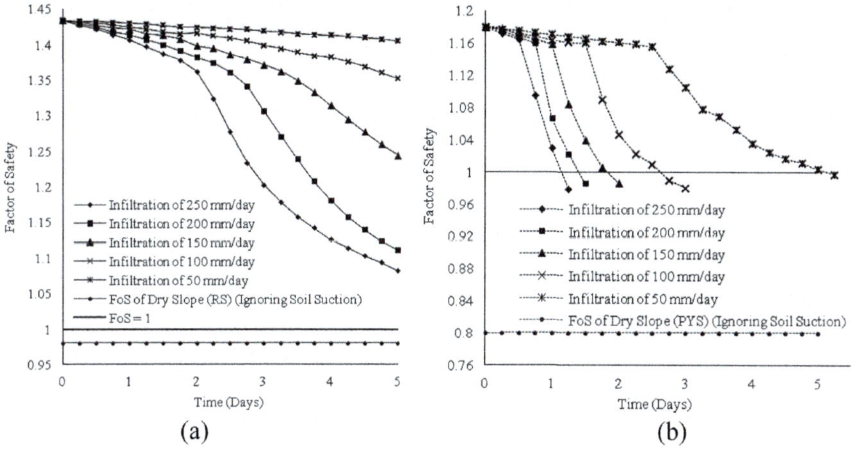

Fig. 7 Temporal degradation of FoS for **a** RS slope **b** PYS slope

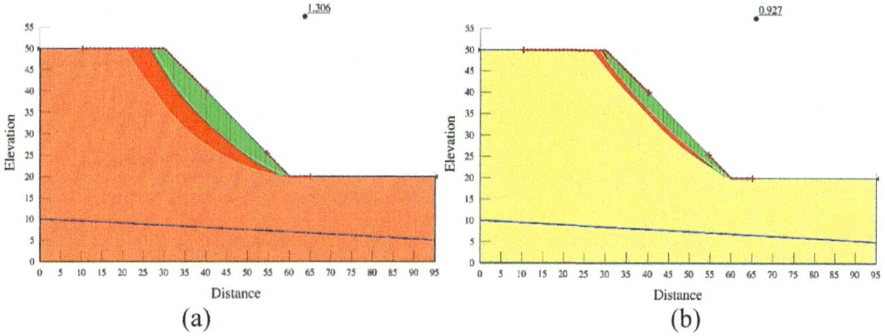

Fig. 8 FoS after an infiltration of 200 mm/day for a duration of 3 days **a** RS slope, **b** PYS slope

As a follow-up, antecedent rainfall infiltration was then used as input to assess the response of the local-scale models for RS and PYS slopes. Figure 10a exhibits rainfall intensity versus time in hours that occurred from the afternoon of 4th October to the midnight of 7th October 2004; the same is used in the simulation. The temporal variation of FoS of the slope is assessed and plotted against time at intervals of 2 h, as shown in Fig. 10b. It can be noted that the PYS slope attained marginal stability after 76 h of rainfall, and underwent collapse by the end of 80th h. Under the same scenario, the RS slope did not exhibit any signs of failure. Such analyses can serve as the basis for the prediction of rain-triggered landslides and develop early warning where data based on actual observation of landslide occurrences and coupled rainfall storm histories are not available.

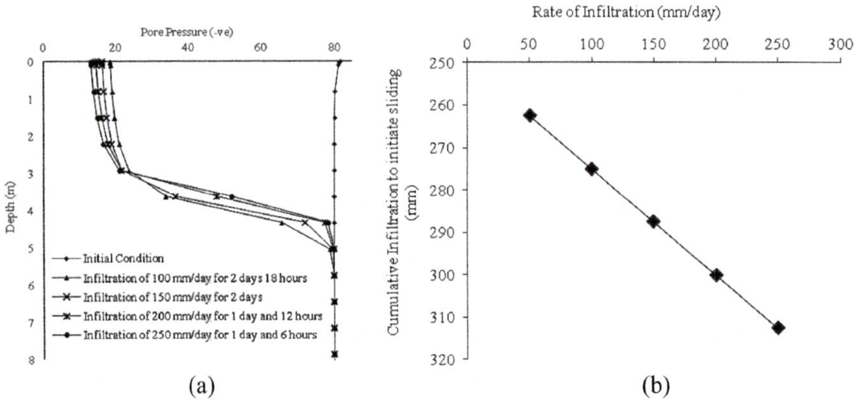

Fig. 9 Early warning protocol for a PYS slope based on infiltration. **a** Pore pressure within the slope at the moment of its failure. **b** Total cumulative rainfall required for initiating slippage within the slope against the rate of applied infiltration

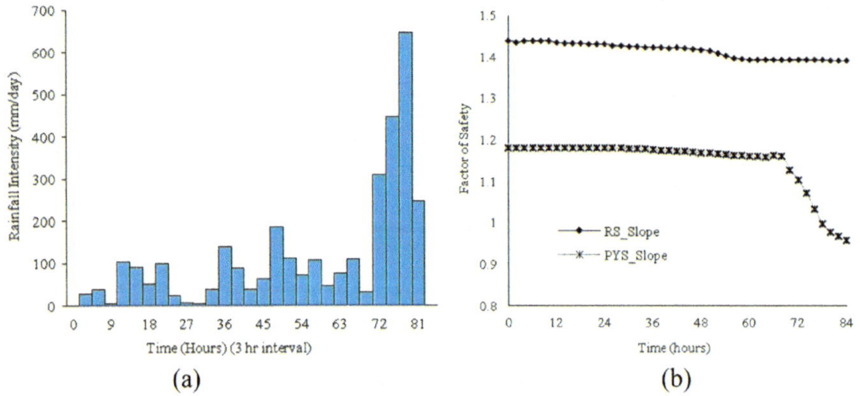

Fig. 10 **a** Rainfall storm event history of 4th October to 7th October 2004. **b** Temporal degradation of FoS for RS and PYS slopes subjected to actual antecedent rainfall infiltration

Outcomes from Regional Scale Analysis: As mentioned earlier, all three physically-based models have been utilized to decipher the landslide zonation across the region. Figure 11 gives the values of the "steady-state specific recharge" required to initiate instability at different locations within the study area. The SHALSTAB model was successful in identifying the effect of the topographical features of the hillslopes in causing greater potential to landsliding. Lower recharge was required to initiate instability in regions of topographical convergence (i.e., the gully regions where the accumulation of flow takes place during a storm event). The SHALSTAB results appropriately represented the topographic and hydrological factors that control landslide occurrences. However, the SHALSTAB model gives the output in the form

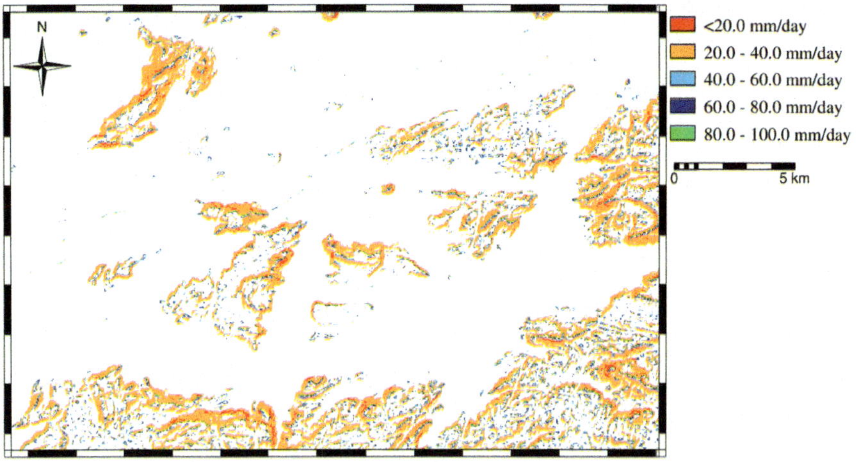

Fig. 11 Steady-state recharge map required to initiate instability in hillslopes of Guwahati city (SHALSTAB output)

of "steady-state specific recharge". Comparing the values to the rainfall precipitation that the study area is exposed to (Fig. 3), it becomes extremely difficult to interpret the stability conditions of the region solely based on the results obtained from SHALSTAB. Based on the outcomes, if the rainfall is represented through the steady-state recharge values, then the entire region would be unrealistically prone to landslides. Thus, a better correlation among rainfall, infiltration characteristics, climatological factors, vegetation cover and subsurface hydrogeological factors need to be established to define the threshold values required to cause instability.

As an outcome of using SINMAP, Fig. 12 provides the Stability Index (SI) and the minimum Factor of Safety (FoS_{min}) if the SI exceeds 1 at that particular position. Some steeper hillslopes have SI values ranging from 0.75 to 1.0, a majority of the hillslopes is found to have a FoS_{min} value ranging from 1.0 to 1.25, while still few areas of gentler slopes are in the range of 1.25–1.5. Based on the outcomes, the entire region is identified as unconditionally stable, and there is no chance of landslides unless triggered by external causal factors. As per the classification based on Pack et al. [18], the hillslopes represent a quasi-stable to moderately stable state, and the slightest disturbance could lead to catastrophic landsliding. This assessment is a much closer approximation to reality, especially as the anthropogenic activities in the region are also considered to be one of the potent causal factors of landslides in this region.

TRIGRS caters to the application of different storm events and rainfall scenarios to simulate the stability condition of the study region. Based on the rainfall history depicted in Fig. 10a, the change in the stability for a pre- and post storm period is analyzed. A very stable condition is assessed over the region for the pre-storm period as shown in Fig. 13, with most of the hillslopes exhibiting a FoS > 1.5, along with a few highlighting marginal stability. This prediction is in line with the high stability

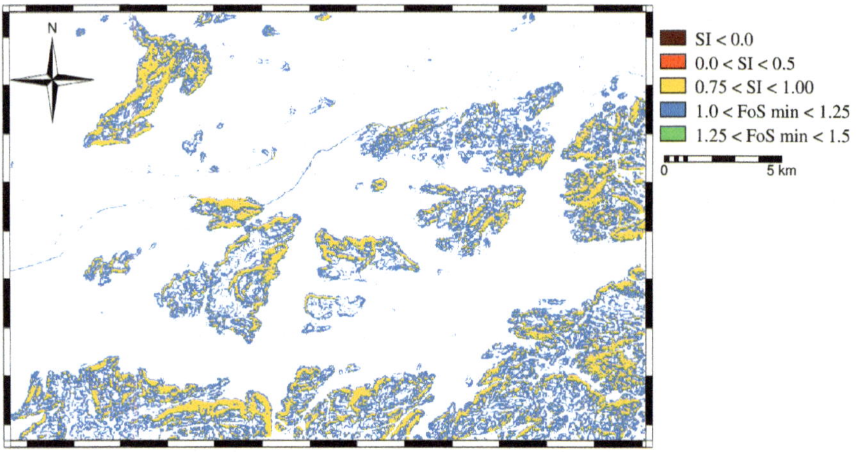

Fig. 12 Stability Index (SI) Map for Guwahati city (SINMAP output)

manifested during the dry season when the unsaturated condition and matric suction induced high shear strength to sustain even the vertical slopes. The post-storm period exhibits a drastic change in regional stability. The FoS drops to marginally stable condition (1.0–1.25) over most of the hilly regions as shown in Fig. 14. However, yet, FoS was observed to remain above 1.0, thereby ensuring limiting failure condition is not reached.

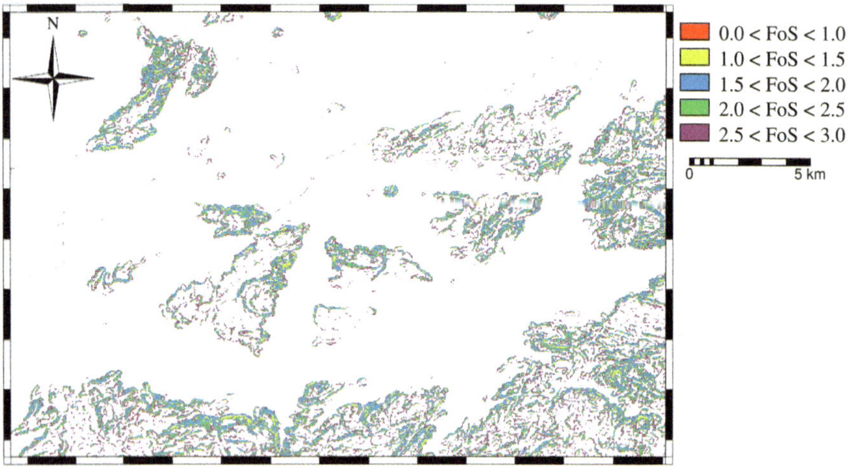

Fig. 13 Factor of safety (FoS) map for Guwahati city in the pre-storm period (TRIGRS output)

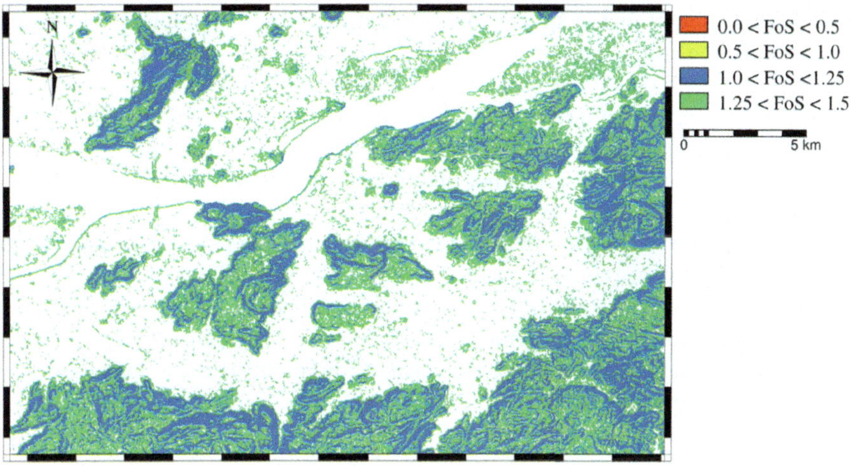

Fig. 14 Factor of safety (FoS) map for Guwahati city in the post-storm period (TRIGRS output)

5 Conclusions

Based on the present study, the following important conclusions can be drawn:

- Suction within a soil layer plays an instrumental role in sustaining the stability of a natural or cut slope. Ignoring the suction and unsaturated soil mechanics fail to provide the evolution of rainfall-induced landslides in a local and regional scale. The influence of rainwater infiltration in gradually depleting the shear strength plays a pivotal role in such widespread failure mechanism.
- It is not necessary for a hillslope to fail only if it is saturated. Failure can occur even in the unsaturated state, especially if the soil has high permeability and minimal cohesion such as pale yellow silty soils. On the other hand, reddish silty clay soils can exhibit much higher resilience to failure that is contributed by the cohesion component.
- Coupling of finite element-based transient seepage analysis and limit equilibrium-based slope stability analysis, a predictive model for local-scale rain-induced slope instability can be developed to assess the stability degradation and failure of the natural residual soil slopes. The success of such a model would largely depend on the input parameters determined from extensive investigations and data collation.
- For various types of soils, SWCC and UHCC play a pivotal role in governing the porewater pressure distribution within the layer, which, in turn, influences the evolution of the failure mechanism in such slopes.
- The regional-scale study that depending upon the local-scale soil typologies and their response to infiltration, the hillslopes within Guwahati city are highly potent to undergo rainfall-induced landsliding. Dedicated effort would be required to develop the region-specific analytical understanding while emphasizing the physics of the problem.

- The study shows that in spite of the inherent differences in the hydrogeological concepts, the adopted physically-based models provide a prediction that is quite similar to each other. The study proves that rainfall infiltration is a significant controlling factor, yet it cannot be considered as the single most important factor in the widespread landslides experienced in the region. More attention is required to incorporate land use and other anthropogenic activities as causal factors in destabilizing the hill slopes of the region.
- The study proves that the application of physically-based models (SHALSTAB, SINMAP, TRIGRS) can take into account the hydrogeological factors in a quantitative manner and is capable of describing the regional rainfall-induced landslides when expressed in the GIS framework. The results of such activity can assist various agencies involved with technical and community services to the region in case of an impending disaster.

References

1. Kalita UC (2001) A study of landslide hazards in North Eastern India. In: Proceedings of the fifteenth international conference on soil mechanics and geotechnical engineering, Istanbul, Turkey, vol 1–3, pp 1167–1170
2. Sarma AK, Bora PK (1994) Influence of rainfall on landslide. In: International conference on landslides, slope stability and the safety of infra-structures, Malaysia
3. GSI: Geological Survey of India, Landslide Home, Post-disaster studies (2013)
4. Phukon P, Chetia D, Das P (2012) Landslide susceptibility assessment in the Guwahati City, Assam using analytic hierarchy process (AHP) and geographic information system (GIS). Int J Comput Appl Eng Sci 2(1):2231–4946
5. Bhusan K, Kundu SS, Goswami K, Sudhakar S (2014) Susceptibility mapping and estimation of rainfall threshold using space based input for assessment of landslide hazard in Guwahati city in North East India. In: The international archives of the photogrammetry, remote sensing and spatial information sciences, vol XL-8, pp 15–19
6. Selby MJ (1993) Hillslope materials and processes. Oxford University Press Inc., New York
7. Glade T, Crozier MJ (2005) The nature of landslide hazard impact. In: Glade T, Anderson M, Crozier MJ (eds) Landslide hazard and risk. Wiley, pp 43–74
8. Glade T, Crozier MJ (2005) A review of scale dependency in landslide hazard and risk analysis. In: Glade T, Anderson M, Crozier MJ (eds) Landslide hazard and risk, Wiley, pp 75–138
9. Wesley LD (2010) Geotechnical engineering in residual soils. Wiley
10. Das UK, Saikia BD (2010) Shear strength of unsaturated residual soils of the hills in Guwahati. In: Proceedings of Indian geotechnical conference, GEOtrendz, pp 679–682
11. Das UK, Saikia BD (2011) Evaluation of a prediction model for shear strength of unsaturated soils. In: Proceedings of Indian geotechnical conference, Kochi, pp 643–646
12. Gasmo JM, Rahardjo H, Leong EC (2000) Infiltration effects on stability of a residual soil slope. Comput Geotech 26:145–165
13. Tsaparas I, Rahardjo H, Toll DG, Leong EC (2002) Controlling parameters for rainfall-induced landslides. Comput Geotech 29:1–27
14. Huat BBK, Ali FH, Rajoo RSK (2006) Stability analysis and stability chart for unsaturated residual soil slope. Am J Environ Sci 2(4):154–160
15. Rahardjo H, Ong TH, Rezaur RB, Leong EC (2007) Factors controlling instability of homogeneous soil slopes under rainfall. J Geotech Geoenviron Eng ASCE 133(12):1532–1543

16. Thiebes B, Bell R, Glade T, Jäger S, Anderson M, Holcombe L (2013) A WebGIS decision-support system for slope stability based on limit-equilibrium modelling. Eng Geol 158:109–118
17. Montgomery DR, Dietrich WE (1994) A physically based model for the topographic control on shallow landsliding. Water Resour Res 30:1153–1170
18. Pack RT, Tarboton DG, Goodwin CN (1998) The SINMAP approach to terrain stability mapping. In: Proceedings of international congress of the international association for engineering geology and the environment, Balkema, Rotterdam, Netherlands, pp 1157–1165
19. Baum RL, Savage WZ, Godt JW (2002) TRIGRS—a FORTRAN program for transient rainfall infiltration and grid-based regional slope stability analysis. U.S. Geological Survey Open-File Report 2002–0424
20. Savage WZ, Godt WJ, Baum RL (2004) Modeling time–dependent slope stability. In: Proceedings of IX international symposium on landslides, Rio de Janeiro, Brazil, pp 23–38
21. Guwahati Metropolitan Development Authority: Master Plan for Guwahati Metropolitan Area - 2025 (Part - 1), Guwahati (2009)
22. Sarma CP, Dey A, Murali Krishna A (2020) Influence of digital elevation models on the simulation of rainfall-induced landslides in the hillslopes of Guwahati, India. Eng Geology 268(105523):1–13
23. Maswood M, Goswami DND (1974) Basic rocks from the Precambrian terrain around Guwahati. Assam, the Indian mineralogist. J Mineral Soc India 15:55–62
24. Shukla RC (1989)Study of granite rocks around Kamakhya hill and adjoining area. In: Geological survey of India, Assam, record–122, IV, pp 72–73
25. Fredlund DG, Xing A, Fredlund MD, Barbour SL (1996) The relationship of the unsaturated soil shear strength to the soil-water characteristic curve. Can Geotech J 33:440–448
26. GeoSlope: Manuals of Geostudio 2007 Software Suite, GEO-SLOPE International Ltd. (2007)
27. Wu WS, Sidle R (1995) A distributed slope stability model for steep forested basins. Water Resour Res 2097–2110 (1995).
28. Claessens L, Heuvelink GBM, Schoorl JM, Veldkamp A (2005) DEM resolution effects on shallow landslide hazard and soil redistribution modeling. Earth Surf Process Landforms 30:461–477
29. Dietrich WE, Bellugi D, Real de Asua R (2001) Validation of the shallow landslide model, SHALSTAB, for forest management. In: Wigmosta MS, Burges SJ (eds) Land use and watersheds: human influence on hydrology and geomorphology in urban and forest areas. Water Science Application, AGU, Washington DC, vol 2, pp 195–227
30. Gardner WR (1958) Some steady-state solutions of the unsaturated moisture flow equation with application to evaporation from a water table. Soil Sci 85:228–232
31. Krahn J, Fredlund DG (1972) On total and osmotic suction. Soil Sci 114(5):339–348
32. Fredlund DG, Morgenstern NR, Widger RA (1978) Shear strength of unsaturated soils. Can Geotech J 15:313–321
33. Fredlund DG, Rahardjo H (1993) Soil mechanics for unsaturated soils. Wiley
34. Sarma CP, Murali Krishna A, Dey A (2015) Landslide early warning based on geotechnical slope stability model for the Guwahati region. In: Indian geotechnical conference, Pune, India, pp 1–10
35. Bromhead EN (1992) The stability of slopes, 2nd edn. Taylor & Francis
36. Baum RL, Savage WZ, Godt JW (2008) TRIGRS—a FORTRAN program for transient rainfall infiltration and grid-based regional slope stability analysis, version 2.0. U.S. Geological Survey Open-File Report 2008–1159

Structural and Geotechnical Approach of Characterization of the Causes of Meriema Landslide Along the Kohima–Wokha Road Section of the National Highway No. 61, Near Nagaland University-Kohima Campus Approach Road Junction, Nagaland, India

Chabungbam Mangi Khuman, Keneisazo Nagi, Soibam Ibotombi, Heisnam Sanatomba, Arenlong Aier, and K. Notoka

Abstract The Meriema Landslide is a kind of circular failure where unconsolidated materials comprising mostly of broken rock masses of shale are found moving downslope *en masse*. The bedrock where this landslide occured comprised of shale belonging to the Disang Group. To be able to decipher the causes of such kinds of slumping down of pieces of broken rocks *en masse* geological, structural and geotechnical investigations are carried out. It could be established that the majority of the joints are radial to the length of curvature surface of the crown of the landslide. It is found plausible that the thrusting activity near the slide area caused the already developed joints particularly those perpendicular to the length of the failure surface are widen. It is also observed that the majority of the joints also dip at high angles; mostly clustering between 60–80°. From the circular surface of the crown, the maximum thickness of overburden above the failure surface at the deepest point (before slide) is also found with the help of a total station survey. The maximum vertical pressure the rocks suffered at the deepest point is then calculated which comes out to be 0.701 MPa. Comparing this value with an average shear strength of 8.32 MPa for samples of the bedrock found out from geotechnical investigations, it is learnt that the rock didn't fail on account of the load. In the rainy season during incessant rain, the rocks are saturated with water; filling all the open spaces of the joint planes with water making the friction reduced considerably, and hence the

C. M. Khuman (✉) · K. Nagi · A. Aier · K. Notoka
Department of Geology, Nagaland University, Kohima Campus, Meriema, India
e-mail: mangikhuman@nagalanduniversity.ac.in

S. Ibotombi · H. Sanatomba
Department of Earth Sciences, Manipur University, Imphal, India

S. Mitra et al. (eds.), *Disaster Management and Risk Reduction: Multidisciplinary Perspectives and Approaches in the Indian Context*,
https://doi.org/10.1007/978-981-99-6395-9_2

body of jointed rock mass above the failure surface could not withstand the weak shear stress. The broken pieces of rock masses could slump down over the circular surface of failure developed because of a mosaic of closely spaced and interconnected numerous joint planes dipping at various angles and directions. The Meriema Landslide is thus considered to have been controlled by lithology and structure.

Keyword Meriema landslide · Disang shale · Joints · Circular surface · Weak shear stress

1 Introduction

Gravity tugs constantly downward on every mass of material everywhere on earth, causing a variety of mass wasting or mass movements, whereby geological materials are moved downward, commonly downslope, from one place to another. The downslope pull (shearing stress) tending to cause mass movements is related to the mass of the material and the slope angle. Counteracting the shearing stress is friction for loose materials or, shear strength in the case of a coherent solid. When shearing stress exceeds frictional resistance or the shear strength of the material, sliding occurs [1]. Therefore, factors that increase shearing stress, decrease friction, or decrease shear strength tend to increase the likelihood of sliding, and vice versa. Basically, mass movements occur whenever the downward pull of gravity overcomes the resisting forces. Landslide is a general term for the results of rapid mass movements. The material moved can range from unconsolidated, fairly fine material (for example, soil or snow) to large, solid masses of rock. Landslides are responsible for the destruction of natural resources and property, disruption of communication and transportation systems, and even loss of life.

In the hilly region of Nagaland state of northeast India, where the bedrocks are generally not only incompetent, but also highly disturbed and deformed, the mass failure is not, in most cases, because of the flow of soils only. Landslides occur commonly on account of the slumping down of broken pieces of soft bedrocks. The Meriema Landslide along the Kohima–Wokha Road section of the National Highway No. 61, near Nagaland University, Kohima Campus approach road junction, is a kind of mass movement where the broken rock masses slide down *en masse* during incessant rain in the monsoon. The bedrock where this landslide occurs is comprised of shale belonging to the Disang Group. The general tendency of downslope movement of the unconsolidated masses of bedrocks of the state of Nagaland is obvious from widespread occurrences of extensive creeping and subsequent slumping down of rock masses in the hill slopes across various sections of the roadways of the state, thereby sometimes causing problems of landslide particularly in the rainy season. This clearly shows that slope and certain bedrock features are susceptible to such kind of mass movements. But mass movement doesn't occur everywhere similar rock types and slopes are exposed. This indicates that, in addition to slope and certain rock type features, there are some other factors that control the causes of such landslides in which broken rock masses slump downslope.

It is strongly felt that there must be some other additional factors other than slope and rock type by virtue of which mass movement of the Meriema Landslide has taken place. Thus, for deciphering the causes of this landslide, an approach is made in which geological, structural and geotechnical features are examined. To be able to retrieve the role of litho-character and other features, geological and structural mapping of the surrounding area is carried out. From the prepared geological and structural map, a geological profile along a section almost across the strike of the bedrocks and passing through the slide area is worked out. The joints of rocks in the surrounding area are also analysed to examine their correlation with the causes of this landslide. The geotechnical features of the bedrocks are again examined to get an idea about the lithological properties like shear strength and other relevant parameters. The maximum pressure the rocks suffered at the deepest point of the failure surface is also calculated to examine whether the load could make the rocks fail.

1.1 Location and Description of Meriema Landslide

The Meriema landslide near Meriema village of Kohima district occurs along the Kohima–Wokha road section of National Highway No. 61. The slide area is located near the Nagaland University-Kohima Campus approach road junction (Fig. 1) and falls approximately within 25°42′56.5′′N—25°43′01′′N latitudes and 94°05′07.5′′E—94°05′10.5′′E longitudes (Fig. 2). The mass failure and subsequent slumping take place along a circular surface (Fig. 3a, b). Figure 3a shows the crown of the landslide, while Fig. 3c is a photograph of the foot of the landslide. The bedrock, exposed in the circular failure surface that could survive failure and over which failed broken pieces of rock slide down, is found to be shale/mudstone. The adjoining area of this landslide is seen still slowly subsiding down (Fig. 3a, b and d). The location of the photograph shown in Fig. 3d is approximately at 25°43′0.6′′N latitude and 94°04′58.9′′E longitude, which is quite far downslope from the slide area. The photograph (Fig. 3d) shows a road-cut section in the area for the construction of a new roadway (Kohima bypass road). From the photograph (Fig. 3d), it is evident that the rocks are slumping down *en masse* over a very wide circular surface but still could have not been completely failed. Thus, it is very much obvious that the stability of a wide area adjacent to this particular landslide has been disturbed and the bedrocks have been subjected to continued slow downslope subsidence. An attempt is made in this communication to work out the causes of this landslide and the continued slow subsidence of the adjoining area.

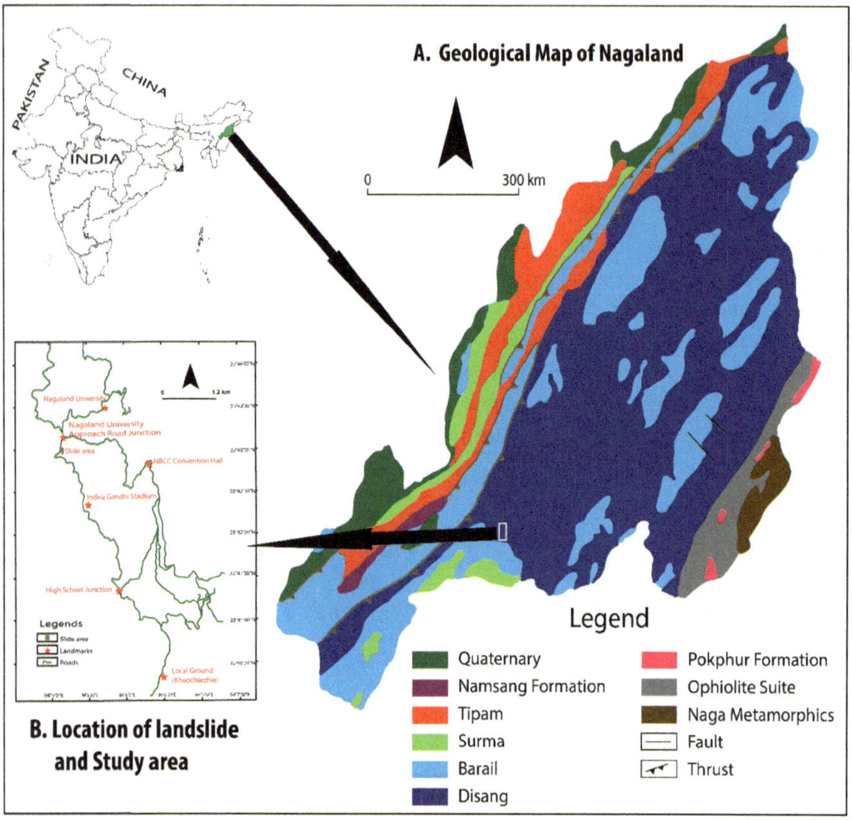

Fig. 1 Geological map of Nagaland and location of the study area

2 Geological and Tectonic Setting

Approximately 1200 km long narrow sigmoidal belt, the Indo-Myanmar Ranges (IMR), linking the Arunachal Himalaya to the north and Andaman–Nicobar Islands to the south, has three distinct segments from south to north, viz., Arakan Yoma, Chin Hills and Manipur–Naga Hills, each of about 400 km length [2]. The Manipur–Naga Hills segment is the northernmost part of the IMR. The state of Nagaland lies in the northern part of the Manipur–Naga Hills segment. Nagaland, for the most part, is hilly, except for the low-lying alluvial tracts bordering the Assam valley. The hilly region of the state is comprised of approximately NE-SW trending Patkai, Barail and associated ranges characterized by their varied distinctive geomorphic features [3]. The constituent ranges of the Naga Hills are characterized by intense folding and thrusting, and these are the result of the various phases of the Tertiary orogenic upheavals.

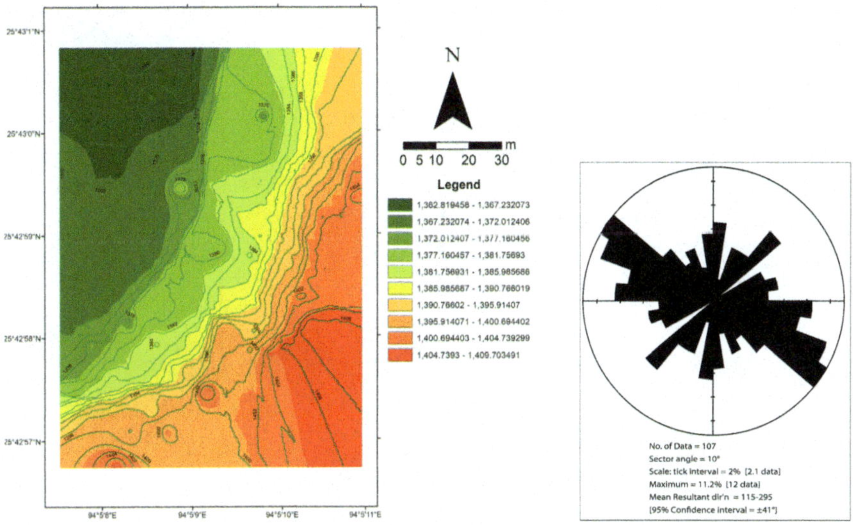

Fig. 2 DEM image of Meriema Landslide. The rose diagram of the joint planes of litho-units of the surrounding area (Fig. 5a) at the inset shows that the radius of curvature is almost parallel to the direction in which most of the joints are concentrated

Based on morphotectonics, the Naga Hills have been longitudinally divided, from west to east, into three distinct units [4, 5], namely—the Schuppen Belt, the Inner Fold Belt and the Ophiolite Belt (Fig. 1). The Schuppen Belt has been defined as a narrow linear belt of about 350 km stretch and a major portion of it passes along the hill ranges bordering the state of Assam. It is postulated that this belt in Nagaland comprises six or possibly more thrust sheets in between the bounding major thrusts—Naga thrust on the west and the Disang thrust on the east [6]. The litho-units in the Schuppen Belt are juxtaposed against each other along various imbricate thrust sheets, the oldest being in the east and the youngest in the west. Thrust sheets of the Tipam Group are overthrusted by those of the Surma Group, which are again overthrusted by thrust sheets of the Barail Group.

The Inner Fold Belt occupies the central part of Naga Hills and extends up to Pangsu Pass in Arunachal Pradesh. A large spread of Disang group of rocks [7] with isolated covers of Barail [4], as well as Disang-Barail Transition sequences, characterize the geological setting of this belt. The Palaeogene rocks have been folded into a series of anticlines and synclines and are confined within two major tectonic zones viz. the Disang thrust to the west and the Ophiolite thrust to the east. The Inner Fold Belt is occupied by two major synclinoria, namely the Kohima synclinorium to the South and the Patkai synclinorium to the North. In Kohima synclinorium, the core is occupied by younger sediments of the Surma Group [8]. The age of Disang sediments is assigned to be Upper Cretaceous to Eocene [7], while that of Barail range from Upper Eocene to Oligocene [4]. The Disang Group comprises mostly of a monotonous sequence of dark grey to black splintery shale in addition to

Fig. 3 Field photographs of Meriema Landslide and evidence of subsidence in the nearby area. **a** The landslide as seen when facing South-Southeast. It shows the crown of circular surface of failure. b The landslide as seen when facing northeast. **c** The foot of the landslide. **d** A road-cut near the slide area for construction of a new roadway. The rocks slumping down *en masse* from both sides are also seen moving downslope over a very wide circular surface

intercalations of siltstone and fine to medium-grained sandstone. The Barail Group consists predominantly of arenaceous sediments characterized by light grey to brown, fine to medium-grained sandstone with minor to considerably thick interbeds of shale. The Barail sediments are overlain by the molasse sediments characterized as the Surma [4] and Tipam [7] Groups. The Surma Group lies unconformably over the Barail Group. The Tipam sediments again lie conformably over those of Surma. The age of the molasse sediments is believed to extend from the Oligocene upto Miocene [4, 7].

The inner fold belt is overthusted by the ophiolite melange zone. The ophiolite belt consists predominantly of tectonic slices representing various units of the sequence of ancient oceanic lithosphere that include the ultramafic suite, mafic dykes and sills, pillow lavas, pelagic sediments like chert and limestone, etc., in a jumbled manner [9, 10]. The ophiolite melange zone in Nagaland has a very thick linear horizon of conglomerate lying over the ultramafics with a tectonic contact. The fossil assemblages from the associated limestone have suggested an Upper Cretaceous to Lower

Eocene age for the Ophiolites [11, 12]. The ophiolite melange zone is again overthrusted by a metamorphic complex known as the Naga metamorphics. The metamorphic complex is comprised of an intensely thrusted group of metasediments, which is predominantly made up of low to medium-grade phyllitic schist, hornblende schist, quartzite, quartz-sericite schist, mylonite, etc. [2]. Brunnschweiler [2] assigned them Pre-Mesozoic or older age but Acharyya et al. [11], however, stated them to be of Proterozoic age.

2.1 Geology of the Study Area

The study area falls within the inner fold belt (Fig. 1). The major principal rock type is shale belonging to the Disang Group. Dark grey to black splintery shale along with intercalations of siltstone and fine- to medium-grained sandstone are the characteristic sediments of the Disang Group. The black shale becomes brown when weathered. The Disang shale is found to be capped by younger sandstone-dominated layers belonging to the Barail Group as outliers in some of the isolated hilltops (Fig. 4b). The arenaceous Barail sediments are predominantly of light grey to brown, fine- to medium-grained sandstone with minor to considerably thick interbeds of shale. The litho-units of the surrounding region (of the landslide area) have been intensely folded and thrusted (Fig. 4a, b). The rocks are also found to have been jointed to a considerable extent. Joints are breaks (planes of broken surfaces) of natural origin in the body of rock that lack visible displacement. They are developed when the rocks have been subjected to a system of tectonic forces during orogenesis.

3 Geological Investigations

Events of geological history and other phenomena are imprinted on the rocks. The geologists unveil them by carrying out certain studies to decrypt those encrypted imprints. Geological mapping is one of the techniques through which those imprints exposed on the surface are translated into geological language to arrive at a rational interpretation of the evolutionary history of the rocks as well as other geological phenomena. It also gives clues to the lithological relationship of the litho-units. Geological mapping in essence is recording and plotting the observations obtained from the rock exposures on a base map and logically interpreting those observations towards unravelling how the litho-units are found exposed and related to each other in the present-day field setting, thereby helping in deciphering their evolutionary history.

a

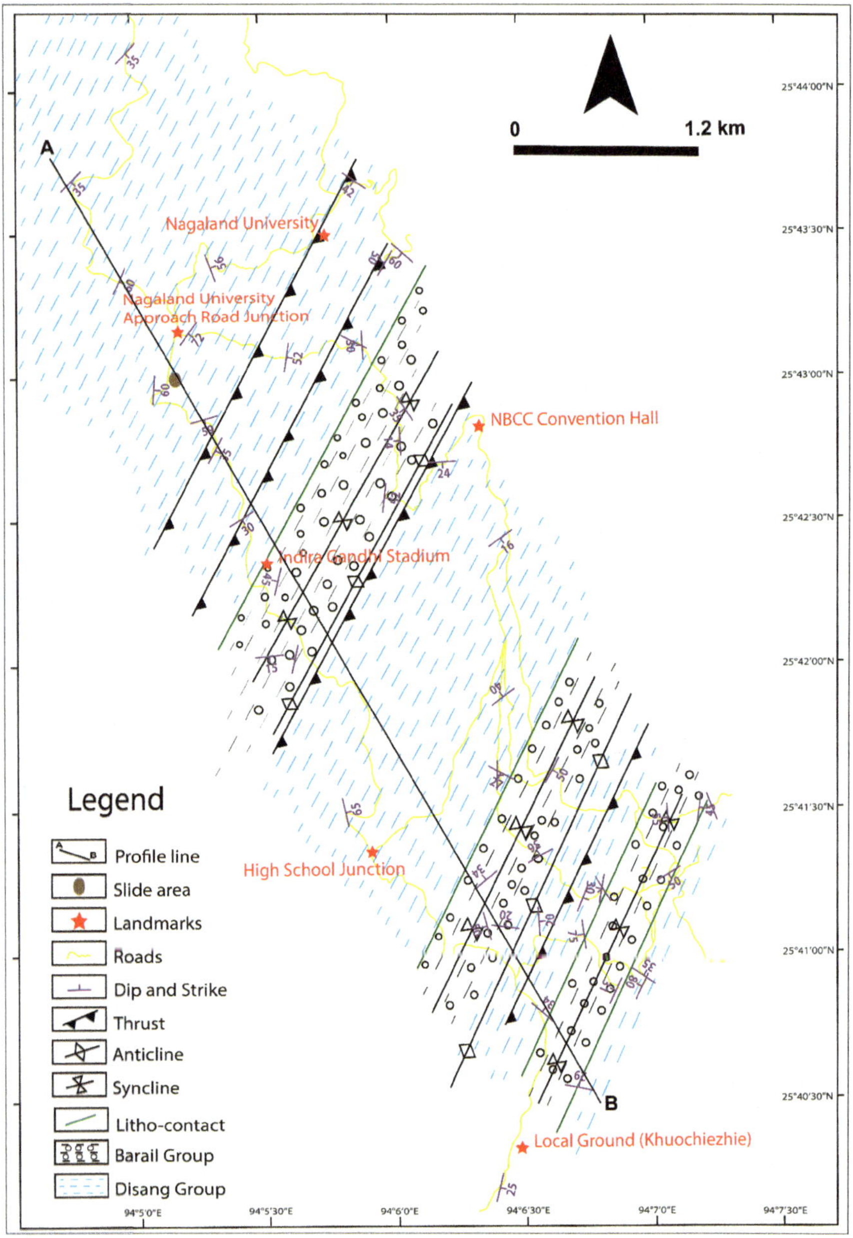

Fig. 4 **a** Geological and structural map (prepared) of the study area. **b** Geological section along the profile line AB

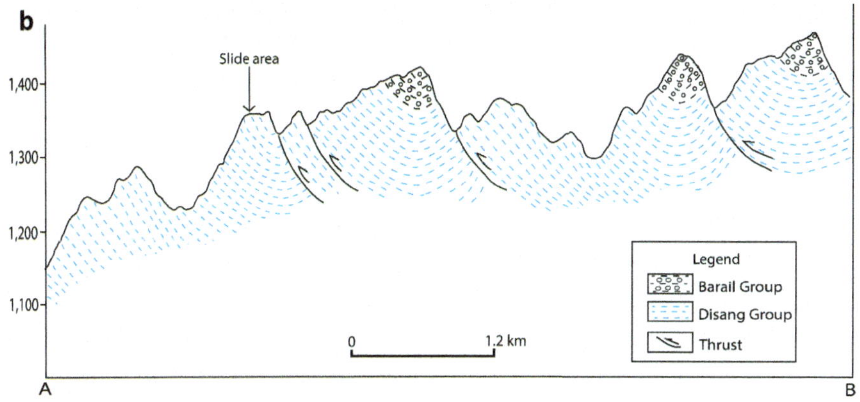

Fig. 4 (continued)

3.1 Geological Mapping and Preparation of Geological Profile

Geological and structural mapping is carried out in this piece of research work with an aim to explore the role of lithology and structure in the causes of the Meriema Landslide. Bedding data of the bedrocks at 53 locations from the surrounding area (covering approximately 20 sq. km.) of the landslide have been collected. The data along with the respective locations as well as other related data derived from them are shown in Table 1. The dip-strike data of the beds along with their respective locations are plotted in the base map using Arc GIS 10.7 software (Fig. 1). Using the orientation and pattern of the dip-strike data of the litho-units, and taking into consideration of the other field data like the locations where evidence of thrusting was observed, the types of rocks found in the different locations, etc., a geological and structural map (Fig. 4a) is constructed. Synclines and anticlines are inferred from the pattern of dip-strike data. As the mapping is carried out in limited sections the boundaries of litho-contacts are drawn using intrapolation and extrapolation techniques. Hence, there is a chance of having little variation of the actual litho-contact boundaries with those portrayed in the map.

A representative profile line AB is chosen in the geological map in such a way that the line is almost perpendicular to the general trend (strike) of the beds and passes through the landslide area as well as in the region where the maximum concentration of stations of observations are clustered. A cross-sectional profile (geological profile) along AB is constructed (Fig. 4b) from the prepared geological map. From the geological profile, it is clearly evident that the Barail sediments are occurring as outliers in the form of synclines covering some of the high peaks. The study area is closely folded and thrusted. A thrust is daylighted very close to the landslide area.

Table 1 Bedding attitude data of the litho-units of the surrounding area of Meriema Landslide

Stations	Coordinates	Strike (degree)	Dip direction (degree)	Dip amount (degree)	Pole value (degree)	Pole direction (degree)
1	N 25° 40′11.0″ E 94° 06′23.8″	194	104	25	65	284
2	N 25° 40′47.8″ E 94° 06′33.6″	128	38	30	60	218
3	N 25° 41′06.2″ E 94° 06′33.0″	(a) 174	84	20	70	264
		(b) 244	154	10	80	334
4	N 25° 41′03.2″ E 94° 06′43.0″	351	261	75	15	81
5	N 25° 40′51.8″ E 94° 06′50.3″	25	295	31	59	115
6	N 25° 41′12.5″ E 94° 06′47.6″	328	238	30	60	78
7	N 25° 41′11.1″ E 94° 06′44.1″	207	117	60	30	297
8	N 25° 41′36.6″ E 94° 06′24.4″	183	93	55	35	273
9	N 24° 41′36.6″ E 94° 06′24.4″	297	207	27	63	27
10	N 25° 41′53.1″ E 94° 06′25.0″	53	323	40	50	143

(continued)

Table 1 (continued)

Stations	Coordinates	Strike (degree)	Dip direction (degree)	Dip amount (degree)	Pole value (degree)	Pole direction (degree)
11	N 25° 42′25.3″ E 94° 06′25.3″	235	145	16	74	325
12	N 25° 42′01.2″ E 94° 05′36.5″	12	282	29	61	102
13	N 25° 42′01.2″ E 94°05′ 30.9″	236	173	15	75	353
14	N 25° 42′29.8″ E 94° 05′25.1″	238	148	30	60	328
15	N 25° 42′44.4″ E 94° 05′19.3″	207	117	35	55	297
16	N 25° 42′50.8″ E 94° 05′16.5″	254	164	59	31	344
17	N 25° 42′57.0″ E 94° 05′05.1″	256	166	44	46	346
18	N 25°43′08.8″ E 94°05′13.2″	217	127	72	18	307
19	N 25°43′ 6.6″ E 94°05′ 51.2″	297	207	35	55	27
20	N 25° 43′06.4″ E 94° 05′52.8″	355	265	30	60	85
21	N 25° 42′52.7″ E 94° 06′02.5″	325	235	35	55	55

(continued)

Table 1 (continued)

Stations	Coordinates	Strike (degree)	Dip direction (degree)	Dip amount (degree)	Pole value (degree)	Pole direction (degree)
22	N 25° 42′46.2″ E 94° 06′01.3″	356	266	14	76	86
23	N 25° 42′34.3″ E 94° 05′58.4"	187	97	52	38	277
24	N 25° 42′41.3″ E 94° 06′11.6″	266	176	24	66	356
25	N 25° 43′03.8″ E 94° 05′36.2″	189	99	52	38	279
26	N 25° 43′23.1″ E 94° 05′18.7″	159	69	56	34	249
27	N 25° 43′40.4″ E 94° 05′51.6″	303	213	42	48	33
28	N 25° 43′24.4″ E 94° 05′58.8″	20	290	55	35	110
29	N 25° 43′25.3″ E 94° 06′02.4″	195	105	50	40	285
30	N 25° 43′29.4″ E 94° 06′03.4″	328	238	11	79	58
	Fault-plane	1	171	29	61	351
31	N 25°40′52.7" E 94°06′56.3"	25	295	80	10	115
32	N 25°40′55.1" E 94°06′57.5"	122	32	35	55	212
33	N 25°41′15.5" E 94°07′03.1"	240	150	50	40	330

(continued)

Table 1 (continued)

Stations	Coordinates	Strike (degree)	Dip direction (degree)	Dip amount (degree)	Pole value (degree)	Pole direction (degree)
34	N 25°41′29.8" E 94°07′13.1"	338	248	20	70	68
35	N 25°41′29.4" E 94°07′11.1"	190	100	45	45	280
36	N 25°41′27.5" E 94°07′02.7"	354	264	55	35	84
37	N 25°41′37.4" E 94°06′37.0"	210	120	50	40	300
38	N 25°41′29.4" E 94°05′48.7"	166	76	65	25	256
39	N 25°42′17.3" E 94°05′32.7"	16	286	45	45	106
40	N 25°43′07.5" E 94°05′04.3"	205	115	50	40	295
41	N 25°43′23.8" E 94°05′58.2"	30	300	50	40	120
42	N 25°43′25.5" E 94°06′02.2"	310	220	60	30	40
43	N 25°43′20.3" E 94°04′57.1"	205	115	60	30	295
44	N 25°43′41.0" E 94°04′45.8"	225	135	35	55	315
45	N 25°44′07.6" E 94°04′58.9"	225	135	35	55	315
46	N 25°42′57.2" E 94°050′5.1"	185	95	60	30	275
47	N 25°41′20.4" E 94°06′33.3"	60	330	36	54	150
48	N 25°41′13.1" E 940 6′47.2"	10	280	30	60	100
49	N 25°41′15.2" E 94°06′20.5"	50	321	34	56	141
50	N 25°41′05.6" E 94°06′25.3"	98	8	20	70	188
51	N 25°41′05.1" E 94°06′20.6"	344	254	50	40	74
52	N 25°40′48.3" E 94°06′33.4"	129	39	34	56	219
53	N 25°40′32.4" E 94°06′42.1"	102	12	29	61	192

3.2 *Joint Analysis*

Joints are defined as failure planes along which there has been no relative displacement. It is the most common type of geological structure and may be seen in any rock exposure and almost in all classes of rocks. As the joints are developed when the rocks have been subjected to a system of tectonic forces during orogenesis, their orientation and pattern could be used to retrieve the principal stress directions of the system of forces that brought about the mountain-forming processes. The joint analysis is carried out in this piece of research work with an aim to understand their pattern and to examine whether these failure planes have any role in bringing about the downslope mass movement of the Meriema Landslide.

Attitudes of 107 joints were collected from the study area. The data of the joints are represented in Table 2. With the dip amount and dip direction data of the joints, a rose diagram (Fig. 5a) is prepared using GEORIENT software. A frequency distribution of the dip amounts of joints is prepared and is shown in Table 3. The histogram of the frequency distribution data of dip amounts of joints is shown in Fig. 5b.

4 Geotechnical Investigations

Mass movement owing to the failure of incompetent shale rock has taken place during the rainy season. Why the shale belonging to the Disang Group has failed causing serious problems of landslides along the National Highway No. 61. To understand the causes of this landslide, geotechnical investigations have been carried out to ascertain how the rocks have failed and slumped down.

Rocks have three mechanical strength parameters such as compressive strength, shear strength and tensile strength in order of decreasing strength. Normally, compressive strength and tensile strength of intact rock samples are determined in the laboratory by using a simple compressive testing machine employing methods such as (i) the Uniaxial compression test, and (ii) the Brazilian test, also known as the diametric compression test. The shear strength can be determined directly in the laboratory using a triaxial compression/shear testing machine. However, in low confining pressure/stress, shear strength can also be evaluated using the strength parameters of uniaxial compressive strength and tensile strength, and by constructing Mohr's stress circles and finding out the common tangent representing the Coulomb's failure equation given by

$$\tau = c + \sigma_n \tan\varphi \tag{1}$$

where τ—the shear strength, c—the cohesion, σ_n—the normal stress and φ—the internal friction angle.

In order to determine the mechanical strength of the rock samples Uniaxial Compression Test for Compressive Strength and the Brazilian Test for Tensile

Table 2 Data of attitude of joints of the litho-units of the surrounding area of Meriema Landslide

Serial no	Joints attitudes		
	Dip amount (degree)	Dip direction (degree)	Strike (degree)
1	80	133	43
2	80	270	180
3	70	207	117
4	80	193	103
5	60	280	190
6	70	263	173
7	90	183	93
8	70	184	94
9	61	296	206
10	69	58	148
11	80	38	128
12	64	94	4
13	83	160	70
14	75	43	133
15	75	191	101
16	69	182	92
17	55	292	202
18	79	314	224
19	65	39	129
20	76	246	156
21	56	301	211
22	81	19	109
23	76	117	27
24	81	99	9
25	66	109	19
26	86	26	116
27	70	267	177
28	50	349	259
29	79	255	165
30	31	48	138
31	60	338	248
32	79	229	139
33	80	27	117
34	70	237	147
35	65	354	264
36	64	338	248

(continued)

Table 2 (continued)

Serial no	Joints attitudes		
	Dip amount (degree)	Dip direction (degree)	Strike (degree)
37	55	224	134
38	70	154	64
39	82	320	230
40	45	201	111
41	85	310	220
42	81	140	50
43	78	23	113
44	70	66	156
45	40	8	98
46	64	355	265
47	35	311	221
48	80	16	106
49	54	343	253
50	65	244	154
51	73	38	128
52	70	5	95
53	74	74	164
54	65	122	32
55	79	203	113
56	71	111	21
57	89	199	109
58	81	264	174
59	50	17	107
60	55	10	100
61	75	320	230
62	80	238	148
63	75	30	120
64	70	32	122
65	80	216	126
66	85	203	113
67	64	213	123
68	65	192	102
69	85	300	210
70	86	271	181
71	80	344	254

(continued)

Table 2 (continued)

Serial no	Joints attitudes		
	Dip amount (degree)	Dip direction (degree)	Strike (degree)
72	74	65	155
73	76	352	262
74	75	55	145
75	51	161	71
76	73	48	138
77	41	178	88
78	70	214	124
79	62	305	215
80	82	179	89
81	32	86	176
82	90	170	80
83	75	133	43
84	75	35	125
85	35	3	93
86	70	185	275
87	80	350	80
88	50	30	120
89	50	105	195
90	60	220	310
91	70	85	175
92	65	225	315
93	60	72	162
94	70	29	119
95	75	301	31
96	75	196	286
97	55	215	305
98	85	94	184
99	85	208	298
100	65	337	67
101	30	217	307
102	80	95	185
103	80	40	130
104	80	280	10
105	70	182	272
106	65	135	225
107	55	243	333

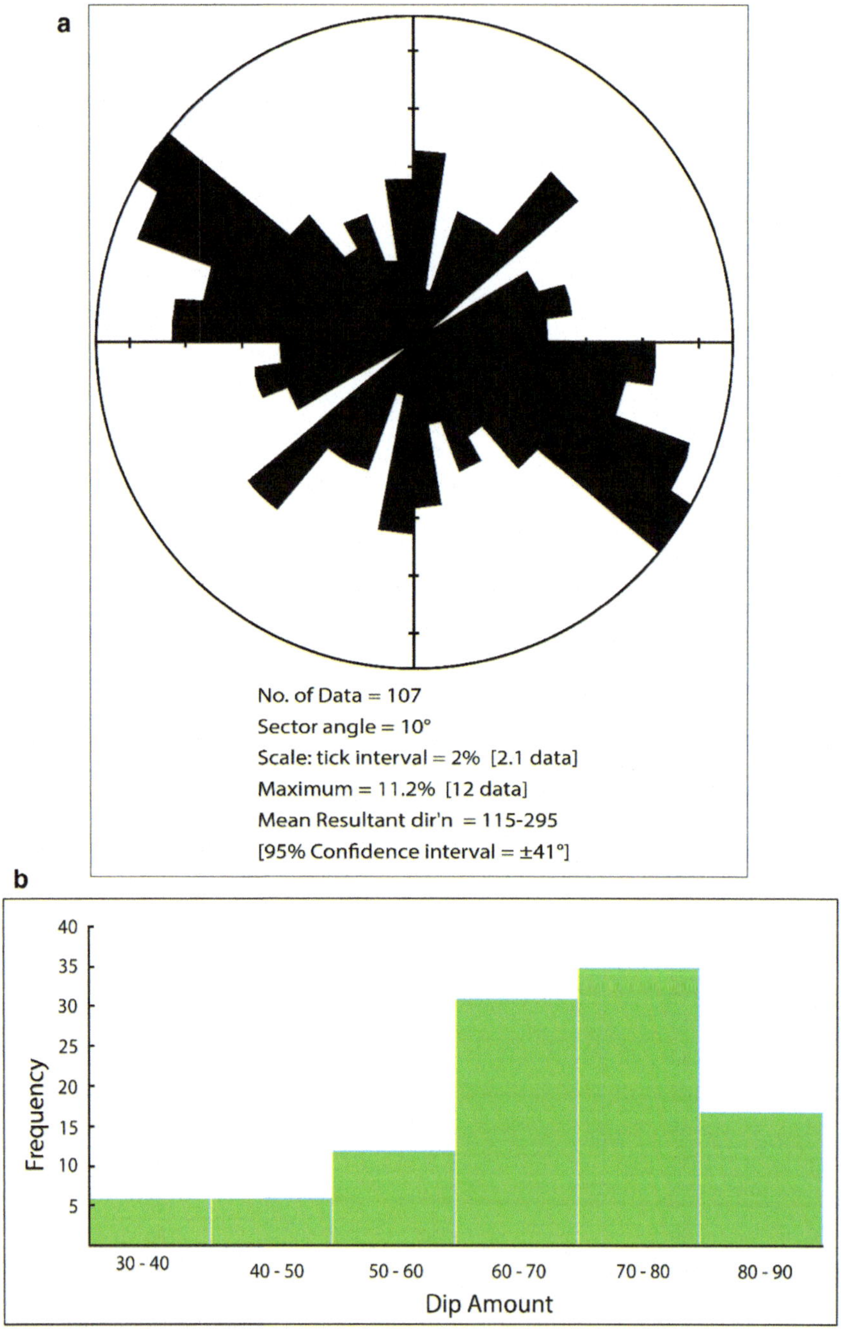

Fig. 5 **a** Rose diagram of joint data of the study area. **b** Histogram showing the relationship between dip amount and frequency of joints

Table 3 Frequency distribution data of dip amount of the joints of the rocks of the surrounding area of Meriema Landslide

Class interval	Frequency
30–40	6
40–50	6
50–60	12
60–70	31
70–80	35
80–90	17
	$\Sigma f = 107$

Strength are employed for this work. The shale/mudstone samples were relatively weak and fragile in nature. So, there were difficulties in preparing a larger number of samples or specimens from the cores. However, with careful cutting and trimming of the samples, six specimens, each of the Uniaxial Compression Test and Brazilian Test were prepared from the core samples. These were tested and the results are shown in Tables 4, 5, 6 and 7.

In order to find out the values of cohesion (c), internal friction angle (φ), etc., Mohr's stress circles have been plotted for samples 1 and 2 from which failure

Table 4 Uniaxial compressive test data of Shale/Mudstone

Sample no	Length (cm)	Diameter (cm)	Load P(KN)	Remarks
1	10	5.1	49	Shale/Mudstone
2	10.2	5.1	47	Shale/Mudstone
3	10.1	5.1	45	Shale/Mudstone
4	10.1	5.1	42	Shale/Mudstone
5	10.2	5.1	50	Shale/Mudstone
6	10.3	5.2	52	Shale/Mudstone

Table 5 Calculation of compressive strength of Shale/Mudstone

Rock sample	Load, P in KN	Mean diameter	Radius, r	Compressive strength, $C_o = P/A\ (P/\pi r^2)$ in MPa
1	49	5.1	2.55	24
2	47	5.1	2.55	23
3	45	5.1	2.55	22
4	42	5.1	2.55	21
5	50	5.1	2.55	24
6	52	5.2	2.6	24

Table 6 Results of Brazilian test of shale/mudstone

Cylindrical Sample N	Length (cm.)	Diameter (cm.)	Load in KN	Remarks
1	5.2	5.1	17	Shale/Mudstone
2	5.1	5.1	16	Shale/Mudstone
3	5	5.1	24	Shale/Mudstone
4	5.1	5.1	18	Shale/Mudstone
5	5.2	5.1	22	Shale/Mudstone
6	5.2	5.1	20	Shale/Mudstone

Table 7 Calculation of tensile strength of Shale/Mudstone

Rock sample	Load, P	Mean diameter, D (cm)	Length, L (cm)	Tensile strength, $T = 2P/\pi DL$ MPa
1	17	5.1	5.2	4
2	16	5.1	5.1	$3.9 \approx 4$
3	24	5.1	5.0	6
4	18	5.1	5.1	4
5	22	5.1	5.2	5
6	20	5.1	5.2	5

envelopes have been obtained as shown in Fig. 6a, b. Similarly, for the other four samples also, Mohr's stress circles have been plotted (but not shown) for the determination of their cohesion (c) and internal friction angle (φ). From the values of compressive strength (Table 5), tensile strength (Table 7), cohesion (c), internal friction angle (φ), shear strengths of the different six samples are calculated and are as shown in Table 8.

4.1 Specific Gravity Determination

In this test, the rock samples are first powdered. Then, the specific gravity is determined by using 50 ml density bottle. The density bottle is the standard method used in the laboratory. For this, the mass M_1 of the empty, dry bottle is first taken. A sample of oven-dried rock, cooled in a desiccator, is put in the bottle, and the mass M_2 is taken. The bottle is then filled with distilled water gradually, removing the entrapped air either by applying a vacuum or by shaking the bottle. The mass M_3 of the bottle, sample and water is taken. Finally, the bottle is emptied completely and thoroughly washed; clean water is filled to the top, and the mass M_4 is taken. Based on these four observations, the specific gravity can be computed as follows:

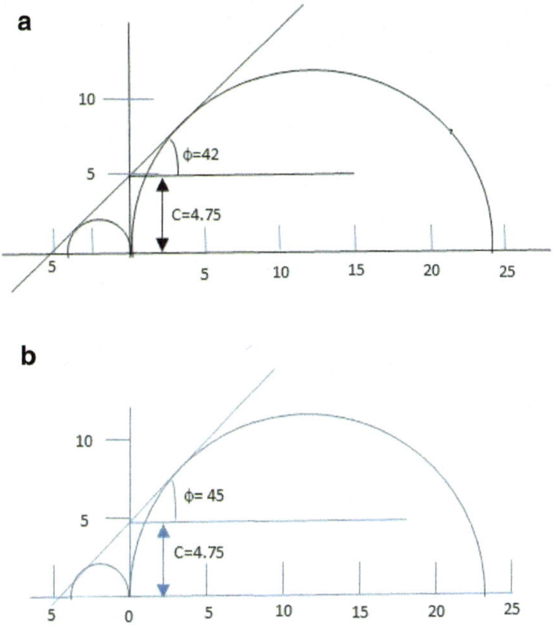

Fig. 6 **a** Cohesive (C) and internal friction angle (ϕ) of sample 1. **b** Cohesive (C) and internal friction angle (ϕ) of sample 2

Table 8 Calculation of shear strength of shale/mudstone

Sample	Compressive (MPa)	Tensile (MPa)	Cohesion C	Internal friction angle (ϕ)	Normal stress (σ_n) (MPa)	Shear strength $\tau = c + \sigma_n \tan\varphi$ (MPa)
1	24	4	4.75	42	4	8.35
2	23	4	4.75	45	3	7.75
3	22	6	6.4	33	5	9.65
4	21	4	4.75	42	3	7.45
5	24	5	5.50	39	5	7.55
6	24	5	5.25	41	4.5	9.16
Average shear strength = 8.32 MPa						

$$\text{Specific gravity, } G = M_2 - M_1/(M_4 - M_1) - (M_3 - M_2) \quad (2)$$

Thus, the specific gravity of the six samples of shale/mudstone is determined and the values are shown in Table 9.

Table 9 Specific gravity of mudstone samples

Rock sample	Wt. of empty bottle in (g) M_1	Wt. of bottle + sample in (g) M_2	Wt. of bottle + sample + water in (g) M_3	Wt. of bottle + water in (g) M_4	Specific gravity G = $(M_2-M_1)/(M_4-M_1)-(M_3-M_2)$
1	18.25	19.64	51.37	50.54	2.48
2	17.98	19.04	42.68	42.01	2.72
3	18.26	19.54	51.35	50.55	2.67
4	18.18	20.18	48.13	47.27	2.69
5	18.19	19.42	46.92	46.17	2.56
6	18.05	19.38	51.16	50.35	2.72
Average specific gravity = 2.64					

5 Calculation of Maximum Load of Overburden

The circular surface of failure is not completely exposed as some debris mass is found still covering and sitting above the surface at the lower level (Fig. 3a, b). Thus, the maximum thickness of the rock column above the circular failure surface (before being moved down of rock mass) could not be surveyed directly in the field (from the crown). This maximum thickness is found indirectly with the Total Station Survey. The sketch of the methodology of determining this maximum thickness of overburden (rock) is shown in Fig. 7. If the slid down rock masses are brought back to the original position, the column of rock (before sliding) occupying the maximum thickness (at the deepest point above the failure circular surface) will fall in the line dividing the circular surface into two similar mirror halves. Had there not been still some debris the thickest column could have been easily found out. The debris at the bottom of the surface of curvature made recourse to the adoption of an indirect method. The surface elevation of the column of rock (before sliding) that was occupying the maximum thickness standing above the failure circular surface turned out to be 1394.6 m above the m.s.l. The deepest point of the circular surface where the thickest column of rock (before being slid) was occupied is not exposed but covered by debris. The vertical line representing the occupational position of the thickest column (in the figure) meets the debris at 1372 m. As the circular surface at the lower level is not exposed, it is assumed to be as deep as the level at which the second step of circular failure initiates. The elevation at which the second step of failure starts is 1367.5 m. Thus, the total thickness of the maximum column of rock above the circular surface is taken as the difference between the surface elevation of 1394.6 m and elevation of 1367.5 m at which the second step of failure starts.

Thus, the vertical thickness of the rock column going down the deepest point of the circular surface, VTdp is given by

$$\begin{aligned} \text{VTdp} &= 1394.6 - 1367.5 \text{ m} \\ &= 27.1 \text{ m} \end{aligned}$$

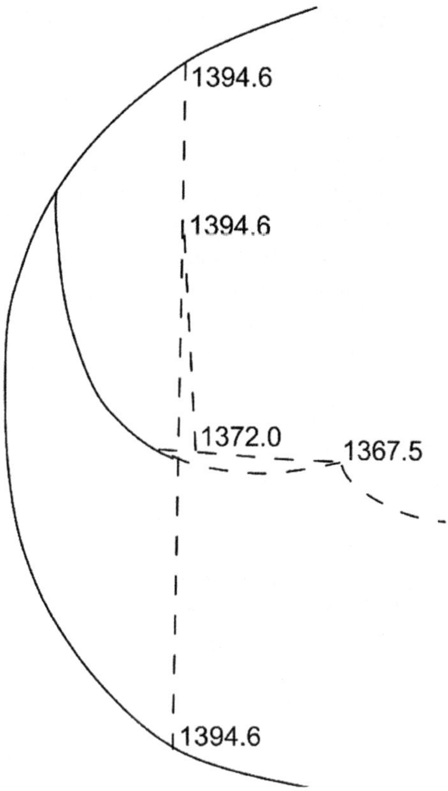

Fig. 7 Sketch of the slide (facing south west) depicting the maximum thickness of overburden of rocks above the failure surface at the deepest point (before slide). The figures in the diagram are in meter (the details are explained in the text)

And the average sp. gravity of the rock (Table 9)

$$= (2.48 + 2.72 + 2.67 + 2.69 + 2.56 + 2.72)/6$$
$$= 2.64$$

Therefore, the maximum pressure the rock suffered at the deepest point

$$= (2.64 \times 1000\ \times 9.8 \times 27.1)/10^6\text{MPa}$$
$$= 0.701\text{MPa}$$

(which is extremely less than average Shear strength, τ of 8.32 MPa).

6 Discussion and Conclusion

The study area belonging to Indo-Myanmar Range (IMR) orogen is characterized by the presence of intense folding and thrusting. The orogenesis of the (IMR) is attributed to the interaction of the adjoining three plates of Eurasia, India and Myanmar (microplate) in which the relative motion of the Indian plate with respect to the Myanmar plate took the key role in the mountain building process [13]. It has been shown that the motion of the Indian plate with respect to the Myanmar plate induces a dextral shear couple deformation mechanism (Fig. 8). The deformation mechanism depicted in Fig. 8 explains the tectonic features of the southern portion of the Manipur–Nagaland segment of the IMR. In the northern portion, the deformation ellipse will be rotated by a few degrees clockwise on account of the arcuate nature of the IMR. Such a deformation mechanism caused the rocks of the region to have been compressed in the WNW-ESE (but almost NW–SE in Nagaland) and extended in the NNE-SSW (almost NE-SW in Nagaland) directions. According to the deformation mechanism and the trend of mountain ranges of the study area, the thrusts and the fold axes are oriented along NNE-SSW to almost NE-SW; the synthetic strike-slip faults along ENE-WSW to almost E-W, and the antithetic strike-slip faults along SSE-NNW to almost N-S; and the normal faults along ESE-WNW to almost SE-NW are in conformity with the deformation mechanism of the region. Consequently, all the lineaments and the joints are found concentrated mostly along the specified directions as postulated by the deformation mechanism, i.e., parallel along trends of thrusts, strike-slip faults and normal faults.

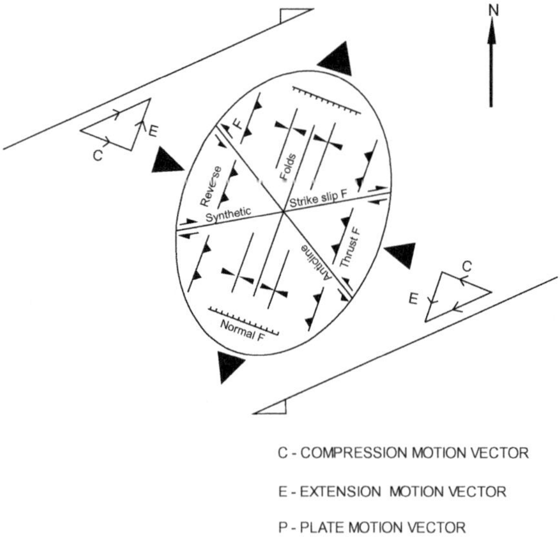

Fig. 8 Dextral shear deformation mechanism of the IMR (after Soibam, 2006)

Figure 5a shows that the joints in the study area are mostly concentrated along the NW–SE direction, which is parallel to the compression direction even though clustering to a smaller extent is also observed in other directions defined by the deformation mechanism. It is also understood from Fig. 2 that the radius of curvature of the failure surface is almost parallel to the direction having maximum joint plane concentration. Thus, it appears that the joints are responsible for making the rock masses fail. There is considerable extension roughly along NE-SW parallel to the elongation direction of the failure surface as well as perpendicular to the majority of joints. This could have made the opening up of the joint planes wide. It is also seen in Fig. 4a, b that a thrust runs very close to the slide area. Thus, the nearby thrusting activity could have caused the already developed joints, particularly those striking across the thrust and parallel to the radius of curvature of the failure surface, to widen more in the defined directions of the joint planes. As the thrust runs nearly along the NE-SW direction, the resultant force of compression in the area must be directed roughly along NW–SE parallel to the radius of curvature of the failure surface. The compressive force and consequently, the resultant extension could have been concentrated more in the area where the landslide occurred.

The maximum pressure the rock suffered at the deepest point of the failure surface of the Meriema Landslide is found to be 0.701 MPa. On the other hand, the average shear strength of the bedrock is 8.32 MPa (Table 8). If we compare the shear strength (τ) values for the failure of rocks at low confining pressures, these values are found to be extremely small to make the rocks (shale/mudstone) fail even if the weight of the saturated water is included in the mass of the overburden. It is understood from Fig. 9 that the load (gravity) is always greater than the shear stress for all the slope angles less than the vertical. The value of the maximum load of 0.701 MPa is the one produced when the gravity tugs down vertically. Thus, the shear strength the rocks actually suffered must have been again lesser than the calculated maximum load of 0.701 MPa. Hence, the rock doesn't fail on account of the shear stress component of the load. Consequently, structures of the rocks play an important role in such kind of landslides in which the rocks move en masse.

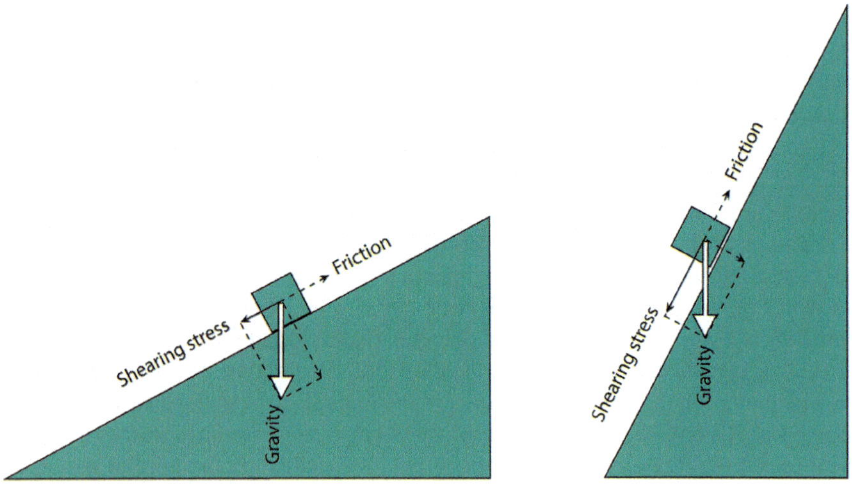

Fig. 9 Relationship of of slope angle and slide potential

Thus, it is established that such a kind of mass movement is controlled by structure. Near this particular slide area, a thrust passes through. It is inferred that the thrusting activity caused the already developed joints particularly those perpendicular to the length of the failure surface are widened. It is also observed that the majority of the joints dip at high angles (Fig. 5b); mostly clustering between 60–80°. In the rainy season during incessant rain, the rocks are saturated with water; filling all the open spaces of the joint planes with water. The saturation of water reduces the friction considerably and hence the body of jointed rock mass above the failure surface could not withstand the weak shear stress. The planes of widened interconnected joints and dipping at various angles (mostly high angled) could create a circular surface above which the broken rock pieces could move down *en masse*. So, the already broken (jointed) rock pieces failed and slid down along the failure surface developed by the mosaic of widened and interconnected joints dipping at various angles and orientations. Thus, the Meriema Landslide is considered to have been controlled by incompetent lithological character and geological structure.

Acknowledgments This research article is an embodiment of a part of the outcome of the research project entitled "*Role of Lithology and Geological Structure in causing Landslide in and around Kohima along the Road Section of NH-27 (Dimapur-Imphal Highway) and Landslide Hazard Mitigation*" funded by the NRDMS, DST, Government of India vide its project sanction No. *NRDMS/ 02/57/017(G)*. The authors therefore deeply acknowledge NRDMS, DST, Government of India for the benevolent support.

References

1. Montgomery CW (2011) Environmental Geology, 9th edn. McGraw Hills, New York
2. Brunnschweiler RO (1966) On the geology of Indo-Burman Ranges. J Geol Soc Australia 13:137–195
3. Supongtemjen, Thong GT, Walling T (2021) Landslide susceptibility mapping of Kohima town, Nagaland. In: Rao BV (ed) Recent advances in earth science research in North East India, South Eastern book agencies, Guwahati, pp 22–51
4. Evans P (1932) Explanatory notes to accompany a table showing the Tertiary succession in Assam. Trans Mining Geol Int India 27:155–260
5. Directorate of Geology and Mining (1978) Unpublished report. Govt. of Nagaland
6. Mathur LP, Evans p (1964) Oil in India, a brochure in 22nd I.G.C. New Delhi, pp 1–86
7. Mallet FR (1876) On the coal fields of Naga hills bordering the Lakhimpur and Sibsagar districts Assam. Geol Sur India Memoirs 12(2):166–363
8. Oldham RD (1883) Report on the geology of parts of Manipur and Naga Hills. Geol Soc India Mems 19:216–242
9. Khuman CM, Soibam I (2010) Ophiolite of Manipur—its field setting and petrotectonic significance. Mem Geol Soc India 75:255–290
10. Soibam I, Khuman CM, Subhamenon SS (2015) Ophiolitic rocks of the Indo-Myanmar Ranges, NE India: relicts of an inverted hyper-extended continental margin basin? In: Gibson GM, Roure F, Manatschal G (eds) Sedimentary basins and crustal processes at continental margins: from modern hyper-extended margins to deformed ancient analogues, Geological Society, London, Special Publications, vol 413, pp 239–267
11. Acharyya SK, Roy DK, Mitra ND (1986) Stratigraphy and palaeontology of the Naga Hills ophiolite belt. Geol Sur India Memoirs 119:64–74
12. Vidyadharan KT, Srivastava RK, Bhattacharyya S, Joshi A, Jena SK (1986) Distribution and description of major rock types. Geol Sur India Memoir 119:18–27
13. Soibam I (2006) Relative plate motions in and around Manipur and its implications on the tectonics of Indo-Myanmar Ranges. Himalayan Geol 27(2):111–122

Assessment and Analysis of Geotechnical Properties of Saron Veng Landslide, Aizawl, Mizoram

Lallawmsanga, Christopher Lalthazuala, Lalhmingsangi, Lalhlimpuia, Shiva Kumar, and Laldinpuia

Abstract The study of Saron Veng landslide is situated at the eastern limb of the Aizawl anticline. The landslide occurred on 13th June 2017 at 7:00–8:00 A.M. It swept away one residential building where seven families lived, and a link road between Saron Veng and Tuithiang Veng was blocked for more than a month. The improper drainage system, weak lithology, and high permeability also enhance the triggering factor for a landslide. The landslide covers an area of 3500 m^2, measuring 101.49 m in length and 44.196 m in width. The lithology of the site is composed of siltstones and sandstones, which were overlain by a thick regolith. Geological and geotechnical fieldwork was done based on various parameters like in-situ rock strength, particle size distribution for soil classification, seepage water quality, Atterberg's limit, Direct Shear test, and analyses based on Limit Equilibrium Method (LEM). The primary soil type is coarse to very coarse sand. Water analysis shows that seepage water is fit for drinking. The average in-situ rock strength is 18N/mm^2; soil can be classified as 'slightly plastic' from Atterberg's values. Direct shear shows that the soil has weak cohesion. The safety factor ranges from 0.372 to 0.391, indicating that instability slope. It can be concluded as 'debris slide,' soft plastic type of soil, weak cohesion, and an ultimate bearing capacity of 631.88 shows that unsafe for any structure. The seepage acts as a lubricant for sliding, added by rainfalls. Benching to reduce load and decrease shear strength and proper drainage system are suggested as mitigation measures.

Keywords Debris slide · Atterberg's limit · Direct shear test · Limit equilibrium method

Lallawmsanga · Laldinpuia (✉)
Centre for Disaster Management, Mizoram University, Aizawl, Mizoram, India
e-mail: dinpuiageo@gmail.com

C. Lalthazuala · Lalhmingsangi · S. Kumar
Department of Geology, Mizoram University, Aizawl, Mizoram, India

Lalhlimpuia
Geologist Consultant, Aizawl Municipal Corporation, Aizawl, Mizoram, India

S. Mitra et al. (eds.), *Disaster Management and Risk Reduction: Multidisciplinary Perspectives and Approaches in the Indian Context*,
https://doi.org/10.1007/978-981-99-6395-9_3

1 Introduction

A landslide is the movement of slope materials of rock mass, soil/debris, vegetation, and earth down the slope under the earth's gravity [10]. The landslide can be of anthropogenic and natural factors. Most of the landslides in Mizoram occur during the monsoon season when the soil gets saturated and is induced naturally or due to anthropogenic activities [7, 20], Laldinpuia et al. 2014; [8–10]. The study area Saron Veng landslide is situated in the eastern limb of the Aizawl anticline, which coordinates between 23°43′50.47″ N and 92°43′10.01″ N at the height of 1010 m. The landslide occurred on the 13th of June 2017 at around 7:00–8:00 A.M (Fig. 1).

In this paper, the assessment of Saron Veng landslide is done based on the geotechnical investigations and LEM analysis to determine the factor of safety for appropriate mitigation measures suggestion.

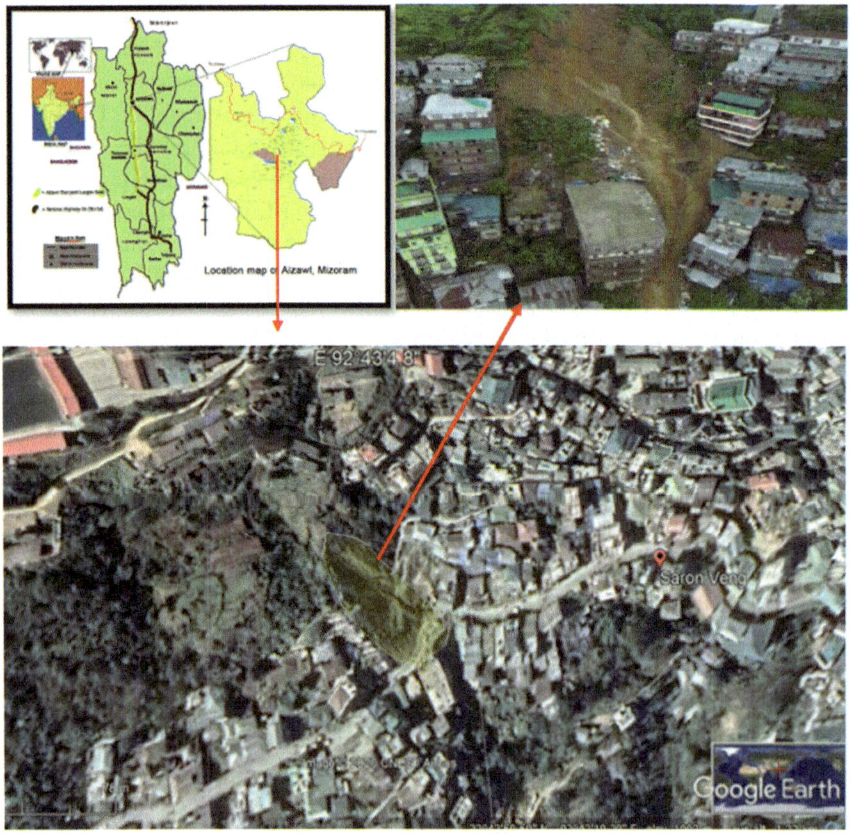

Fig. 1 Location map of study area, Saron Veng Landslide

2 Materials and Methods

2.1 Field Study

The study area was geologically investigated with the help of toposheets, Brunton compass and Field Move clino, hand-held GPS, digital camera, drone (Phantom 5.0, 4 k resolution), and measuring tape for field investigation.

2.2 Geotechnical Investigation

The geotechnical investigation includes Schmidt's Hammer (N/NR type), where the in-situ strength of the outcrop rock bed is determined, and the rebound values are converted to the respective value of uniaxial compressive strength of the rock per ASTM D5873– 14 [1, 3]. Particle Size Distribution using Sieve analysis, Atterberg's limits include Plastic limit, Liquid limit, Liquid Index, and Consistency Index, which is used to determine the soil texture, moisture content, and size distribution [2, 4, 19]. The direct shear test includes collecting the soil sampling for analysis done at three different points to determine the ultimate bearing capacity [5]. Water sampling for investigation was done at three other points, viz., Spring point, seepage-I, and seepage-II, respectively, in which physical–chemical properties were tested, and rainfall data also collected.

2.3 LEM Analysis

The study areas were analyzed in a LEM package, using the Ordinary/ Fellenius method, Bishop simplified method, Janbu's simplified method, Spencer's method, and GLE/Morgenstern Price's Method using LEM package Slide 6.0 software.

The LEM analysis is the solution for equations that provide deterministic information relating to the stability of a slope. The factor of safety (FOS) can also be expressed as a ratio between the slope's actual cohesion or friction angle and the cohesion or friction required for the slope to be stable. After several iterations, it is possible to locate the most crucial position for the potential slip surface, which is the surface that yields the minimum factor of safety for the slope in conventional terms and is theoretically the critical slip surface [13].

3 Results and Discussion

The young and immature geology of Mizoram has been a problem as it is vulnerable to landslide hazards [7, 20], (Laldinpuia et al. 2014). Due to high rainfall during the monsoon and human activities, many landslides have been in Mizoram [10].

The Saron Veng landslide analysis through geotechnical and LEM analysis results can be shown with logistic view inappropriate headings.

3.1 Field Study

The landslide area covers an area of about 3500 m^2; 101.49 m in length and 44.196 m in width (Fig. 2). The lithology of the area is mainly composed of sandstone and siltstones overlain by a thick regolith.

Fig. 2 Field study using a drone to know the geometry and effect of landslide

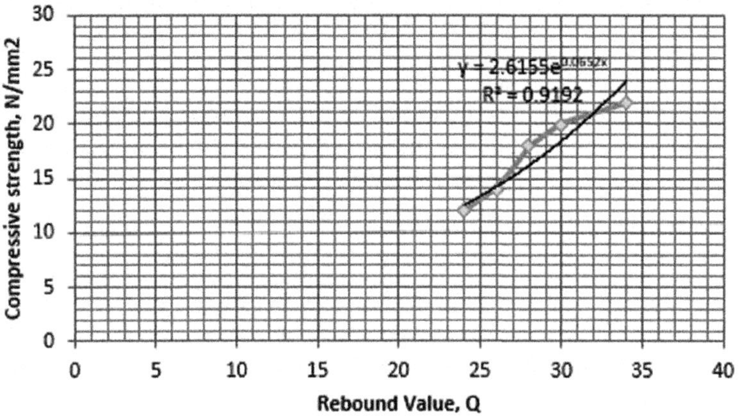

Fig. 3 Hammerlink curve for Saron Veng landslide (Schmidt Hammer, N/NR)

3.2 Geotechnical Investigations

3.2.1 Schmidt Hammer

The in-situ rock strength is determined using a Schmidt hammer (N/NR type).

The observed average rebound value is 28 and is equivalent to 18 N/mm^2 (Fig. 3).

3.2.2 Grain Size Analysis of Soil

The main soil type is ***coarse sand to very coarse sand*** from sieve analysis.

3.2.3 Atterberg's Limit

The plasticity index (Ip) is calculated from Figs. 4, 5 and 6, respectively. From the Plasticity Index values, the nature of soil can be classified as 'slightly plastic' [6] (Table 1).

3.2.4 Liquidity Index and Consistency Index

See Tables 2 and 3.

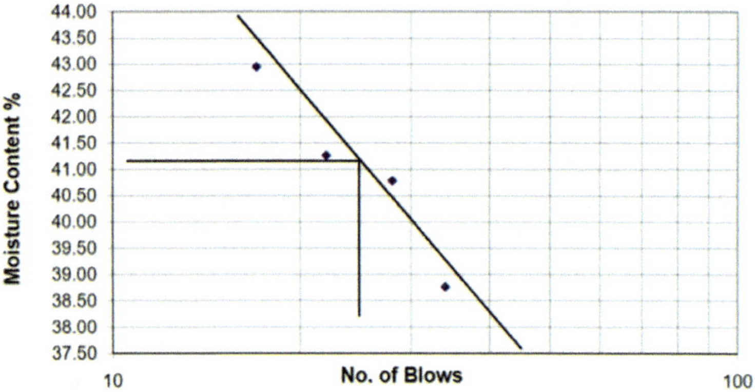

Fig. 4 Liquid limit graph for sample-1

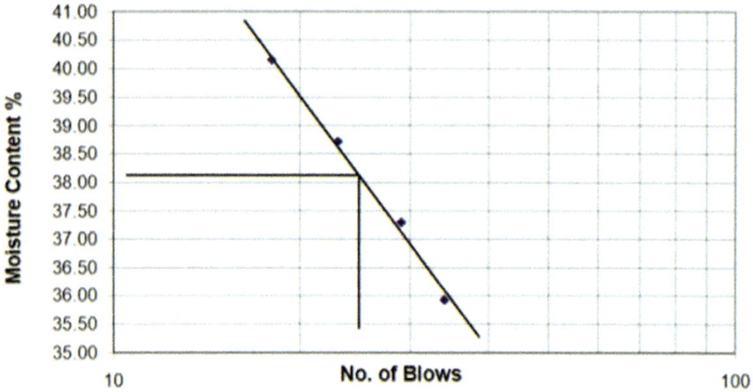

Fig. 5 Liquid limit graph for sample-2

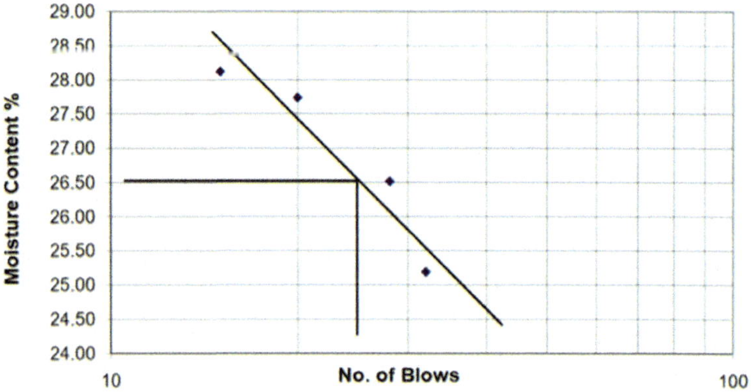

Fig. 6 Liquid limit graph for sample-3

Table 1 Atterberg's limit values of Saron Veng

Sample no	Location	W_L	W_P	I_P
1	N23° 43.753′ and E092° 43.338′	41.20	30.92	10.28
2	N23° 43.793′ and E092° 43.363′	38.10	28.77	9.33
3	N23° 43.800 and E092° 43.37′	26.50	21.22	5.28

[W_L = Liquid limit; W_P = Plastic limit; I_P = Plasticity index]

Table 2 Liquidity index and consistency index

Sample no	Location	Moisture content	W_L	W_P	I_P	Liquidity index, I_L	Consistency index, I_C
1	N23° 43.753′ and E092° 43.338′	36.65	41.20	30.92	10.28	0.557	0.442
2	N23° 43.793′ and E092° 43.363′	34.05	38.10	28.77	9.33	0.566	0.434
3	N23° 43.800 and E092° 43.37′	24.46	26.50	21.22	5.28	0.613	0.386

Table 3 States of soil consistency based on consistency Index, IC (BS 5930)

Sample no	Location	I_c	State
1	N23° 43.753′ and E092° 43.338′	0.442	Soft (Plastic)
2	N23° 43.793′ and E092° 43.363′	0.434	Soft (Plastic)
3	N23° 43.800 and E092° 43.37′	0.386	Soft (Plastic)

3.2.5 Direct Shear Test

The observed angle of friction ranges are 23.43°, 29.54°, and 30.68° with cohesion of 2.4 kN/m^2, 1.3 kN/m^2, and 1.0 kN/m^2, and the average ultimate bearing capacity is 631.88 (Table 4).

3.3 Hydrogeological Data

3.3.1 Water Analysis

The three prominent water seepage at the rupture surface was collected and tested in the laboratory. Three prominent water seepages are physically and chemically

Table 4 Direct shear test data of Saron Veng

Pit no	Location	γ_d	ϕ	c	q_{ult}
1	N23° 43.753′ and E092° 43.338′	14.40 kN/m^3	29.54°	2.40 kN/m^2	619.61
2	N23° 43.793′ and E092° 43.363′	17.40 kN/m^3	30.68°	1.00 kN/m^2	752.80
3	N23° 43.800 and E092° 43.37′	18.30 kN/m^3	23.43°	1.50 kN/m^2	524.03

[ϕ = angle of friction; c = cohesion; qult = ultimate bearing capacity]

normal for drinking purposes. But the seepage acts as a lubricant for sliding, added by rainfalls. The water analysis reports are in Tables 5 and 6, respectively.

3.3.2 Aizawl Rainfall Data

According to the Meteorological Centre, Directorate of Science and Technology, Aizawl received 774 mm of rainfall before 12th June 2017, and 180 mm on 13th June 2017, i.e. landslide occurred on this day. 349 mm rainfall within three days before the landslide occurred. The soils are saturated, influenced seepages and act as lubricant for sliding [12, 14] (Laldinpuia et al. 2014) (Fig. 7 and Table 7).

3.4 Lem Analysis

The study areas were analyzed in a LEM package Slide 6.0 using the Ordinary/Fellenius method, Bishop simplified method, Janbu simplified method, Spencer method, GLE/Morgenstern Price Method [13]. Based on the three parameters, viz., unit weight, cohesion (c), and angle of friction (φ) given in Table 2.2, LEM was plotted in the Ordinary/Fellenius method, Bishop's simplified method, Janbu's simplified method, Spencer's method, and Morgenstern and Price's method (Fig. 8). The study area is an active landslide. The factor of safety is less than unity, ranging from 0.372 to 0.391, which is unstable nature (Table 8; [8, 10]).

Table 5 Physical properties of spring water and seepage

Physical properties	BIS specification for drinking water		Spring point	Seepage-I	Seepage-II	
	Requirement limit	Permissible limit	N23°43′ 810″ and E092°43′ 350″	N20°43′ 793″ and E092°43′ 363″	N20°43′ 802″ and E092°43′ 360″	
pH	6.5–8.5		6.21	6.12	6.08	
Odour	Unobjectionable	–	Odourless	Odourless	Odourless	
Taste	Agreeable	–	Tasteless	Tasteless	Tasteless	
Colour	Unobjectionable	–	Colourless	Colourless	Colourless	
Turbidity (in NTU)	1.0	5.0	0.54	1.37	2.16	
Electrical Conductivity (in micro mhos/cm)	500.0	–		115.6 @ 25.2 °C	108.2 @ 25.3 °C	130.6 @ 25.3 °C
Total Dissolved (in mg/l)	500.0	2000.0		57.70	54.10	65.40

Table 6 Chemical properties of spring water and seepage

Chemical properties (*in mg/l*)	BIS specification for drinking water		Spring point	Seepage-I	Seepage-II
	Requirement limit	Permissible limit	N23°43′ 810″ and E092°43′ 350″	N20°43′ 793″ and E092°43′ 363″	N20°43′ 802″ and E092°43′ 360″
P-Alkalinity	–	–	–	–	–
M.-Alkalinity	–	–	40.0	40.0	44.0
Total Alkalinity	200.0	600.0	40.0	40.0	44.0
Total Chloride	250.0	1000.0	10.0	10.0	11.0
Total hardness	300.0	600.0	36.0	36.0	52.0
Total iron	0.3	0.3	0.0	0.2	0.1

Table 7 Aizawl rainfall data in 'mm'

Date	Jan	Feb	Mar	Apr	May	June	July	Aug	Sept	Oct	Nov	Dec
1	0.0	0.0	0.0	0.0	1.1	6.0	3.0	0.0	25.0	0.0	0.0	0.0
2	0.0	0.0	0.0	1.1	2.1	28.0	1.0	0.0	3.0	4.0	0.0	0.0
3	0.0	0.0	0.0	3.5	0.0	24.0	3.0	0.0	3.0	32.0	0.0	0.0
4	0.0	0.0	0.0	64.9	0.0	56.0	22.0	33.0	9.0	0.0	0.0	0.0
5	0.0	0.0	0.0	25.8	0.0	11.0	8.0	4.0	0.0	0.0	0.0	0.0
6	0.0	0.0	0.0	11.7	0.0	0.0	0.5	6.0	0.0	0.0	0.0	0.0
7	0.0	0.0	0.0	0.0	25.0	0.0	12.0	0.0	11.0	12.0	0.0	0.0
8	0.0	0.0	0.0	0.0	0.0	0.0	0.0	1.0	3.0	23.0	0.0	0.0
9	0.0	0.0	0.0	0.0	0.0	4.0	2.0	4.0	41.0	5.0	0.0	5.0
10	0.0	0.0	0.0	0.0	0.0	**46.0**	2.0	1.0	1.0	0.0	0.0	19.0
11	0.0	0.0	18.4	0.0	0.0	**61.0**	10.0	13.0	7.0	0.0	0.0	11.6
12	0.0	0.0	32.4	0.0	0.0	**62.0**	2.0	66.0	2.0	0.0	0.0	2.0
13	0.0	0.0	0.0	0.0	0.3	**180.0**	6.0	3.0	1.0	1.0	0.0	0.0
14	0.0	0.0	0.0	0.0	0.0	2.5	72.0	0.0	22.0	0.0	0.0	0.0
15	0.0	0.0	0.0	0.0	17.0	0.5	38.0	14.0	0.0	0.0	0.0	0.0
16	0.0	0.0	0.0	2.5	39.0	39.0	11	2.0	0.0	0.0	5.0	0.0
17	0.0	0.0	0.0	0.0	0.0	3.5	0.0	0.0	7.0	13.0	1.0	0.0
18	0.0	0.0	0.0	6.2	0.0	15.0	1.0	54.0	7.0	4.0	2.0	0.0
19	0.0	0.0	5.1	0.0	0.0	19.0	2.0	6.0	22.0	15.0	0.0	0.0
20	0.0	0.0	0.0	2.4	14.0	6.0	13.0	3.0	2.0	31.0	0.0	0.0
21	0.0	0.0	11.1	2.2	68.0	0.0	2.0	32.0	0.0	2.0	0.0	0.0
22	0.0	0.0	0.0	1.5	3.0	26.0	1.0	12.0	1.0	41.0	0.0	0.0
23	0.0	1.5	0.0	8.1	4.0	29.0	16.0	13.0	1.0	0.0	0.0	0.0
24	0.0	17.2	0.0	2.5	0.0	13.0	7.0	32.0	0.0	0.0	0.0	0.0
25	0.0	0.0	0.0	3.4	0.0	46.0	15.0	67.0	0.0	0.0	0.0	0.0
26	0.0	0.0	8.7	0.9	0.0	24.0	0.0	40.0	12.0	0.0	0.0	0.0
27	0.0	0.0	0.0	0.0	0.0	24.0	15.0	13.0	0.0	0.0	0.0	0.0
28	0.0	0.0	0.0	0.0	4.0	4.0	2.0	21.0	28.0	0.0	0.0	0.0
29	0.0		0.0	8.0	18.0	13.0	8.0	27.0	18.0	0.0	0.0	0.0
30	0.0		1.1	0.3	3.0	1.0	34	3.0	13.0	103.0	0.0	0.0
31	0.0		5.8		18.0		66.0	9.0		43.0	0.0	0.0
Avg	0.0	0.7	3.1	4.8	7.0	24.8	12.1	15.5	8.0	10.6	0.3	1.2
Total	0.0	18.7	95.9	145.0	216.5	**743.5**	374.5	479.0	239.0	329.0	8.0	37.6

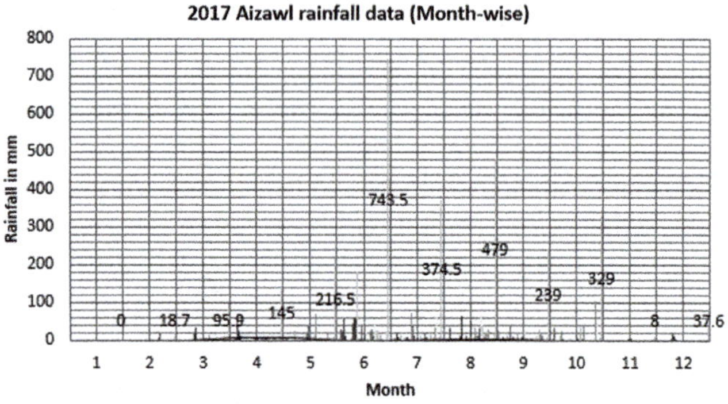

Fig. 7 Graph of Aizawl rainfall data (2017)

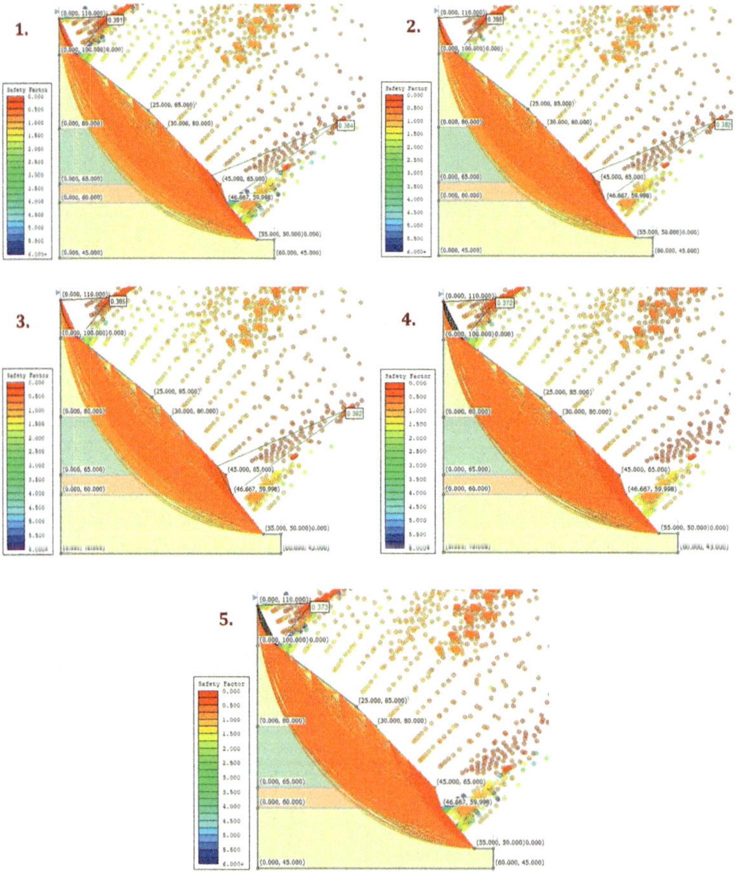

Fig. 8 LEM Analysis of Saron Veng landslide using different methods: 1. *Bishop Method*), 2. *Spencer Method*, 3. *GLE/Morgenstern Method*, 4. *Janbu Simplified Method,* and 5. *Ordinary/ Fellenius Method*

Table 8 Factor of safety for Saron Veng landslide using LEM analysis

Factor of safety (FoS): deterministic value				
Ordinary/Fellini's	Bishop's simplified	Janbu's simplified	Spencer's method	GLE/Morgerstern
0.373	0.384–0.391	0.372	0.382–0.388	0.382–0.385

4 Conclusion

Saron Veng landslide can be classified as a 'debris slide'. The representative soil is coarse sand, soft, and slightly plastic type. It has weak cohesion and an ultimate bearing capacity of 631.88, which shows that it is unsafe for any type of structure. The seepage may act as a lubricant for sliding, which is added by rainfalls. Benching to reduce load and decreasing shear strength, and a proper drainage system are suggested as mitigation measures.

Acknowledgements The authors are grateful to the concerned government department, local authorities, and people of Saron Veng and Tuithiang Veng for their cooperation during field studies and damage assessment of the affected area.

References

1. Aydin A, Basu A (2005) The Schmidt hammer in rock material characterization. Eng Geol 81:1–14
2. ASTM Standard D6913–04 (2009) Standard test methods for practice size distribution (gradation) of soils using sieve analysis. Annual book of ASTM standards, 04.08. Philadelphia (PA): ASTM
3. ASTM Standard D5873- 14 (2014) Standard test method for determination of rock hardness using rebound hammer. Annual book of ASTM standards, 04.08. Philadelphia (PA): ASTM
4. BIS (1985) Indian standard methods of test for soils, part V determination of liquid and plastic limit. IS: 2720 (Part 5)
5. BIS (1996) IS: 2720 (Part 13)—1986 Indian standard methods of test for soils, Part 13 direct shear test (Second Revision), UDC 624: 131.439.5, p 12
6. Coduto DP (1999) Geotechnical engineering: principles and practices. Prentice Hall, New Jersey, pp 125–130, 136–155
7. Kumar S, Singh TN, Sawmliana C (1996) On the instability problems in South Hlimen Landslide, Mizoram. Indian J Eng Mat Sci 4:78–80
8. Laldinpuia (2019) Cut slope stability analysis of Rangvamual landslide along Aizawl airport road, Northeast India: proceedings of the 2nd GeoMEast international congress and exhibition on sustainable civil infrastructures, Egypt 2018—the official international congress of the soil-structure interaction group in Egypt (SSIGE). https://doi.org/10.1007/978-3-030-01941-9_16
9. Laldinpuia (2014) Geological investigation of slumping localities in Saiha Town, Southern Mizoram, North-East India. In: Sassa K, Canuti P, Yin Y (eds) Landslide science for a safer geo-environment, volume 3: targeted landslides. Springer International Publishing Switzerland, pp 407–411

10. Laldinpuia (2015) Geological studies of the rocks at landslide prone localities in Aizawl, Mizoram; Mizoram University (Ph.D thesis)
11. Petrucci O, Gulla G (2010) A simplified method for assessing landslide damage indices. Nat Hazard 52:539–560
12. Raju M (2001) Role of seismicity and rainfall as causative factors of extensive landslides in Mizoram. National seminar on geotechniques and geological hazards in the Indian context, April 24–26, 2011, GSI NER, Pre-seminar volume, pp 165–167
13. RocScience Inc. (2010). Slide v. 6.0- 2D limit equilibrium slope stability analysis. Toronto. Ontario: Canada. www.rockScience.com
14. Shimoma S, Orense RP, Maeda K, Towhata I (2004) Experimental study on landslides caused by heavy rainfall. J Natural Disaster Sci 26(1):15–26
15. Sardana S, Verma AK, Verma R, Singh TN (2019) Rock slope stability along road cut of Kulikawn to Saikhamakawn of Aizawl, Mizoram, India. Nat Hazards 99:753–767
16. Singh TN, Singh R, Singh B, Sharma LK, Singh R, Ansari MK (2016) Investigation and stability analyses of Malin village landslide of Pune district, Maharashtra, India. Nat Hazard 81:2019–2030. https://doi.org/10.1007/si11069-016-2241-0
17. Singh CD, Singh J (2013) Landslides caused due to ignorance- case studies from North-East Inida. J Geol Soc Ind 82:91–94
18. Stead D, Coggan J (2012) Numerical modeling of rock- slope instability. In: Clague JJ, Stead D (eds) Landslide types, mechanism and modeling. Cambridge, New York, pp 144–158
19. The Soil Engineering Sectional Committee (2004) Indian standard classification and identification of soils for general engineering purposes (1st revision, IS: 1498–1970 Edition 2.2). BIS: New Delhi
20. Tiwari RP, Sharma BL, Singh B (1996) Geotechnical appraisal of Bawngkawn landslide, Aizawl, Mizoram. In: Proceedings of international conferences on disaster and mitigation. Madras 1:A4 124–A4 131
21. Verma AK, Sardana S, Sharma P, Laldinpuia, Singh TN (2019) Investigation of rockfall prone road cut slope near Lengpui Airport, Mizoram, India. J Rock Mech Geotech Eng 11(1):146–158
22. Zairemmawii, Lalhlimpuia H, Lalhmingsangi, Vanthangliana V, Laldinpuia, Kumar S (2021) Rockfall analysis of state highway along the Southern Vicinity of Aizawl, Mizoram. Sci Technol J 9:100–112. https://doi.org/10.22232/stj.2021.09.02.13

Impact of Heavy Precipitation on Landslide Due to Climate Change and Probable Remedial Measure

Joyita Golder, Sudipto Halder, and Gupinath Bhandari

Abstract Landslides are the most frequent worldwide natural disaster. In landslide prone area, repetition of landslides may be observed; however, landslides may also occur in a fresh area. In the present study, a fresh landslide was observed during March 2022, at Pangthang, East Sikkim. Analysis through Geospatial Technology, ground survey and soil investigation were carried out for the estimation of landslide susceptibility at concerned location, and it was obtained as non-susceptible; however, the landslide occurred after a prolonged untimely precipitation with high intensity, which may be the Impact of Climate Change. The analysis of the concerned slope was carried out by Bishop (1955) Method, with and without considering the effect of precipitation, and obtained the failure of slope with the effect of precipitation, because of pore water pressure. An attempt was made to provide an eco-friendly and low-cost remedial measure. To reduce the pore water pressure, sand drain has been proposed to arrest the subsurface water and to drain out through the bottom of the slope. The surface of the slope was dressed up and mixed with the vetiver grass seeds, on top of which a layer of Jute Geotextile has been placed, to arrest the surface scouring. The reinforcement of the slope has been done by using Sal-Bullah, as indigenous technology and material, which led to the sub-division of the slope height and reduction in surcharge load.

Keyword Fresh landslide · Untimely precipitation · High intensity · Climate change · Pore water pressure · Sand drain · Sal-bullah · Eco-friendly · Surface vegetation · Scouring · Stable slope

J. Golder (✉) · G. Bhandari
Department of Civil Engineering, Jadavpur University, Kolkata 700032, India
e-mail: joyitag.civil.rs@jadavpuruniversity.in

G. Bhandari
e-mail: gupinath.bhandari@jadavpuruniversity.in

S. Halder
School of Water Resources Engg, Jadavpur University, Kolkata 700032, India
e-mail: sudiptoh.swre.rs@jadavpuruniversity.in

S. Mitra et al. (eds.), *Disaster Management and Risk Reduction: Multidisciplinary Perspectives and Approaches in the Indian Context*,
https://doi.org/10.1007/978-981-99-6395-9_4

1 Introduction

Movement of soil mass in the downward direction, either in case of natural or manmade slopes, is termed as landslides [20]. Landslides are the most frequent worldwide natural disaster. These refer to various types of downward mass movement like, slippage of soil, flow of mud, debris flows, planner block slides, rotational slides, in rugged terrain rock avalanches takes place by catastrophic rock slope failure. These catastrophe causes huge social and economic losses and a large-scale disaster [10, 6, 19, 11, 16]. The hazards and risk assessment related to landslides become a worldwide concern nowadays [1, 10]. The landslides may occur due to various reasons like, seismic action, slope undercutting, putting of surcharge load, and lack of appropriate surface and subsurface drainage system. The landslides may occur due to natural phenomena, anthropogenic activity, and climate change [9]. Due to climate change variation in natural hydrologic system has been observed (IPCC 2007), which is responsible for landslides in hilly terrain. The landslide-prone area may be identified, in general, where the repetition of landslides may be observed,however, landslides may also occur in a fresh area, apart from the prone area, which may be due to climate change. The landslide at Mirik, during July 2015, may be an example of fresh landslide and it was reported that before this specific landslide, the earthquake at Nepal occurred and then the precipitation with high intensity took place in Mirik area. The precipitation during July (Monsoon) is a very usual phenomenon,however, it may be noted that a large-scale earthquake occurred before the above-mentioned precipitation with high intensity, and the landslide occurred after that. It appears that due to earthquake some faults might have been generated within the rock and soil body and due to precipitation with high intensity, huge amount of water could percolate within the rock or soil body through those faults, which have created the pore water pressure. It is well known that the pore water pressure is one of the important responsible parameter for landslide [28].

1.1 Impact of Climate Change

There are three major hydrological factors related to climate change, which have an impact on landslides (Crozier, 2010). The discussed factors are Prolonged and High Intensity Precipitation, Increase in Temperature, and Increase in Wind Speed. With the increase of total precipitation capillary suction of soil decreases and subsequently the effective stress also decreases. Water table may get raised due to high precipitation in totality, causing the reduction of shear strength, and increase in bulk density of soil. The changes in soil characters due to higher precipitation increase the probability of landslides [9]. Due to seepage, the subsurface pore water pressure increases and causes the failure of slopes [8], having different geometry and lithology or soil stratification. The water flow through these natural soil pipes (macro pores) is known as pipe flow. With the increase of rainfall magnitude pipe flow becomes stream flow which can develop pore water pressure and cause instability of slope [27]. With the increase of total precipitation a wetter antecedent conditions will be expected to arise which will have various negative effects on stable slope, like less requirement of rainfall to reach critical condition, development of pore water pressure, rise of water table within the slope which is responsible for reduction of effective normal stress, shear strength, soil suction, and cohesion and besides these it will cause the increase of unit weight of slope material. So it's clear that all these factors will enhance the instability of slope [15, 23]. The pressure exerted by the water molecules, due to the presence of subsurface water after occurrence of the precipitation, by considering the finite element as a tool, was estimated by Cai and Ugaui [4]. High permeability soil may fail within shorter duration rainfall. But in their study they have considered only steady state of seepage and subsequent pore water pressure. When the rain fall intensity increases subsurface drainage rate also increases, seepage, drag forces, particle detachment and piping occurs. Bank scour takes place and removal of lateral and basal support from slopes happens. Chiang and Chang [7] had prepared a model to obtain the Factor of Safety for predicted landslide and tried to assess the worst scenario for the occurrence of landslide, in a mountainous watershed. They have attempted to predict the landslide till the end of the century, by considering the trend of change in precipitation pattern.

Changes in Precipitation. According to Intergovernmental Panel on Climate Change (IPCC) 2007, Global Warming is very clear with the increase of Sea Surface Temperature (SST) and the water vapor in lower troposphere. These two factors are the reason for the occurrence of tropical cyclones (Typhoons). Depending on the historical records and simulation records, it is clear about the relationship between Global Warming and increased typhoon activity and intensity. Typhoons (Tropical Cyclones) and increased rainfall intensity are the result of climate change in the western north Pacific, including Taiwan. With the continuation of Global Warming, landslides are expected to be increased [7]. Intergovernmental Panel for Climate Change (IPCC) is principally working on Climate Change. They have assessed the impact of Climate Change on different physical atmospheric parameters. The large-scale variations in climate events have the potential to cause very large impacts. Changes in precipitation and temperature due to climate change and increase in frequency of heavy precipitation events, increased flood risk, lead to higher rate of soil erosion were mentioned in the climate change assessment of IPCC 2007. The study by Saez et al. [23] discussed about the impact of temporal changes of climatic parameters on the occurrence of landslide, and also predicted the future incidence of landslides, due to the expected climate changes, but the remedial measures were not provided by them. Dixon and Brook [12] discussed about the potential effect of climate change by considering the changing pattern of rainfall, which is responsible for landslide. In their study they had indicated that greater variability of precipitation may have a negative effect on stability of landslide.

Impact of Precipitation Intensity. The excessive precipitation, with high intensity and low duration or low intensity and high duration, the critical value of permeability exceeds and the conditions of slopes may become at higher risk. In case of surface and subsurface drainage systems designed for a lower rainfall intensity, the water flow may take the path on its own, if the actual rainfall intensity becomes excessive, which are partially dangerous and undesired causes various reasons like surface and subsurface erosion, rising of water table, piping, which all causes landslides [15]. The debris flow may also happen due to climate change, which has been studied by Turkington et al. [26]. High-intensity prolonged precipitation or snow melting can cause increased number of landslides [5] in the Mediterranean area. Galeandro et al. [14] have presented a dual permeability model, which simulates the flow of water, due to prolonged low-intensity precipitation, through a fractured swelling soil, overlaying a more permeable soil. The results showed that prolonged low-intensity rainfall may be more harmful than short high-intensity rains, for occurrence of landslides.

1.2 Hydro-geological Studies

The experimental results of Tsaparas et al. [25] said that saturated coefficient of permeability and the precipitation patterns might have significant effect on seepage within a slope of unsaturated soil. Lu and Gobt [17] had presented a framework to study the stability of a slope, under steady unsaturated seepage conditions. The analysis of this study indicated that failure may occur above the water table under steady infiltration. Rounia et al. [22] proposed a model of hydrological and geotechnical interface, which can predict the changes of pore water pressure within embankment and cutting slope. They considered the effects of climate and vegetation on slope deformation and stability. Shan et al. [24] collected various metrological data to investigate the relationship between landslides, pore water pressure at the tailing edge, and the ground temperature. At the tailing edge, a relationship between the slip rate and the pore water pressure, could be identified in case of landslide. Arya et al. [2] studied the effect of crack in a slope and the presence of water within the slope. When water starts percolating through the cracks, it causes pore water pressure. Greater the pore water pressure smaller will be Factor of Safety (FS). If the depth of crack from the surface of slope increases, the FS will decrease. The pore water pressure may be developed due to seepage of precipitated water from the surface to the subsurface, and the pore water pressure is one of the important triggering parameters for slope instability and landslides. Construction in hilly terrain is another important issue while focusing on landslides. The new construction imposes additional surcharge load in the hill slope, which exerted force on vertical as well lateral direction, and the lateral force may increase the probability of landslides. As on date the consideration of impact of climate change has become essential, because direct or indirect impact of the same is responsible for devastating landslides.

2 Study Area

A fresh landslide was observed at Pangthang, East Sikkim (Fig. 1), and hence the present study focuses on this hilly terrain. The geographical coordinates of the location near toe and top are 27.3727°N, 88.5838°E and 27.3724°N, 88.5841°E respectively. Newly constructed buildings at the top of the slope were observed during field visit at the location and a small Hilly Stream locally called Jhora or Nala, is passing through this location, which carries huge amount of water after every heavy precipitation. Field photograph in Fig. 1 shows that an amount of surcharge load coming from the newly constructed buildings, which may be termed as anthropogenic activity, and is imposing a lateral load to initiate the sliding of the land. Also the Jhora have a specific carrying capacity and the excessive flow of water through the Jhora is also a major concern.

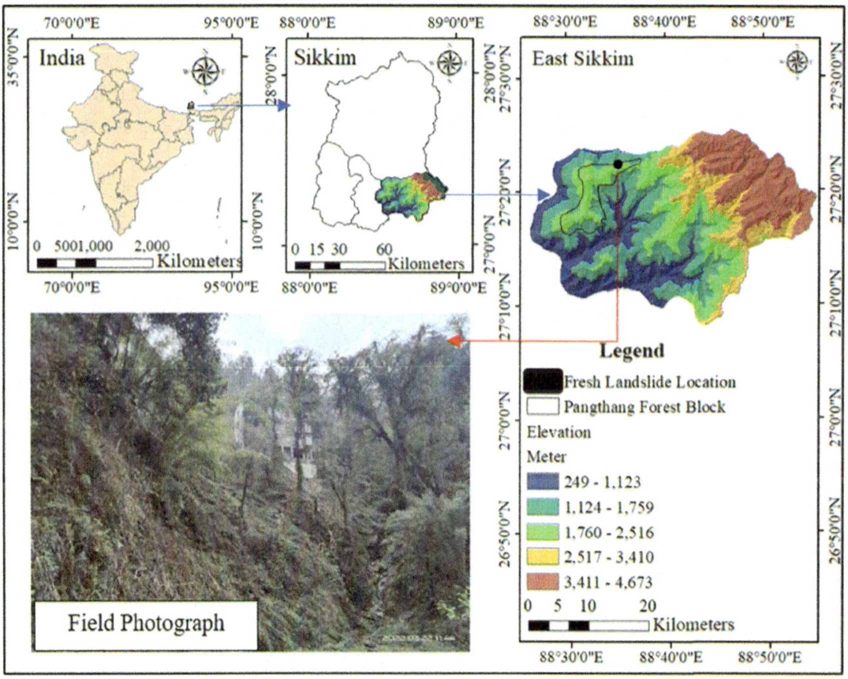

Fig. 1 Study area at pangthang, East Sikkim

3 Field Observation and Data Collection

It has been observed during the visit at field, that the foundation of a recently built super superstructure has become exposed, and a soil mass moved downward in another point, at the middle of the slope, due to landslide, as shown in Fig. 2. The slope under study has an inclination of 48.33°, with a height 26.00 m, as obtained from Contour Map (Fig. 2).

A subsoil investigation was carried out at this study area, to understand the subsoil characteristics. The subsoil profile and the characteristics of subsoil, as presented in Fig. 3, were found to be as follows.

There are two stratums available upto 12.0 m depth, Stratum I (0.00–3.00 m) having the Organic Clay, N Value has observed as 6, having 100% water loss. N value is increasing with depth of Borehole. The parameters of subsoil was obtained as, Grain Size Distribution: Gravel: 24%, Course Sand: 14%, Medium Sand: 34%, Fine Sand: 17%, Silty Clay: 11%; Bulk Density: 15 kN/m^3; NMC: 25%; C: 0.00kN/m^2; $\phi = 27^0$; and Stratum II (3.00–12.00 m) having the Sandy soil with Weathered Granite Gneiss having 100% water loss, and the ϕ value is 32°.

It has also been noted during field visit that before the occurrence of the discussed landslide, a heavy precipitation was noted and huge amount of water started falling from the roof of the buildings which may cause surface scouring.

The trend analysis of 12 years (2009–2020) precipitation data obtained from Indian Meteorological Department, Government of India (IMD), has been carried out, which shows that the rate of untimely precipitation is getting increased (Table 1), which may indicate the Climate Change.

It has already been mentioned that heavy precipitation occurred before the landslide. There is no presence of proper drainage system. After heavy precipitation and improper drainage system water starts percolating through the ground. Additionally

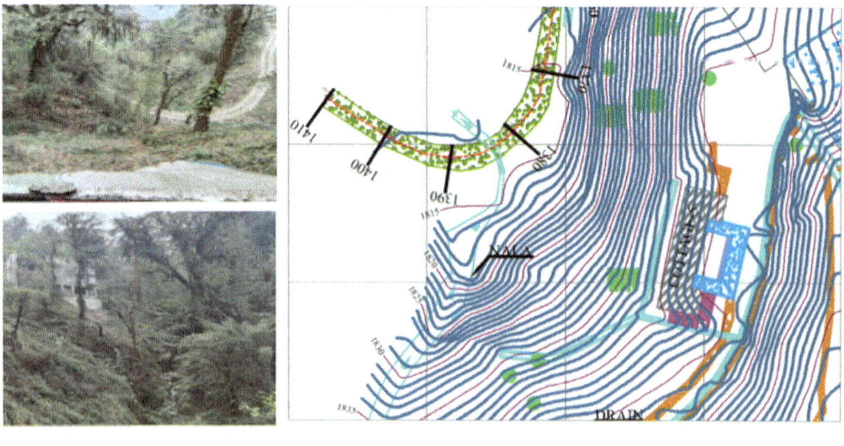

Fig. 2 Field observation **a** Exposed foundation; **b** Landslide at the middle of the slope; **c** Contour map

SITE: Pangthang, East Sikkim								Date: March 25, 2022
DRILLINNG EQUIPMENT: Diamond Drilling with Single tube Barrel								
DRILLING HOLE NO.: 01		**DRILLING DIAMETER:** NX Size		**CASHING DIAMETER:** NX Size			**ORIENTATION:** Vertical Downward	**RL. (m):** 1898m AMSL
RUN (m)		**Depth (m)**	**Log**	**Type of Sample**	**Core**		**Water loss in %**	**Description of Formation**
From	**To**				**Recovery (%)**	**Numbers**		
0	0.5	0.5		Slush	–		100	Organic soil
0.5	1.5	1		Slush	–		100	A horizon, Soil with organic content (Yellowish clay) supplemented by SPT and sample collected from various depth
1.5	3	1.5		Slush	–		0	A horizon, Soil with organic content (Black Soil) supplemented by SPT and sample collected from various depth
3	4.5	1.5		Slush	–		0	Sands (Weathered Lingtse Granitic Gneiss)
4.5	6	1.5		Slush	–		100	Sands (Weathered Lingtse Granitic Gneiss)
6	7.5	1.5		Slush	–		100	Sands (Weathered Lingtse Granitic Gneiss)
7.5	9	1.5		Slush	–		100	Sands (Weathered Lingtse Granitic Gneiss)
9	10.5	1.5		Slush	–		100	Sands (Weathered Lingtse Granitic Gneiss)
10.5	12	1.5		Slush	–		100	Sands (Weathered Lingtse Granitic Gneiss)

Depth	Bulk Density (kN/m^3)	MC (%)	Liquid Limit (%)	Plastic Limit (%)	Grain Size Distribution					Shear Strength Parameters		SPT $(N_1)_{60}$
					Gravel	Sand			Silty Clay	c (kN/m^2)	φ (°)	
						Coarse Sand	Medium Sand	Fine Sand				
1-1.45	17.00	33.4	–	–	30.4	20	37.4	11	1.2	0	27	7
2-2.45	17.34	21.8	–	–	25.6	11.6	38.8	22	2	0	32	64

Fig. 3 Detail results of subsoil investigation

Table 1 Trend of precipitation at East Sikkim, as obtained from IMD for the Period of 2009–2020

Precipitation	2009	2020	% Increase
Average annual precipitation (mm)	190.93	237.53	24.41
Average monsoon precipitation (mm)	311.10	448.23	44.08

a hilly Jhora present just beside the slope was spilled over and the water percolated toward the subsurface. This percolated water may cause the development of pore water pressure. It was observed from the subsoil investigation that the organic clay is prevailing up to 3 m, hence the effect of pore water pressure may create more vulnerable situation, which might be an important reason for the landslide and exposure of foundation. Apart from the impact of pore water pressure, the anthropogenic activity like surcharge load by construction of superstructure may accelerate the landslide.

4 Application of Geospatial Technology

The application of remote sensing and GIS has been carried out to check the landslide susceptibility of the identified location. The parameters studied through Geospatial Technology are Elevation Map, Slope Map, Aspect of Slope, Lithology, Terrain Ruggedness Index and Fault Lines.

4.1 Elevation Map

The most important morphometric attribute for a better understanding of the landforms of the study area is the Elevation Map [21]. The most common methods of landslide hazard assessment using remote sensing heavily depend on 3-dimensional Terrain visualization. An analysis and study on the Digital Elevation Model (DEM), on the Shuttle Radar Topography Mission (SRTM) data (https://earthexplorer.usgs.gov/), having 30 m resolution, was carried out. The passive factors of landslide represent the physical features of the study area associated with landslide occurrences, as those found in the topographical setting. In that sense, given the important role played in the DEM, as data input. A DEM of the study area was built through remotely sensed data obtained from the SRTM. The Elevation map represents variation of altitude of the study area. The whole landscape of study area shows wide variations in elevation, ranging between 1145 m and 2619 m. The Pangthang Landslide was observed in the north eastern part of the study area (Fig. 4).

4.2 Slope Map

The next important morphometric attribute is Slope Map. Slope is defined as the rate of change of elevation at a surface location, measured as an angle in degrees. Development of slope includes all the combined effects of drainage, topography, or surface undulation. A slope angle map has been prepared to analysis of slopes in the study area. SRTM DEM (https://earthexplorer.usgs.gov/) was used to delineate slope in the study area. The varied nature of the topographic expression of the study area has well been quantified by this parameter, which ranges between $< 13°$ and $41.3°$. The map has been classified into five (5) Zones: Zone 1 ($<13°$) having a moderate to low-level steep slope covering alluvial plain, Zone 2 ($13-20°$) having moderate level steep slope covering valleys, Zone 3 ($20.1-27.1°$) having moderate to high-level steep slope covering moderately dissected hill and valleys, Zone 4 ($27.2-34.2°$) having high-level steep slope covering moderate to high-level hilly areas, and Zone

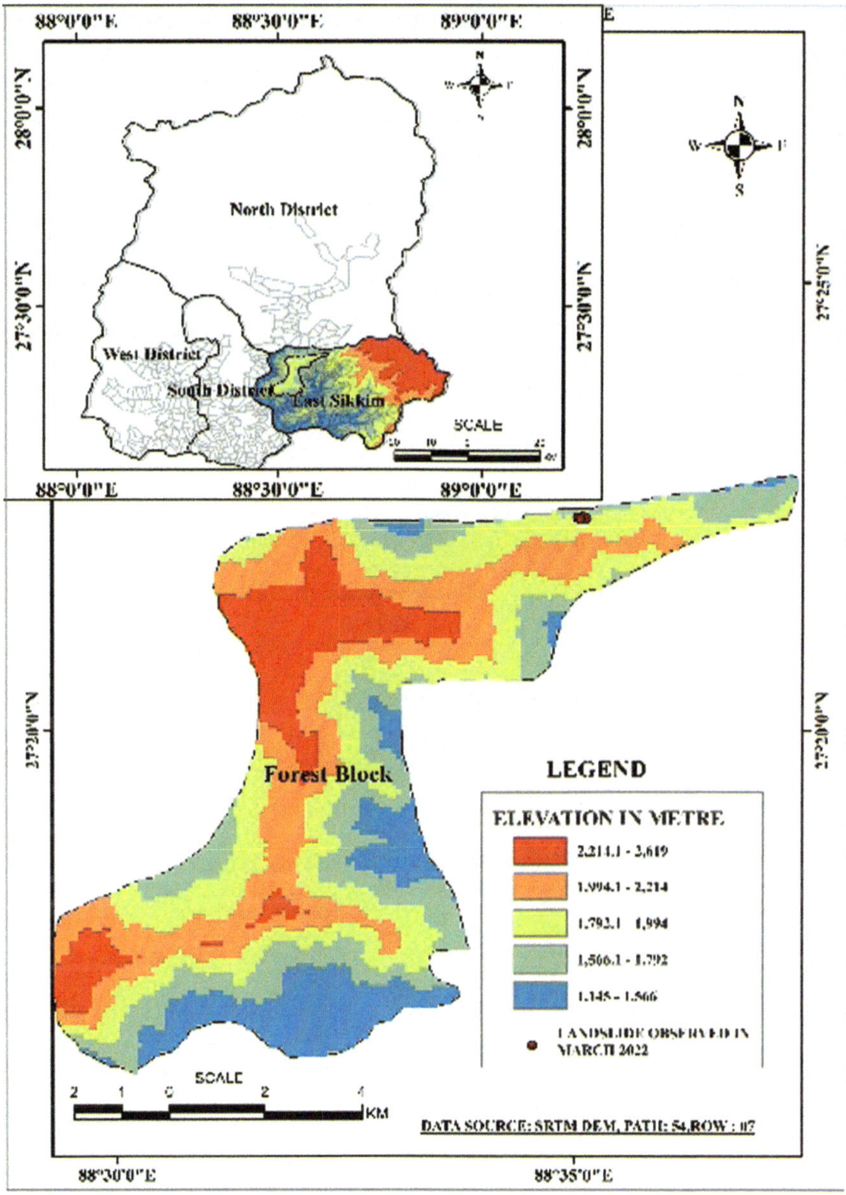

Fig. 4 Elevation map at East Sikkim

5 (34.3–41.3°) having very steep slope covering highly dissected hilly areas. In the north-western part of East Sikkim particularly in uninhabited forest block, the mean figure of average slope is around 30° while the minimum is below 13°, the variability being almost above 70%. Slope is steep on the escarpments. Most of the undulating low lands slope are found in lower slope region along river valley, as shown in Fig. 5.

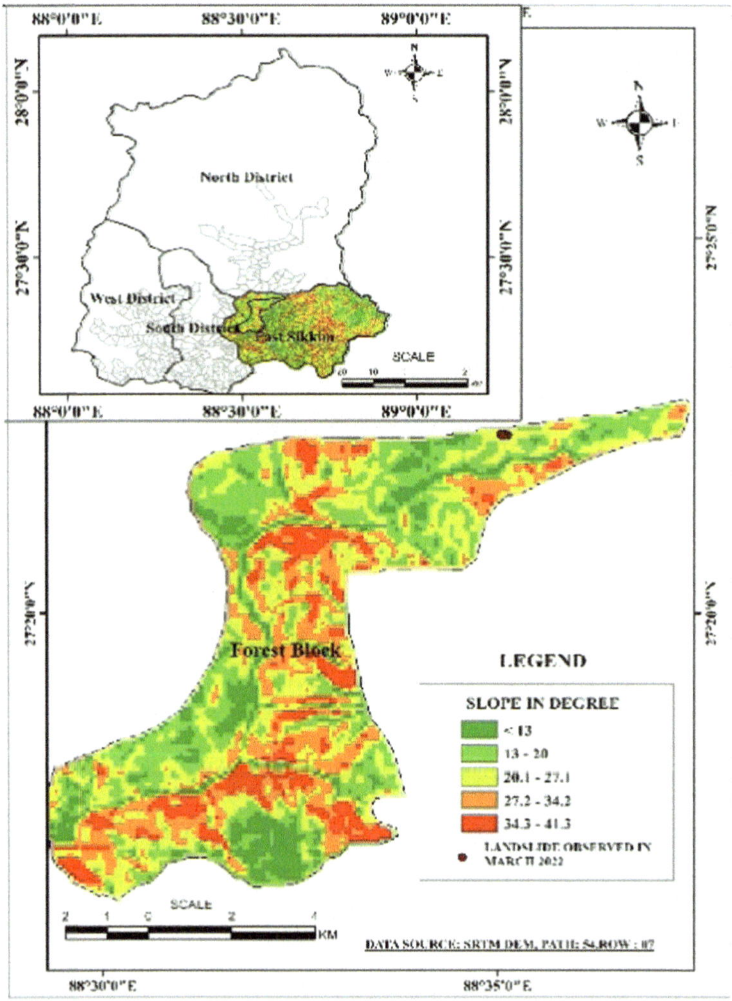

Fig. 5 Slope map at East Sikkim

4.3 Aspect of Slope

The Aspect of Slope can make very significant influences on its microclimate [21]. In the Himalayan region, south facing slopes are warm, wet, and forested but north facing slopes are cold, dry, and glaciated. SRTM DEM (https://earthexplorer.usgs.gov/) was used to delineate slope aspects in the study area. The slope facing in the study area is: North facing 337.5–360, North West facing 292.5–337.5, West facing 247.5–292.5, South West facing 202.5–247.5, South facing 157.5–202.5, South East facing 112.5–157.5, East facing 67.5–112.5, North East facing 22.5–67.5, North facing 0–22.5, as shown in Fig. 6. The Pangthang Landslides were observed in the north facing slope of the study area. This particular area under East Sikkim is essentially controlled by low slope angles, about 30 degrees or below and no evidence of slope instability is observed. Therefore it appears to be free from present and future landslide hazard in normal weather.

4.4 Terrain Ruggedness Index

This deals with detailed surface configuration. It involves more numbers of contours to bring out the details of micro landform development that relate to the present erosional agency. SRTM DEM (https://earthexplorer.usgs.gov/) was used to delineate Terrain Ruggedness Index (TRI) in the study area, and presented in Fig. 7. It is the product of Drainage density and maximum basin relief. An extremely high value of TRI indicates dominancy of both variables and slope is not only steep but also long, considered highly susceptible to landslides, on the other hand lower value of TRI indicates the controlling factors are unlikely to have adverse pressure on the slope stability, and hence the chance of slope failure is minimized by low slope angle. Minimum value of TRI of the study area is 5 and maximum TRI is 84. The Pangthang Landslide is observed in very low TRI zone, and hence this area appears to be non-susceptible to landslide.

4.5 Lithology

The Lithological study for the present study area was carried out by using Bhukosh Data (https://bhukosh.gsi.gov.in/). This study shows the hard strata in the study area, as shown in Fig. 8, which needs very high level of external forces for landslide. The natural external forces may be Seismic force or the subsurface pore water pressure, it may be noted that the anthropogenic activities may also offer external forces. In the present study area there are two forces, like anthropogenic activity and the pore water pressure became active.

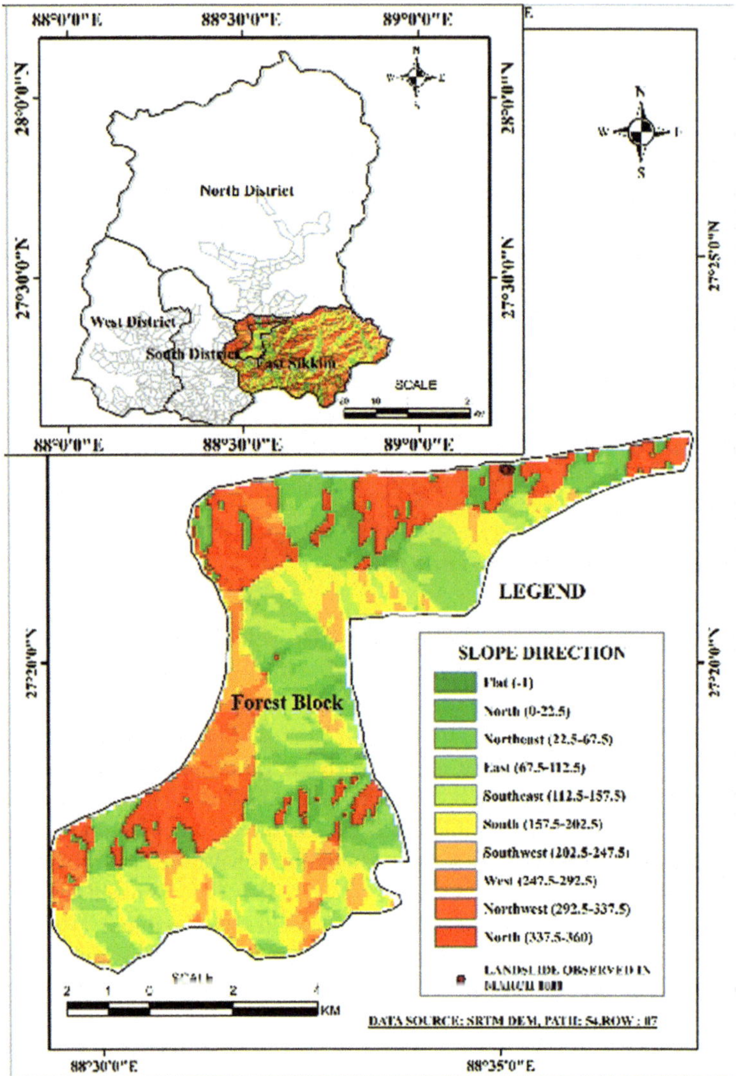

Fig. 6 Aspect of slope at East Sikkim

4.6 Fault Line

The presence of faults has high influence on various mechanical behaviors of rock and soil masses like strength and deformation. Due to the presence of faults, rocks and soils become weak and destabilized by developing the structural discontinuity [13].

Figure 9 shows the map of fault type (https://bhukosh.gsi.gov.in/) in the study area, which indicates that the various fault lines are passing through different locations,

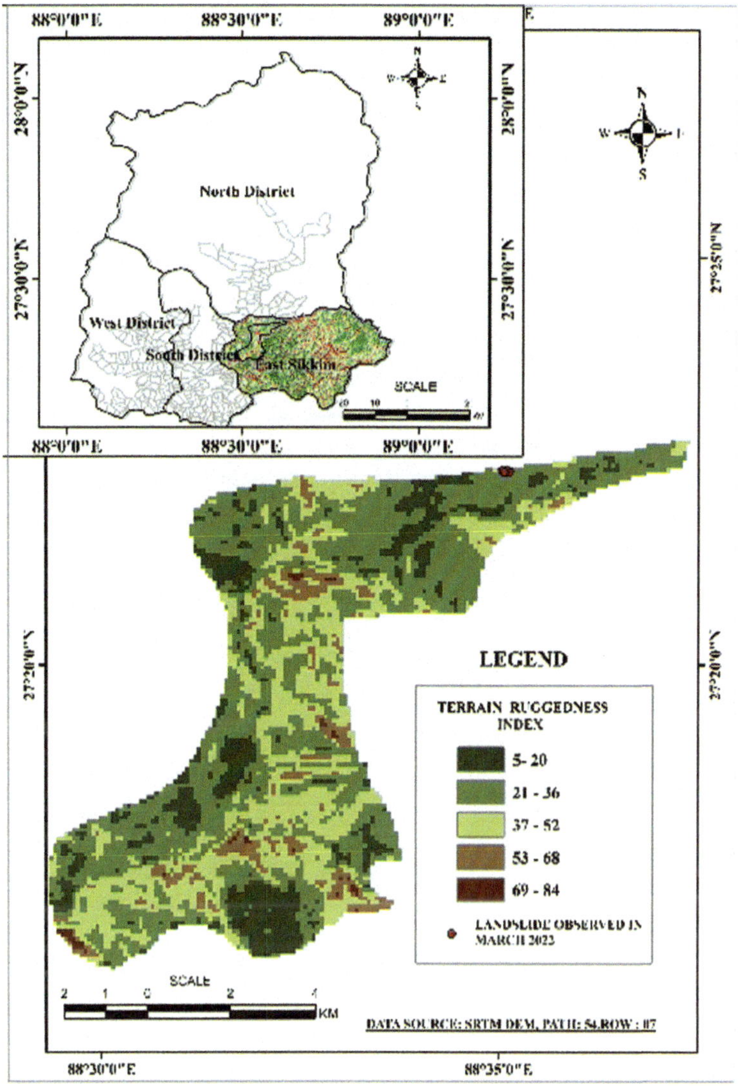

Fig. 7 Terrain ruggedness index map

which are far away from the location under study. One structural discontinuity line passes beside the study area but not passing through the study area. So from the Fault point of view it may be concluded that there is no susceptibility of landslide, in fair weather condition, in the study area.

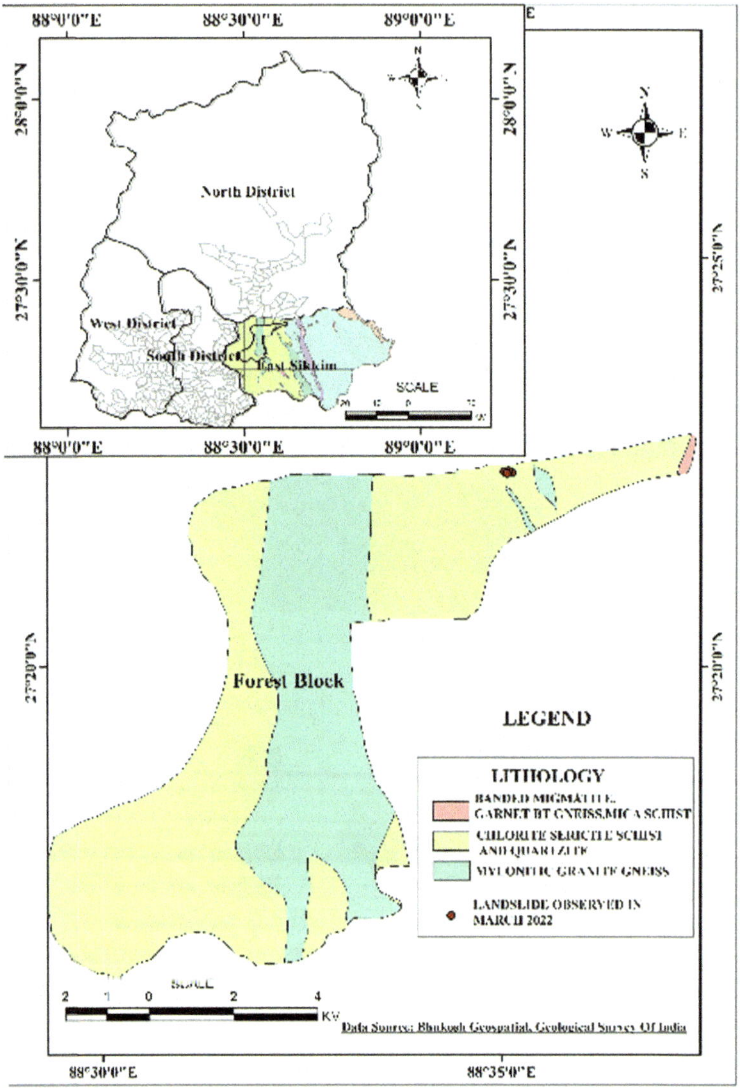

Fig. 8 Lithological map

5 Objectives

The objectives of this study are to stabilize the slope by providing implementable eco-friendly and cost-effective remedial measures by reducing the surcharge load and by considering all the boundary conditions observed during field visit. Human-induced superstructure load and developed pore water pressure appears to be the external parameters, for this landslide, impact of which has been attempted to reduce.

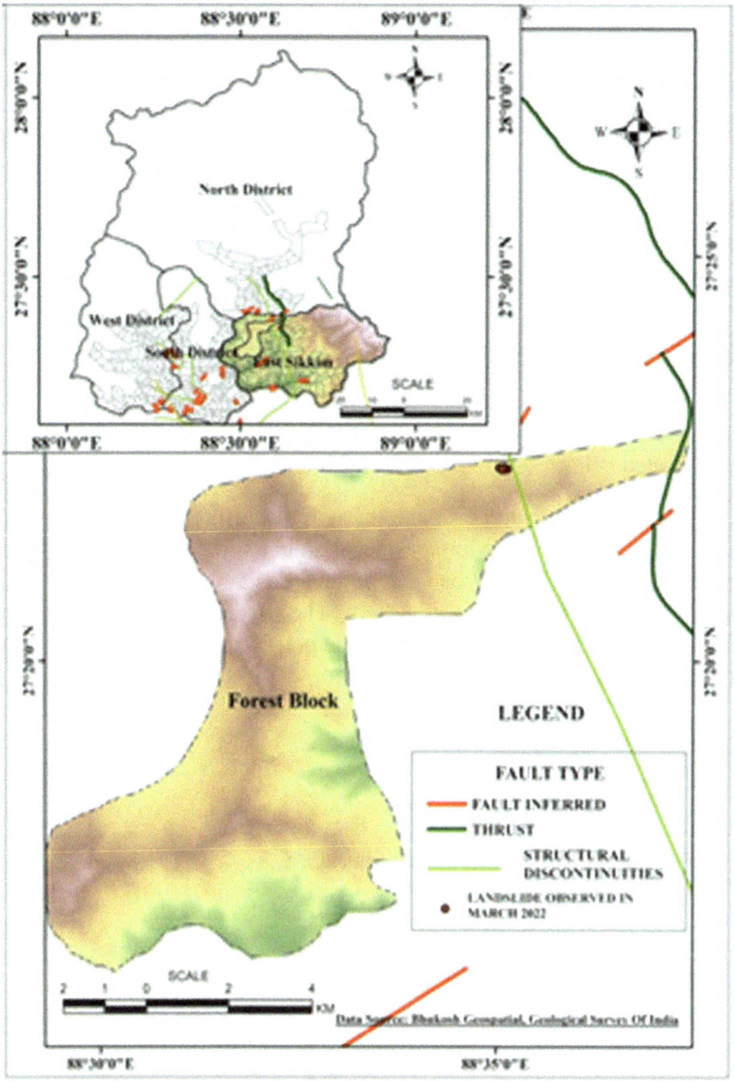

Fig. 9 Fault mapping in the study area

6 Methodology

Study on different landslide causative factors has been carried out to study landslide susceptibility of the concerned location. Contour survey has been carried out and corresponding contour map has been prepared. Geotechnical report has been prepared by carrying out the subsoil investigation. A Remote Sensing analysis has been carried out by taking recent time Satellite Imagery of that particular location, which has shown that the area under study is not susceptible to landslide. Field visit was carried out to observe the fresh landslide. At first analysis of the slope has been carried out through Bishop [3] Method by considering fair weather. As heavy precipitation occurred before the landslide, hence the analysis repeated after considering the effect of higher precipitation in terms of pore water pressure. A further analysis on slope stability was carried out by consideration of implementable, eco-friendly, cost effective remedial measure, to reduce the impact of surcharge load and the pore water pressure.

7 Proposed Remedial Measures

There are three directions of the remedial measure, as shown in Fig. 10, which have been enumerated below:

(1) Attempt has been made to reduce the pore water pressure, and accordingly a remedial measure, by providing the sand drain up to a depth of 3.00 m, has been proposed, which could not be extended beyond that depth, because stratum of Weathered Granite Gneiss starts there. The sand drain has been proposed to arrest the subsurface water to reduce the pore water pressure, and drained out at the bottom of the slope.

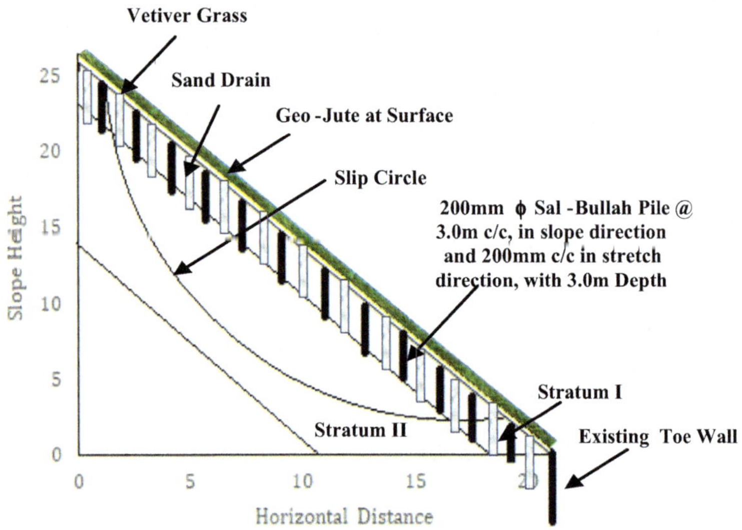

Fig. 10 Slope stabilization by Eco-friendly & Indigenous technology

(2) The surface of the slope was dressed up and mixed with the seeds of vertebra grass, on top of which a layer of Jute Geotextile has been placed, to create the environment for growing up the surface vegetation, required to arrest the surface scouring.
(3) The reinforcement of the slope has been done by using Sal-Bullah, as indigenous technology and material, which led to the sub-division of the slope height and reduction in surcharge load, as well. During pilling of Sal-Bullah, a compaction effect may increase the strength of the prevailing soil strata. Also this reinforcement will be effective to retain the soil against lateral loading and hence the lateral thrust due to anthropogenic activity by construction of superstructure load may be restrained.

8 Results and Discussion

A stable slope could be obtained in extreme weather, while providing the remedial measure as discussed above. The slope was analyzed before and after providing the remedial measure in consideration of impact of Climate Change in terms of heavy precipitation, by using Bishop [3] Method and it was observed that after providing the remedial measure slope became safe. (1) In fare weather condition the Factor of Safety became more than 1.3. It may be noted that the precise value of Factor of Safety, as obtained was 1.39, hence it can be concluded that in fare weather condition the existing slope at Pangthang, East Sikkim, appears to be stable enough. (2) After considering the effect of higher precipitation the factor of safety reduced to 0.93, and slope failure occurred at field also, as shown in Fig. 2. The value of reduced factor of safety is clearly indicating that, for the occurrence of this landslide porewater pressure has huge importance. (3) After providing the remedial measures the slope appears as stable enough because the Factor of Safety increases to 1.48. Surcharge load has been taken by the reinforcement and pore water pressure has been arrested by sand drains, that may be the reason behind the increased factor of safety.

9 Conclusion

The study may be summarized and concluded as follows.

This study focused on a landslide in a hilly terrain. Contour Survey, Subsoil Investigation and study on some landslide causative factors were carried out. It was obtained through Bishop [3] Method, that before the construction of superstructures the slope was stable, in fair weather condition. The existing slope, without anthropogenic activities, like imposing of surcharge load by constructing superstructures, was found to be stable in fair weather condition, and the Factor of Safety was obtained as 1.39.

But after heavy precipitation landslides occurred, a foundation of a building was exposed. A Jhora present just besides the building, and just beside the Jhora at the middle of the slope, it became fail. The stability analysis, through Bishop [3] Method, in consideration with the impact of heavy precipitation, the Factor of Safety was obtained as 0.93. It may be noted that the seismic effect was not considered, because of non-presence of fault line within the study area.

Trend analysis of IMD precipitation data was carried out and it is very clear that the increase of precipitation has taken place in recent years.

Due to heavy precipitation subsurface flow starts and development of pore water pressure occurred, which is the main reason for the landslide, additionally the huge load coming from the superstructure is also very harmful after the landslide.

It is clear after all analysis that this landslide is a fresh landslide and the effect of heavy precipitation or it's an impact of climate change.

The reinforcement of the slope has been done by using Sal-Bullah, as indigenous technology and material, which led to the sub-division of the slope height and reduction in surcharge load, as well.

While the Sal-bullah pilling would be done the soil mass would be compacted, hence the effective shear strength may increase, may be in smaller scale.

Sand drain has been provided to reduce the pore water pressure.

Vetiver grass, on top of which a layer of Jute Geotextile has been placed to arrest surface scouring.

After providing the remedial measures the slopes become completely stable, with a Factor of Safety 1.48.

The probability of further landslides becomes reduced, however a periodic observation and maintenance, has become essential, because the Climate Change is non-linear and so far unpredictable, due to absence of an appropriate model.

Acknowledgements The authors are indebted to the AICTE for providing the Fellowship to carry out the Research Work. The Authors didn't forget to extend their heartiest thanks to Jadavpur University Authority for providing the research amenities, and related infrastructures. Thanks are extended to Classical Paradise Hotels & Resort Ltd., Pangthang, Gangtok, Sikkim, for providing the scope of work in scientific way, for their project.

References

1. Aleotti P, Chowdhury R (1999) Landslide hazard assessment: summary review and new perspectives. Bull Eng Geol Env 58(1):21–44
2. Arya IW, Wiraga IW, Dwipa IGAGS, Pramana IMW (2020) Effect of pore water pressure on soil crack against safety factor of slope stability. In: Journal of Physics: Conference Series 2020, vol 1450, no 1, pp 1–9. IOP Publishing
3. Bishop AW (1955) The use slip circle in the stability analysis of slopes. J Geotech 5:7–17
4. Cai F, Ugai K (2004) Numerical analysis of rainfall effects on slope stability. Int J Geomech 4(2):69–78

5. Cardinali M, Ardizzone F, Galli M, Guzzetti F, Reichenbach P (2000) Landslides triggered by rapid snow melting: the December 1996–January 1997 event in Central Italy. In: Proceedings 1st plinius conference on mediterranean storms 2000. Bios: Cosenzas
6. Chau KT, Sze YL, Fung MK, Wong WY, Fong EL, Chan LCP (2004) Landslide hazard analysis for Hong Kong using landslide inventory and GIS. Comput Geosci 30(4):429–443
7. Chiang SH, Chang KT (2011) The potential impact of climate change on typhoon-triggered landslides in Taiwan, 2010–2099. Geomorphol 133(3–4):143–151
8. Collins BD, Znidarcic D (2004) Stability analyses of rainfall induced landslides. J Geotech Geoenvironmental Eng 130(4):362–372
9. Crozier MJ (2010) Deciphering the effect of climate change on landslide activity: A review. Geomorphol 124(3–4):260–267
10. Dai FC, Lee CF, Ngai YY (2002) Landslide risk assessment and management: an overview. Eng Geol 64(1):65–87
11. Dikshit A, Sarkar R, Pradhan B, Segoni S, Alamri AM (2020) Rainfall induced landslide studies in Indian Himalayan region: A critical review. Appl Sci 10(7):1–24
12. Dixon N, Brook E (2007) Impact of predicted climate change on landslide reactivation: case study of Mam Tor, UK. Landslides 4(2):137–147
13. Duggal SK, Pandey HK, Rawal N (2017) Engineering geology, 1st edn. McGraw Hill Education (India) Private Limited, New Delhi
14. Galeandro A, Doglioni A, Simeone V, Šimůnek J (2014) Analysis of infiltration processes into fractured and swelling soils as triggering factors of landslides. Environ Earth Sci 71(6):2911–2923
15. Gariano SL, Guzzetti F (2016) Landslides in a changing climate. Earth Sci Rev 162:227–252
16. Hewitt K (2009) Rock avalanches that travel onto glaciers and related developments, Karakoram Himalaya, Inner Asia. Geomorphol 103(1):66–79
17. Lu N, Gobt J (2008) Infinite slope stability under steady unsaturated seepage conditions. Water Resour Res 44:1–13
18. Parry ML, Canziani O, Palutikof J, Van der Linden P, Hanson C (eds.) (2007) Climate change 2007—Impacts, adaptation and vulnerability: Working Group II, contribution to the fourth assessment report of the IPCC, vol 4. Cambridge University Press
19. Patton AI, Rathburn SL, Capps DM (2019) Landslide response to climate change in permafrost regions. Geomorphology 340:116–128
20. Ranjan G, Rao ASR (2007) Basic and applied soil mechanics, 3rd edn. New Age International (P) Limited Publishers, New Delhi
21. Rawat MS, Joshi V, Sundriyal YP (2016) Slope stability analysis in a part of East Sikkim, using remote sensing & GIS. In: 2nd International conference on next generation computing technologies 2016. IEEE, pp 51–61
22. Rounia M, Davies O, O'Brien T, Glendinning S (2009) Numerical modeling of climate effects on slope stability. Eng Sustain 162(2):81–89
23. Saez JL, Corona C, Stoffel M, Berger F (2013) Climate change increases frequency of shallow spring landslides in the French Alps. Geology 41(5):619–622
24. Shan W, Xu Z, Guo Y, Zhang C (2015) The impact of climate change on landslides in Southeastern of high-latitude permafrost Regions of China. Front Earth Sci 3(7)
25. Tsaparas I, Rahardjo H, Toll DG, Leong EC (2002) Controlling parameters for rainfall-induced landslides. Comput Geotech, 29(1):1–27
26. Turkington T, Remaître A, Ettema J, Hussin H, Westen CV (2016) Assessing debris flow activity in a changing climate. Clim Change 137(1):293–305
27. Uchida T, Kosugi K, Mizumaya T (2001) Effects of pipe flow on hydrological processes and its relation to landslide: a review pipe flow studies in forested headwater catchments. Hydrol Process 15:2151–2174
28. Wang S, Idinger G, Wu W (2021) Centrifuge modelling of rainfall-induced slope failure in variably saturated soil. Acta Geotech 16(9):2899–2916

Landslide Susceptibility Mapping Using Machine Learning Algorithm

Meghanadh Devara, M. V. Vishwajith, V. K. Maurya, and Ramji Dwivedi

Abstract Landslides are one of the noteworthy threats to human lives and induce huge loss in terms of economical particulars. Landslide is the mass movement of debris, rocks or a slope failure, which occurs due to heavy rainfall, snow melting, earthquakes, etc. So, it is most important to scrutinize and explore the landslide susceptible zones. Accurate Landslide Susceptibility Mapping (LSM) is a pre-requisite for the development of landslide mitigation strategies. Uttarakhand, Sikkim, and Darjeeling in India are witnessing the highest landslide-prone region due to its geographical and topographical conditions. In this research paper, LSM for Sikkim and Darjeeling regions was generated with the help of Support Vector Machine (SVM) learning algorithm. For accurate LSM generation, fourteen landslide causative factors (Elevation, Slope, Aspect, Lithology, Geology, Geomorphology, NDVI, Rainfall, Soil type, Land use Land cover, Distance to roads, rivers, faults and lineaments) have been considered and respective layers were used for the LSM generation of Uttarakhand region. This generated result using Analytic hierarchy process (AHP) is fed to the machine learning model for the prediction of susceptibility mapping. The predicted LSM model is found to be 95% accurate. The proposed technique would be useful for rapid generation of LSM in highly landslide-prone regions.

Keywords Landslides · Weighted overlay analysis · AHP · Landslide susceptibility map · SVM

M. Devara (✉) · M. V. Vishwajith · V. K. Maurya · R. Dwivedi
GIS Cell, MNNIT Allahabad, Prayagraj, Uttar Pradesh, India
e-mail: rgi1703@mnnit.ac.in

M. V. Vishwajith
e-mail: vishwajith.2020gi22@mnnit.ac.in

R. Dwivedi
e-mail: ramjid@mnnit.ac.in

S. Mitra et al. (eds.), *Disaster Management and Risk Reduction: Multidisciplinary Perspectives and Approaches in the Indian Context*,
https://doi.org/10.1007/978-981-99-6395-9_5

1 Introduction

Landslides are found as most risky natural event, causing high amount of losses of lives and huge impact on neighborhood infrastructure in all over the world [1]. The improvement of progressed measures for landslide mitigation and damage control is predicated on the quantitative evaluation of nearby landslide threat that is the potential degree of private and material loss due to detrimental landslide activities [2]. However, this evaluation is complex because of the various variety of environmental and anthropogenic factors concerned, and because the tactics underlying landslide initiation and dynamics are nevertheless poorly understood [3]. Mountainous regions are heavily affected by means of the natural catastrophe of landslides. Like other devastating natural hazards, landslides are vital environmental failures which results damage and destruction to infrastructure, gardens, grasslands, and agricultural fields.

The present study focuses on the Uttarakhand, a state of India. From the survey reports, it is found that in 1998, more than 150 people died due to landslides and a number of villages were affected in the state and more than 4000 causalities were recorded in June 2013 [4]. Landslide threat zones, if accurately highlighted, could be a key to reduce landslide associated losses. There are three types of landslide maps that could be used for better land management, these are landslide inventory map, landslide susceptibility map (LSM) and landslide hazard map. Landslide inventory maps contain all the occurred landslides, LSM highlights landslide potential zones and landslide hazard maps provide detailed information on occurred landslides, types, extent, and failure details.

Landslide susceptibility map (LSM) is considered as an essential measure of landslide management in a landslide-prone region [5]. LSMs could be implemented for minimizing effect of landslide on human kind, infrastructures, road networks, and land-use management [6]. Landslide susceptibility is defined as the probability of landslide occurrences by considering previous cases under the analogous geo-environmental conditions [7]. LSMs could be prepared by various methods, mainly qualitative, quantitative, direct, and indirect. Qualitative methods utilize expert judgment, quantitative used statistical methods to get probability of landside events, direct method considers distribution of unstable spatially, and indirect method utilizes potential landslide causative factors. The chance of landslide prevalence relies upon various conditioning factors as opposed to a single factor. For preparing the LSM, two steps are crucial: Firstly, landslide inventory data which is considered the dependent variable and secondly, landslide conditioning factors which are considered independent variables. Along with above discussed methods, machine learning techniques are highly exploited for LSM preparation. Support vector machine (SVM), random forest (RF), logistic regression (LR), neural networks (NN), etc., conducted a comparison of seven machine learning algorithms and concluded SVM to be in top two best performers along with RF [8].

The objective of this research paper is to prepare a GIS-based dataset of landslide causative factors and to generate landslide susceptibility map using Support Vector Machine (SVM) machine learning model. However many researches have been carried out in the generation of landslide susceptibility map using AHP method by taking not more than 8–10 causative factors. In order to prepare a more robust LSM and its effectiveness in deriving susceptibility of whole state, an increased fourteen thematic layers were used as landslide causative factors for the weighted overlay analysis for the preparation of LSM using AHP method.

2 Study Area and Dataset Collections

2.1 Study Area

In the present study, Uttarakhand, Sikkim, and Darjeeling region are selected as the study area which is the prominent landslide-prone regions of India. Uttarakhand area is characterized by high peaks and deep valleys of elevation range 181–7811 m gradually increasing from south direction to north direction (Fig. 1). It lies between 28° 36′ 57.6″ and 31° 29′ 16.8″ North latitudes and between 77° 3′ 43.2″ and 81° 36′ 57.6″ East longitudes, comprises an area of 53483 km^2. Its extent from east to west of the state is 385 km and its width from north to south is 320 km. This state is covered by various rivers of the Ganges (Ganga) system. It has different varieties of soil which are susceptible to soil erosions. The Northern soil ranges from gravel to stiff clay and the area is highly undulating lesser Himalayan terrain, represented by high ridges/spurs, deep valleys, and abrupt slopes, while southern face is mostly covered by agricultural land.

Sikkim, the mountainous state in India in the North Eastern Himalayas is surrounded in the North and North-East by the Dongkya Range of Tibet, in the South-East by Pangola Range of Bhutan, in the South by Darjeeling Hills, and in the West by the Singalila Range of Nepal. It lies between 27° 1′ 44.4″ and 28° 9′ 46.8″ North latitudes and between 88° 12′ 46.8″ and 89° 0′ 57.6″. Sikkim is about 64 km in East-West and about 113 km in North-South. The elevation in state ranges from 227 to 8252 m (Fig. 1).

The Darjeeling district is located in eastern Himalaya region which is covered by hilly and mountainous terrain having elevation range of 61–3611 m (Fig. 1). It lies between 26° 25′ 15.6" and 27° 14′ 6″ North latitudes and between 87° 51′ 10.8″ and 88° 33′ 46.8″ East longitudes, comprises an area of 2092 km^2. Its extent from east to west of the state is 26 km and its width from north to south is 29 km. Along with triassic rocks, enough variation in soil, huge amount of rainfall, the characteristics of both plain and mountainous topographies exist in this study area.

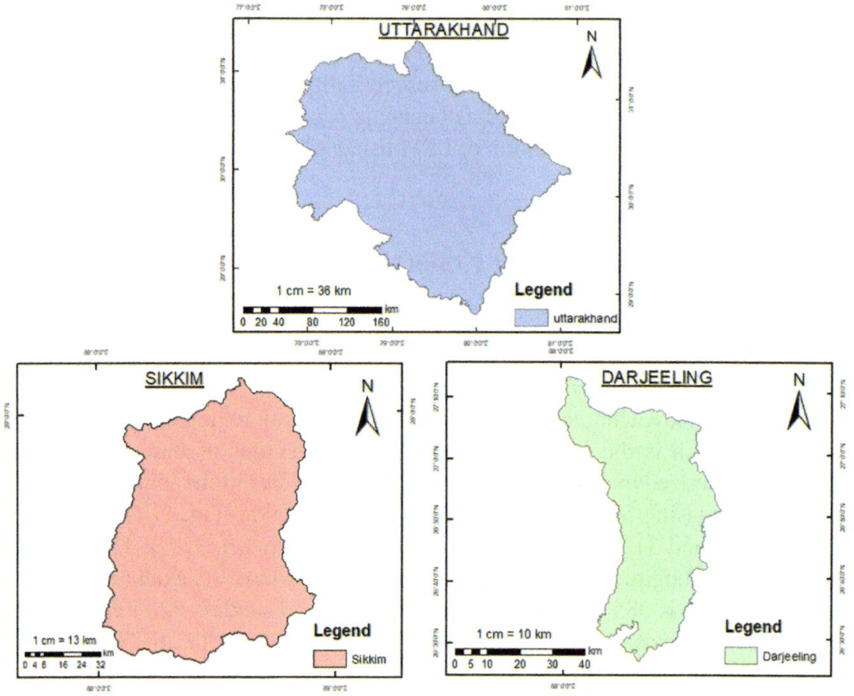

Fig. 1 Boundary map of Uttarakhand, Sikkim and Darjeeling

2.2 *Dataset Collections*

To accomplish the objective of the study, fourteen potential landslide factors are utilized for LSM preparation (Figs. 2, 3, and 4). All the factors along with their source, data type, and resolution are provided in Table 1.

3 Methodology

3.1 *Analytical Hierarchy Process (AHP)*

AHP is a popular Multi-Criterion Decision Analysis (MCDA) technique which is introduced by Saaty in 1980, and has been widely applied successfully during the past 20 years in decision-making systems. Decomposition, comparative judgment, and synthesis of priorities are three principles of the method [10]. These are assigned weights on a nine point ordinal scale (Table 2). Factors are arranged in the form of square matrix. When the factor on horizontal axis is more important than the vertical, this value ranges from 1 to 9 and conversely, this value varies between the reciprocals

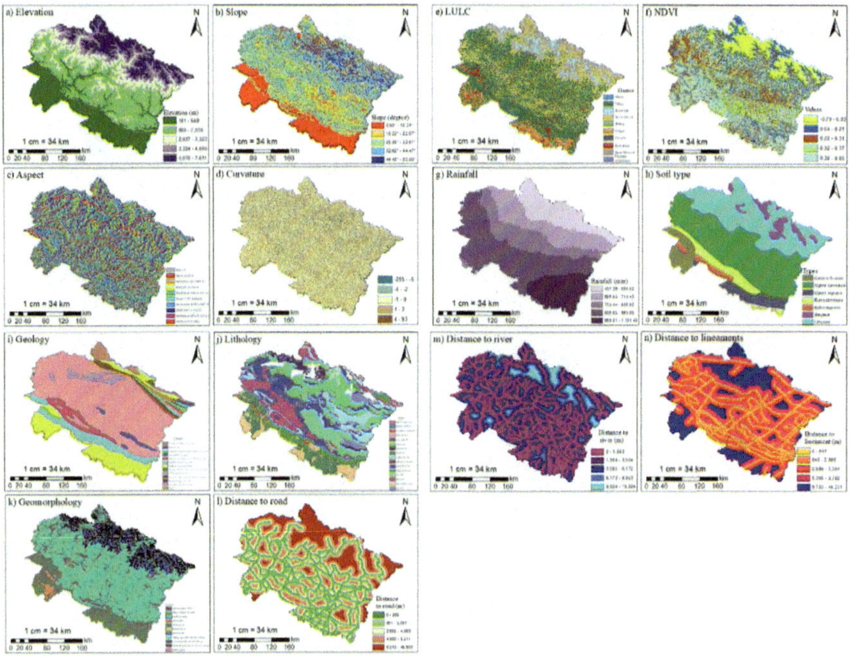

Fig. 2 Fourteen landslide conditioning factors generated for Uttarakhand

1/2 and 1/9 as illustrated in Table 4. Matrix calculation gives weights of factor in terms of eigenvector. The important aspect of the AHP is consistency index (CI) formulated as shown below

$$CI = (\lambda_{max} - n) / (n - 1) \tag{1}$$

where λ_{max} is the maximum eigenvalue and n is the order of the matrix. Calculation of the inconsistency rate is one of the advantages, which makes it possible to reconsider judgments. Consistency ratio (CR) is used to check the consistency comparison matrix, which is calculated as the ratio of consistency index (CI) and random index (RI). RI values depend on number of n factors included as given in Table 3. CR should be <0.1 which is an acceptable level of reliability. Next step is assigning the weights for factors of landslides causative. The weighted overlay is tool available within the GIS environment which is broadly utilized in resolving multicriteria decision-making problems. All criteria weights are multiplied by its allocated factors along with the following equation.

$$LSM = \Sigma(X_i * W_i) \tag{2}$$

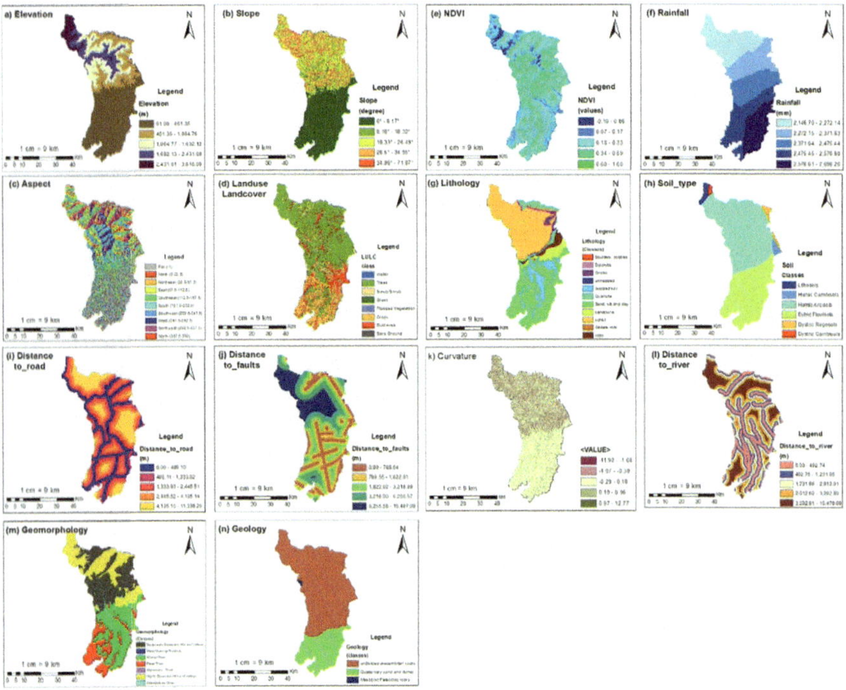

Fig. 3 Fourteen landslide conditioning factors generated for Darjeeling

where; Xi is the level of each rank obtained from reclassification and Wi is the weights of the causative factors obtained from AHP. Landslide susceptibility map is derived by following steps as mentioned in Fig. 5.

3.2 Support Vector Machine (SVM)

SVM, proposed and developed by Vapnik [9], is a supervised learning method that uses an optimal separating hyperplane. SVM splits the information into a high-dimension space and determines optimal hyperplane to classify the data, here landslide and non-landslide. Support vectors are used to generate hyperplane (Fig. 6). SVM utilizes a kernel trick and transforms the data to determine an optimal boundary.

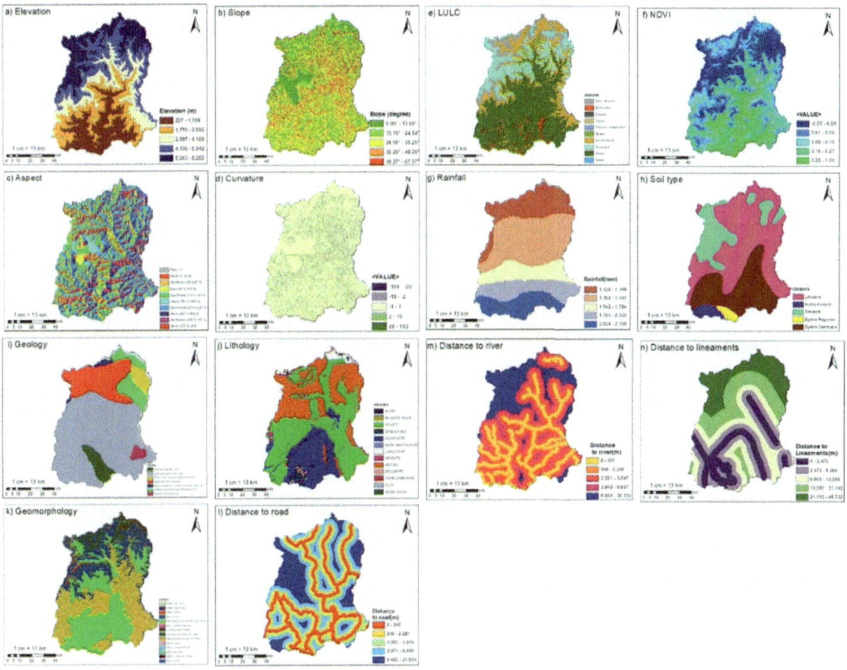

Fig. 4 Fourteen landslide conditioning factors generated for Sikkim

4 Results and Discussions

The proposed methodology has been used for the LSM preparation of Sikkim and Darjeeling. Initially, the LSM for the Uttarakhand is generated using fourteen landslide conditioning factors and AHP is shown in Fig. 7. The criteria weights for all fourteen causative factors are mentioned in Table 5. From this output it can be stated that 15.42% of the total area is at low risk, 56.12% of moderate risk and 28.46% of high risk. The area covered by the low susceptible zone is found to be 7,822.55 km^2, the moderate susceptible zone of 28,469.62 km^2 and high susceptible zone of 14,437.73 km^2. The generated LSM is then validated with NLSM. The pixels from the generated LSM based on AHP of Uttarakhand are further utilized for the training and testing of the SVM model, 80% for training and 20% for testing. The training has been very successful and concluded with 95.7% accuracy, 93.9% precision, and 92.4% recall. Based on notable performance, the trained SVM model has been used to generate LSM for Sikkim and Darjeeling. The output from this SVM model is shown in Fig. 8.

Table 1 Thematic layers used for multi-criteria analysis

Sl. no	Thematic layers	Source	Data type	Resolution/ Scale
a	SRTM DEM	USGS Earth explorer	Raster	30*30 m
b	Slope	Generated from DEM	Raster	30*30 m
c	Aspect	Generated from DEM	Raster	30*30 m
d	Curvature	Generated from DEM	Raster	30*30 m
e	Landuse Landcover (LULC)	ArcGIS ESRI 2020 landcover	Raster	10*10 m
f	Normalized difference vegetation index (NDVI)	Landsat 8 Imagery, USGS Earth Explorer	Raster	30*30 m
g	Rainfall	Climatic research unit data	Raster	0.5*0.5° (55*55 km)
h	Soil type	FAO Digital soil map of world	Vector	1:5 million
i	Geology	USGS	Vector	1:10 million
j	Lithology	Bhukosh GSI	Vector	1:250 K
k	Geomorphology	Bhukosh GSI	Vector	1:250 K
l	Distance to road	Bhukosh GSI	Vector	1:250 K
m	Distance to river	Bhukosh GSI	Vector	1:250 K
n	Distance to lineaments	Bhukosh GSI	Vector	1:250 K

Table 2 Saaty's scale of relative importance

Scales	Degree of preference	Remarks
1	Equally	Two activities contribute equally
3	Moderately	Moderately favor one activity over another
5	Strongly	Strongly or essentially favor one activity over another
7	Very strongly	An activity is very strongly favored over another
9	Extremely	One factor favoured over other in highest degree
2, 4, 6, 8	Intermediate	Compromises between preferences in weights 1, 3, 5, 7 and 9
Reciprocal	Opposite	Used for inverse comparison

Table 3 RI values based on number of included n factors

n	1	2	3	4	5	6	7	8	9	10	11	12	13	14	15
RI	0	0	0.58	0.90	1.12	1.24	1.32	1.41	1.45	1.49	1.51	1.48	1.56	1.57	1.58

Table 4 Pairwise comparison matrix

	B	j	l	e	H	A	d	i	k	g	c	n	m	f
b	1	1	3	2	3	4	5	2	3	9	7	8	9	8
j	1	1	2	2	3	5	3	4	5	6	8	7	8	9
l	1/3	1/2	1	3	3	2	4	6	4	7	2	9	9	8
e	½	1/2	1/3	1	2	3	4	4	3	6	8	5	7	7
h	1/3	1/3	1/3	1/2	1	3	2	2	3	4	5	3	8	6
a	¼	1/5	1/2	1/3	1/3	1	3	1	3	4	5	4	6	8
d	1/5	1/3	1/4	1/4	1/2	1/3	1	2	2	1	3	4	5	6
i	½	1/4	1/6	1/4	1/2	1	1/2	1	3	3	4	3	8	7
k	1/3	1/5	1/4	1/3	1/3	1/3	1/2	1/3	1	2	3	4	2	5
g	1/9	1/6	1/7	1/6	1/4	¼	1	1/3	1/2	1	3	4	3	5
c	1/7	1/8	1/2	1/8	1/5	1/5	1/3	1/4	1/3	1/3	1	2	5	3
n	1/8	1/7	1/9	1/5	1/3	¼	1/4	1/3	1/4	1/4	1/2	1	3	3
m	1/9	1/8	1/9	1/7	1/8	1/6	1/5	1/8	1/2	1/3	1/5	1/3	1	2
f	1/8	1/9	1/8	1/7	1/6	1/8	1/6	1/7	1/5	1/5	1/3	1/3	1/2	1

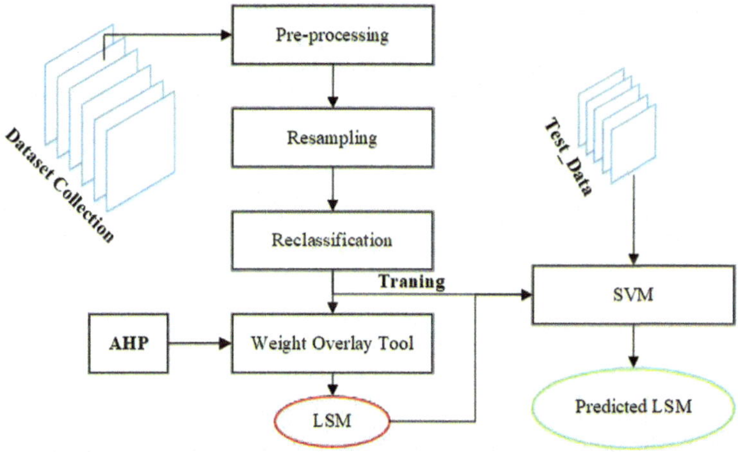

Fig. 5 Flowchart of proposed methodology

5 Conclusions

In this study, we have generated landslide susceptibility maps by machine learning method SVM using fourteen landslide conditioning factors (DEM, slope, aspect, curvature, LULC, NDVI, lithology, geology, geomorphology, rainfall, soil type, distance to river, distance to road and distance to lineaments). The training of the model is done using landslide pixels extracted from one of the most landslide-affected

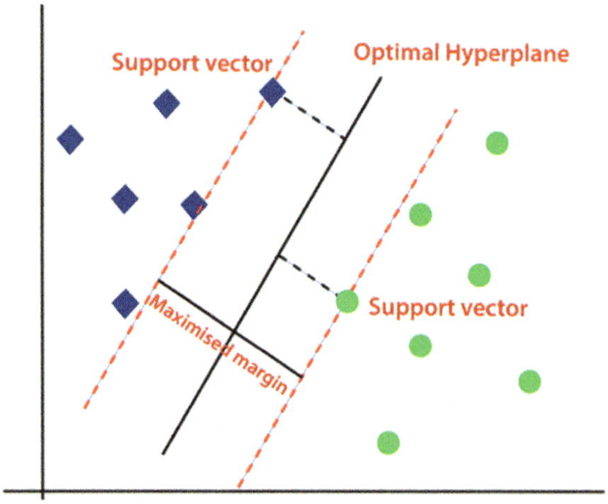

Fig. 6 SVM

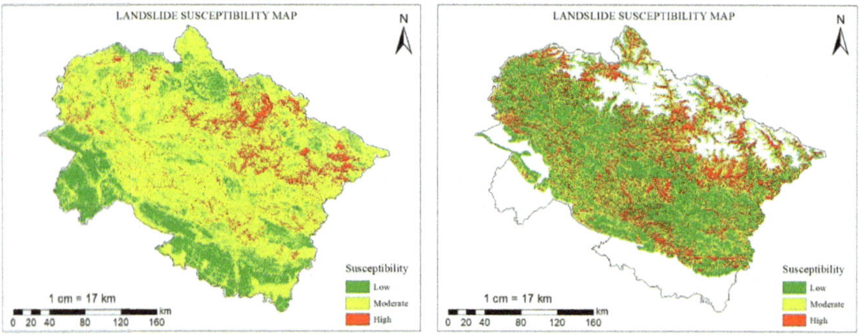

Fig. 7 LSM generated by AHP (left) and NLSM (right)

districts, Uttarakhand. The model achieved 95.7% accuracy, 93.9% precision, and 92.4% recall. Further, the trained model is utilized to predict LSM for Sikkim and Darjeeling. The proposed research work lays a foundation of rapid LSM generation using potential landslide conditioning factors. Further, more machine learning models could be explored to conduct a comparison study delivering the best model for rapid LSM.

Table 5 Criteria weights for thematic layers

Causative factors	Criteria weights
Slope	0.184
Lithology	0.160
Distance to road	0.142
LULC	0.115
Soil type	0.079
Elevation	0.072
Curvature	0.054
Geology	0.058
Geomorphology	0.039
Rainfall	0.030
Aspect	0.025
Distance to lineaments	0.019
Distance to river	0.012
NDVI	0.011

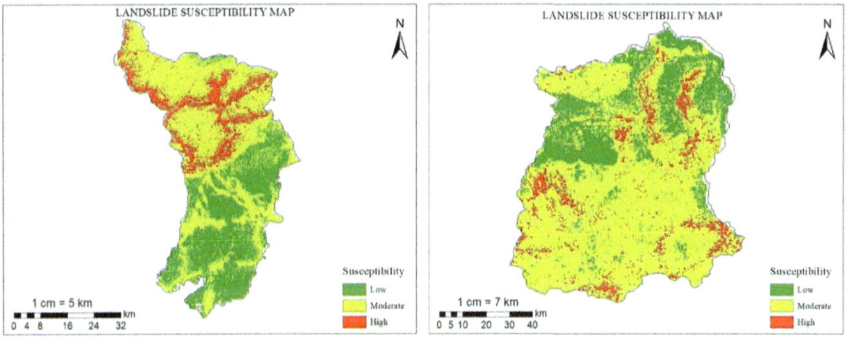

Fig. 8 LSM generated by SVM for Darjeeling (left) and Sikkim (right)

References

1. Guzzetti F, Mondini AC, Cardinali M, Fiorucci F, Santangelo M, Chang KT (2012) Landslide inventory maps: New tools for an old problem. Earth Sci Rev 112(1–2):42–66
2. Kayastha P, Dhital MR, De Smedt F (2013) Application of the analytical hierarchy process (AHP) for landslide susceptibility mapping: A case study from the Tinau watershed, west Nepal. Comput Geosci 52:398–408
3. Sardooi ER, Azareh A, Mesbahzadeh T, Sardoo FS, Parteli EJ, Pradhan B (2021) A hybrid model using data mining and multi-criteria decision-making methods for landslide risk mapping at Golestan Province, Iran
4. Satendra AK, Gupta VK, Naik TK, Roy S, Sharma AK, Dwivedi M (2015) Uttarakhand disaster 2013. National Institute of Disaster Management, New Delhi

5. Roy J, Saha S, Arabameri A, Blaschke T, Bui DT (2019) A novel ensemble approach for landslide susceptibility mapping (LSM) in Darjeeling and Kalimpong districts, West Bengal, India. Remote Sens 11(23):2866
6. Ozer BC, Mutlu BEGÜM, Nefeslioglu HA, Sezer EA, Rouai M, Dekayir A, Gokceoglu C (2020) On the use of hierarchical fuzzy inference systems (HFIS) in expert-based landslide susceptibility mapping: the central part of the Rif Mountains (Morocco). Bull Eng Geol Env 79(1):551–568
7. Roslee R, Mickey AC, Simon N, Norhisham MN (2017) Landslide susceptibility analysis (LSA) using weighted overlay method (WOM) along the Genting Sempah to Bentong Highway, Pahang. Malays J Geosci (MJG) 1(2):13–19
8. Ngo PTT, Panahi M, Khosravi K, Ghorbanzadeh O, Kariminejad N, Cerda A, Lee S (2021) Evaluation of deep learning algorithms for national scale landslide susceptibility mapping of Iran. Geosci Front 12(2):505–519
9. Cortes C, Vapnik V (1995) Support-vector networks. Mach Learn 20:273–297

Design of Horizontal Drains for Chanmari Slope Stabilization at Sikkim

D. Chatterjee, O. S. Charan, and P. Sharma

Abstract Landslides are common in the North-Eastern Himalayan state of Sikkim. Landslide is a serious geo-environmental disaster causing loss of life and property. Short duration high-intensity rainfall is one of the major factors driving these slope instabilities. Out of several such landslides, the Chanmari landslide has been susceptible to major mass movements for many years and rainfall infiltration is the triggering factor for these slope movements. To mitigate landslide disasters, stabilization of the slope is of utmost importance. Installation of horizontal drains within the slope is an effective way of stabilizing the slope by reducing the pore water pressures. Horizontal drains have already been successfully installed in residual soil slopes. A numerical study was carried out to understand the effect of horizontal drains on the Chanmari slope under rainfall infiltration. Parametric studies were conducted to find out the best possible layout of the drains. The slope failed within a few minutes due to the high-intensity rainfall. The drain installed at the middle of the slope kept the minimum factor of safety of the slope higher than the slope without a drain by lowering the water level within the slope. Multiple drains placed at the middle proved to be an effective drainage system.

Keywords Landslide · Rainfall · Factor of safety · Horizontal drains

1 Introduction

Some parts of Sikkim receive heavy rainfall in the monsoon season which leads to landslides causing heavy loss of life and property. When the rainfall infiltrates into the slopes it leads to a decrease in shear strength and matric suction. Heavy rainfalls cause the formation of perched water tables inside the slopes making them prone to failure. Therefore, utmost priority should be given to tackle these disasters. Stabilization of the slope is one way of addressing landslide problems. The stabilization measures that

D. Chatterjee (✉) · O. S. Charan · P. Sharma
Department of Civil Engineering, National Institute of Technology Sikkim, Ravangla, India
e-mail: dooradarshi@nitsikkim.ac.in

S. Mitra et al. (eds.), *Disaster Management and Risk Reduction: Multidisciplinary Perspectives and Approaches in the Indian Context*,
https://doi.org/10.1007/978-981-99-6395-9_6

are carried out commonly in Sikkim include retaining walls and geometry profiling but these methods cannot alone stabilize the slope in the long run. The build-up of excess water within the slope should be minimized as much as possible if not avoided completely. This is where the use of horizontal drains comes in. Horizontal drains have been found to reduce groundwater levels and this has been confirmed by both field and numerical studies. While they are not able to completely drain out all the water from the slope, they can reduce the groundwater level and keep it below a certain depth throughout the duration of the rainfall.

Gjetvaj and Nossan [7] proposed that for the design of horizontal drains for slope stabilization it is necessary to define the location of the drains with respect to the slip surface, i.e., their elevations, length and spacing, as well as the time required for the stabilizing effects to take hold. A numerical study was done by Mukhlisin and Aziz [9] to understand the effect of horizontal drains on slope stability of residual soil slopes. Soils with high permeability proved to be the best material for horizontal drains as they facilitated quick flow of water. It revealed that when horizontal drains were installed within a slope the groundwater levels dropped immediately. Similar findings were also reported by Rajhardo et al. [10]. They have also shown that the drains located at the bottom portions of the slope are more effective because they attract the majority of the groundwater. Drains located at the top regions are not that effective. They have also stated that rainfall infiltration is limited to a certain depth below which the infiltration becomes insignificant. The results of the field measurements done by Rahardjo et al. [11] indicated that horizontal drains were indeed effective in reducing the water table and stabilizing the slope. The numerical models analyzed here indeed show that the groundwater level is reduced after installing horizontal drains. Martin and Siu [8] did field studies and their analysis of piezometric and drain discharge data showed that the drains at the eight selected sites have caused significant lowering of main groundwater levels.

2 Problem Statement

The Chanmari landslide location (Fig. 1) consists of residual soil in Gangtok, Sikkim. In the months of July and August the site receives heavy rainfalls which leads to recurring landslides. A typical slide occurred on June 8 in the year 1997. A 42 mm/hr rainfall intensity led to the disaster killing eight people and considerable loss of property [1]. High-intensity rainfalls of short durations lead to excess build-up of pore water within the slope leading to high groundwater levels which cause instability. The soil type being silty sand facilitates quick infiltration because of its high hydraulic conductivity. The slope also has high deposits of mica which occur as thin sheet structures and have a low frictional resistance which is further reduced due to the presence of water. Thus, the lowering of groundwater levels is of utmost importance to make the slope stable.

Fig. 1 View of the Chanmari landslide [1]

3 Methodology

First the seepage analysis was performed using the finite element method in Seep/W. The pore water pressures obtained from the seepage analysis were used in the stability analysis. The stability analysis was performed utilizing the limit equilibrium method in Slope/W.

3.1 Horizontal Drain Design

There are several ways to design horizontal drains. One method is installing horizontal drains to reduce the pore water pressures to achieve an optimum factor of safety. Here design calculations are carried out based on a model of how the drains will affect the groundwater regime. In another method drains reduce pore water pressure to provide some additional but unquantifiable support to the slope. Normally design calculations do not take into account the effect of drains. The common method of designing involves designing tentative drain locations and varying other parameters such as lengths, spacing based on empirical knowledge and the final layout is determined from assessment of the site conditions encountered during installation [8].

3.2 Seepage Analysis

The equation for a two-dimensional transient water flow is as follows [4]

$$m_w^2 \gamma_w \frac{\partial h_1}{\partial t} = \frac{\partial}{\partial x}\left(-k_{wx}\frac{\partial h_1}{\partial x}\right) + \frac{\partial}{\partial y}\left(-k_{wy}\frac{\partial h_1}{\partial y}\right) + q \tag{1}$$

where m_w^2 is the slope of the soil water characteristic curve, γ_w is the unit weight of water; h_l is the hydraulic head or total head; t is the elapsed time; k_{wx} is the coefficient of permeability with respect to water as a function of matric suction in x direction; k_{wy} is the coefficient of permeability with respect to water as a function of matric suction in the y direction and q is the applied rainfall flux. Equation 1 was solved in Seep/W software using finite element method.

3.3 Stability Analysis

The shear strength equation used for the stability analysis of partially saturated soils incorporating the effect of matric suction is as follows [3]

$$\tau = c' + (\sigma_n - u_a)\tan\phi' + (u_a - u_{\mathrm{w}})\tan\phi^b \tag{2}$$

where τ denotes shear strength of partially saturated soil; c' denotes effective cohesion; ϕ' denotes effective friction angle, $(\sigma_n - u_a)$ denotes net normal stress; σ_n denotes total normal stress; u_a denotes pore air pressure; $(u_a - u_w)$ denotes matric suction; u_w denotes pore water pressure; and ϕ^b denotes the angle representing the rate of increase in shear strength with respect to the matric suction. Equation 2 represents a linear failure envelope. Bishop's simplified method was adopted in the stability analysis on the basis that it takes less time to compute the factor of safety as compared to more rigorous methods. Furthermore, Fredlund and Krahn [2] also showed that this method is capable of obtaining factors of safety almost as accurately as other rigorous methods.

4 Materials

The soil found in Chanmari slope was classified as medium grained sandy soil mixed with boulders and mica gneiss [1]. The soil had a unit weight, γ of 16 kN/m^3, effective cohesion, c' of 0 kPa, effective angle of internal friction, ϕ' of 32.1°, saturated permeability, k_s of 3.625×10^{-6} m/s and ϕ^b of 32.1°. The soil water characteristic curve (Fig. 2) was generated from Seep/W based on the saturated permeability of

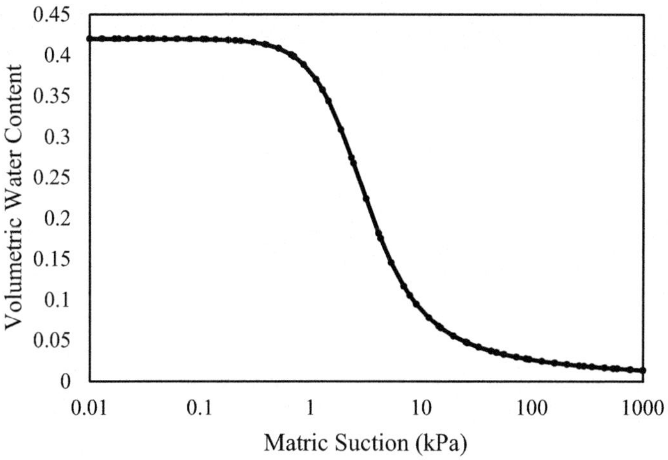

Fig. 2 Soil water characteristic curve

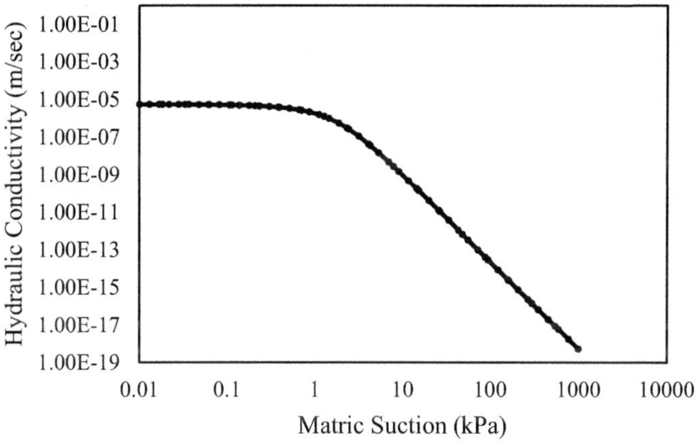

Fig. 3 Permeability function

the sandy soil and the permeability function was established from the soil water characteristic curve as shown in Fig. 3.

5 Numerical Model

The model of the Chanmari slope is shown in Fig. 4. The average inclination of the Chanmari slope is 31° with a total height of 250 m. The models were prepared in GeoStudio software using Seep/W and Slope/W. The finite element model with five

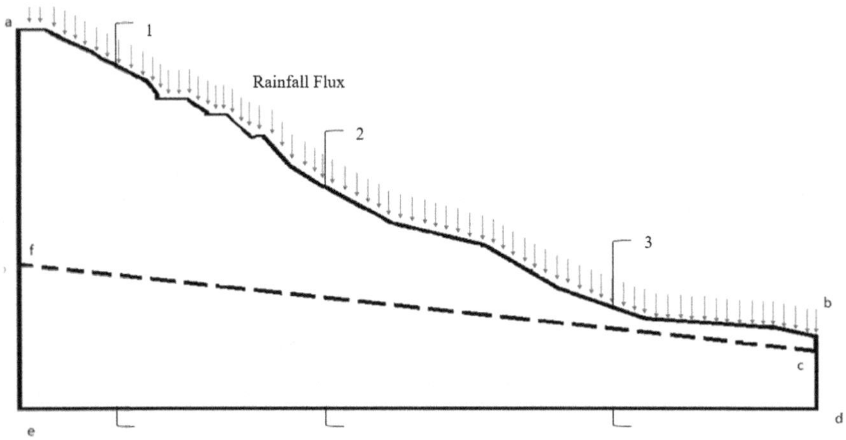

Fig. 4 Chanmari slope configuration [1] showing the different sections

drains at the mid portion of the slope is illustrated in Fig. 5. The rainfall was modeled as a water flux of intensity 0.949 mm/hr and applied to the region face (ab) of the slope. The rainfall flux was applied to the slope surface for 5 h. The rainfall data was collected from the Indian Meteorological Department for the east Sikkim region for the month of June. A no-flow boundary condition was given to the base of the slope and to the sides above the water table (af, bc, and ed). A constant water head of 100 m and 40 m was applied to the left and right (ef and cd) boundaries of the slope to simulate the effect of an initial water table. The drains were modelled with a water rate of 0 m^3/sec. The permeability of the drain material was selected to be 2 $\times$ 10^{-4} m/s. The soils having a high permeability make good material for horizontal drains as the flow is very rapid in these kinds of soils.

6 Results and Discussion

The seepage and stability analysis with both drains and without drains were performed. First the results are presented for the slope without drains. Then the results are represented for the slope with drains with varying parameters like different drain lengths, positions, and number of drains. The results are presented in terms of pore water pressures and factor of safety values obtained from the seepage analysis and stability analysis respectively.

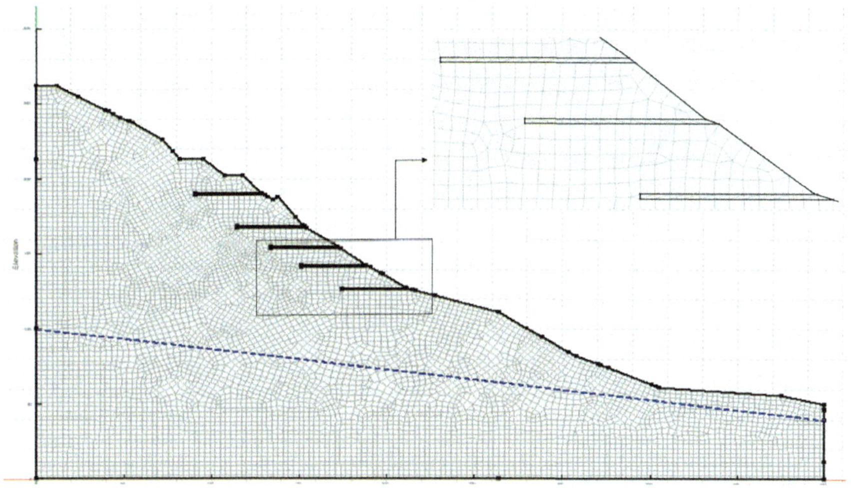

Fig. 5 Finite element model with the horizontal drains

6.1 Stability Without Drains

The Chanmari slope model was analyzed without including drains and the factor of safety values was obtained during the rainfall event. The seepage analysis results showing the pore pressures and the groundwater level after 10 h have been illustrated in Fig. 6. It is observed that the water level falls down considerably when a single drain is installed at the middle of the slope while for the slope without drain the water level follows the slope surface. This suggests that if a proper drainage is designed for the slope, it will be able to effectively stabilize the slope by reducing the groundwater level and thereby the pore water pressures.

Further study confirms the reason for using horizontal drains to stabilize a slope. Figure 7 illustrates the stability of the slope with and without drains under rainfall infiltration. A drain of 40 m length was placed at the middle of the slope.

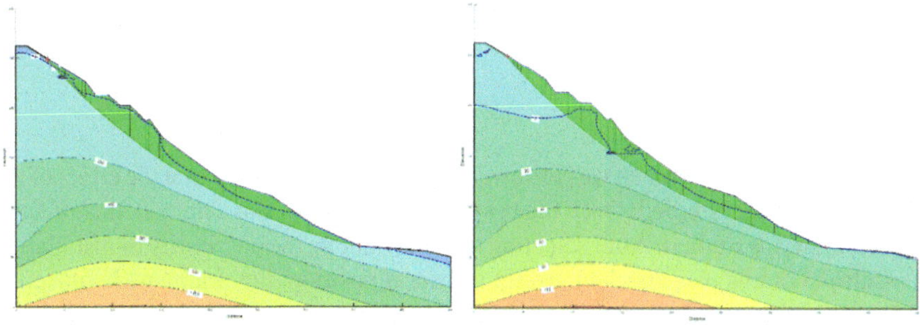

Fig. 6 Groundwater level (represented by dotted lines) without and with drain after 10 h

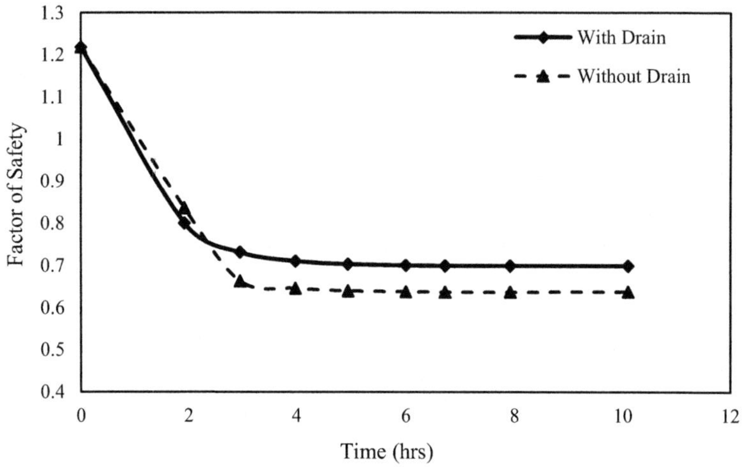

Fig. 7 Change in the stability of the slope with drain (middle, 40 m) and without drain

The minimum factor of safety of the slope decreased by 9.7% more for the slope without a drain. The study clearly shows the reason why drains should be installed in the Chanmari slope to reduce the groundwater level and stabilize it during rainfall periods.

Three different rainfall intensities were applied ($q = k_s$, $q < k_s$, and $q > k_s$ where q is the rainfall flux and k_s is the saturated permeability of the soil). The variation of the factor of safety with time for different rainfall intensities has been shown in Fig. 8. As rainfall starts the factor of safety starts to decrease due to reduction of matric suction as a result of pore water pressure increase. It is observed that the curves for $q = k_s$ and $q > k_s$ coincide with each other which indicates toward a threshold rainfall intensity. Any more infiltration beyond this point is insignificant as the water table is already at maximum level. For all the three cases, the factor of safety goes below one pointing toward failure but stays constant after the stoppage of rainfall. As rainfall stops, the matric suction doesn't decrease anymore and the water level starts to recede.

6.2 *Stability with Drains*

The most common approach was followed to design the drains. At first the effect of the drain position was studied. Once the best possible position was selected, the effect of drain length was studied. After selecting the drain length, the possibility of installation of multiple drains was checked. To fix the drain position, the drains were placed at different positions to compare their factor of safety values. The results showed that drains located in the middle region proved to be most effective in lowering the water table as it intercepted most of the infiltration. The drains located at the top however

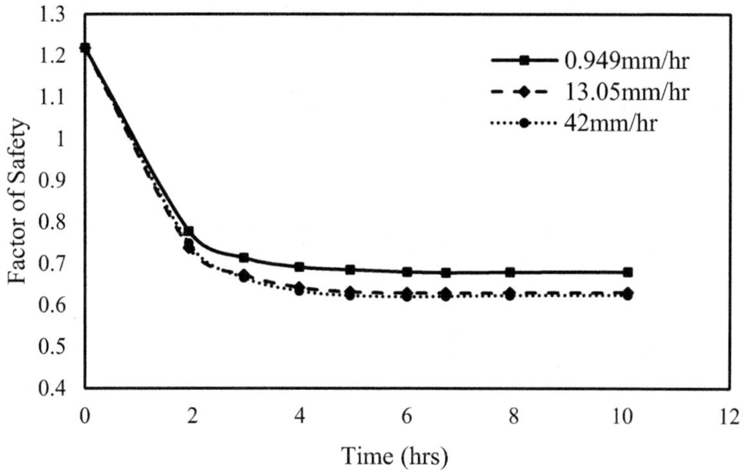

Fig. 8 Variation of factor of safety with rainfall intensity

proved to be ineffective and did not have much impact on the groundwater levels. The pore water pressure graphs after 10 h at Sect. 1 are shown in Fig. 9. It was observed that after a period of 10 h the middle drain has the least pore water pressure corresponding to depth which indicates an effective decrease in groundwater levels. For top and bottom drains, very little distinction could be seen in the pore water pressure variations. However, it could not be concluded that the top and bottom drains are equally effective just from the pore pressure graphs so a factor of safety comparison was done.

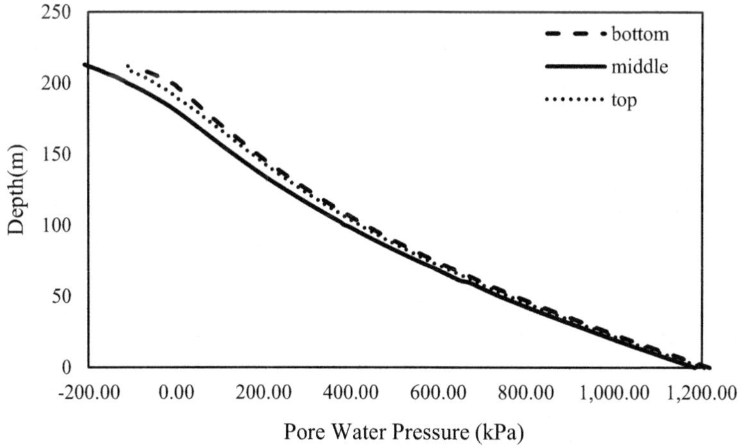

Fig. 9 Pore water pressure variation with depth for different drain positions after 10 h at Sect. 1

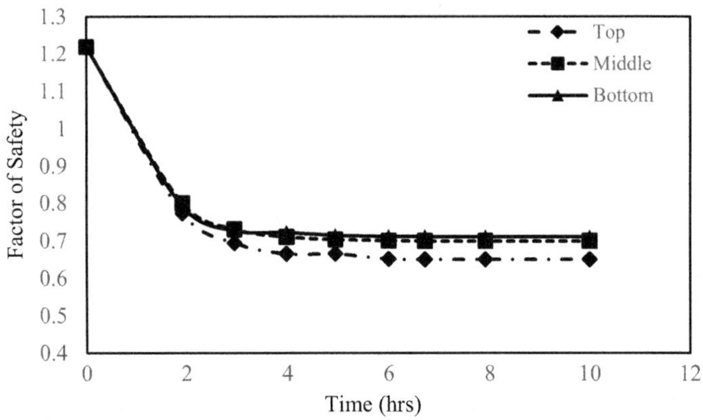

Fig. 10 Variation of factor of safety with time for different drain positions

The variation of factor of safety with time for different drain positions is illustrated in Fig. 10. To study the effect of drain position, three different drain positions (top, middle, and bottom) were selected. The safety factor starts to decrease with rainfall infiltration till five hours. After rainfall stops, the stability of the slope stays constant with time. It is observed that when the drain is placed at the middle or at the bottom of the slope, there is increased stability. The drain when placed in the middle or bottom position, is able to reduce the pore water pressures more than when it is placed at the top position. Therefore, the position of the drain was selected to be placed at the middle of the slope.

After selecting the position, the effect of the drain lengths was studied. Drains of different lengths (20, 30, 40, and 50 m) in the middle region were compared in terms of their factors of safety and the analysis revealed that a length of 50 m was the most optimal. The variation of factor of safety corresponding to different drain lengths has been shown in Fig. 11. The decrease in factor of safety with rainfall infiltration is minimum for the drain length of 50 m indicating a much-reduced water level. Due to higher surface area provided by the 50 m drain, it is able to receive much more water than the other drain lengths. Hence, a 50 m drain length was selected for the design of drains within the slope.

To select the number of drains to be placed a parametric study was performed with a single drain, three drains, and five drains placed at the middle of the slope. The variation of factor of safety for multiple drains versus a single drain has been depicted in Fig. 12. It was observed that the slope with five drains placed at the middle stabilized the slope more than the single drain or the three-drain configuration. The inclusion of multiple drains was found to slow down the reduction of factor of safety with time by allowing more water to flow out of the drains thereby reducing the groundwater level.

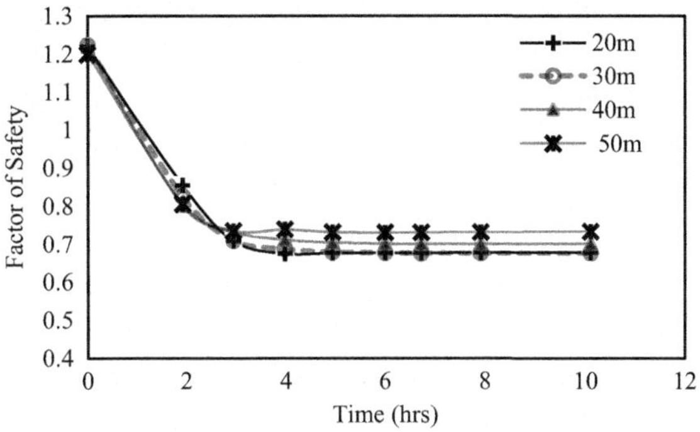

Fig. 11 Variation of factor of safety with different drain lengths in the middle region

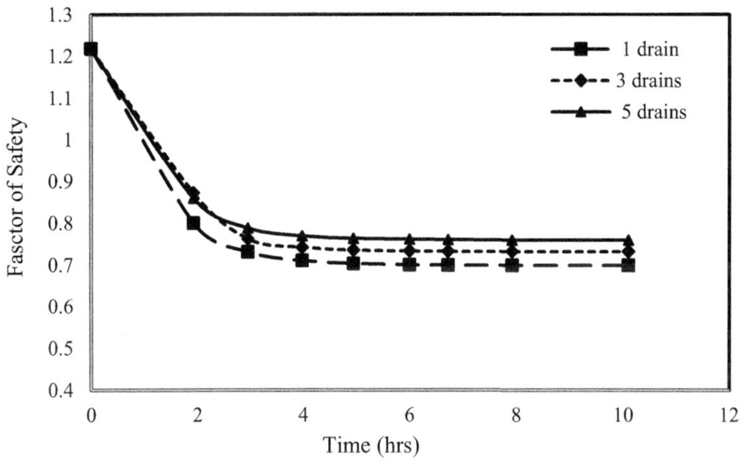

Fig. 12 Variation of factor of safety for multiple drains and single drain in the middle region

7 Conclusion

Solutions to landslide disasters in residual soil slopes under rainfall events need to be found and proper design methodology should be followed. Horizontal drains can prevent the strength of the soil from going below a certain value by maintaining the matric suction by lowering the groundwater level. The focus should be on lowering the main water level and reduction of pore water pressures. It is important to take into account the positioning of the drains such that the drains are able to intercept a significant amount of infiltration. From the study, the following recommendation can be made. The Chanmari slope can be stabilized by installing five 50 m horizontal drains at the middle of the slope.

References

1. Bhasin R, Grimstad E, Larsen JO, Dhawan AK, Singh R, Verma SK, Venkatachalam K (2002) Landslide hazards and mitigation measures at Gangtok, Sikkim Himalaya. Eng Geol 64:351–368
2. Fredlund DG, Krahn J (1977) Comparison of slope stability methods of analysis. Can Geotech J 143:429–439
3. Fredlund DG, Morgenstern NR, Widger RA (1978) The shear strength of unsaturated soils. Can Geotech J 153:313–321
4. Fredlund DG, Rahardjo H (1993) Unsaturated soil mechanics. Wiley, New York
5. Geo-Slope (2021a) Seep/W Users manual. GEOSLOPE International, Calgary, Alberta, Canada
6. Geo-Slope (2021b) Slope/W User's manual. GEOSLOPE International, Calgary, Alberta, Canada
7. Gjetvaj V, Nossan AS (2009) Increase of slope stability with time by drilled drains. In: Proceedings of the 17th International conference on soil mechanics and geotechnical engineering
8. Martin RP, Siu KL (1996) Use of Horizontal Drains for Slope Stabilisation in Hong Kong. HKIE Transactions 3(2):25–36
9. Mukhlisin M, Aziz N (2016) Study of horizontal drain effect on slope stability. J Geol Soc India 87:483–490
10. Rahardjo H, Hritzuk KJ, Leong EC, Rezaur RB (2003) Effectiveness of horizontal drains for slope stability. Eng Geol 69:295–308
11. Rahardjo H, Santoso VA, Leong EC, Ng YS, Hua CJ (2011) Performance of horizontal drains in residual soil slopes. Soils Found 51(3):437–447

Establishing Rockfall Hazards Through the Characteristics of Moving Rock Blocks

K. N. Manohara, Aman Ujjwal, Arindam Dey, and S. Sreedeep

Abstract Rockfall constitutes one of the most disastrous hazards as it involves very rapid movements of rock blocks. Chunks of rock detach from steep slopes and undergo several kinds of motion like falling, rolling, bouncing and sliding due to gravity. Rockfall behavior is greatly influenced by the slope geometry, rock properties and contact properties. These hazards are characterized by certain rockfall parameters like bounce height, maximum runout distance, velocity attained and kinetic energy of the moving rock blocks that can be determined from their simulation. It is therefore prudent to understand the effect of characteristics like shape, size and density of the falling rock blocks on these rockfall parameters. Hence, in the present study, an attempt has been made to establish rockfall hazards through the characteristics of rock blocks. The slope geometry and other properties have been taken from the literature. It has been found that density of the rock blocks does not influence the propensity of the rockfall hazard but their shape and size acts as primary influencing factors. Additionally, sharpness of the edges of the rock blocks was also found to have a marked influence on the rockfall parameters. The impact energy exerted by the falling rocks on the barrier was also studied considering the effect of their shape and size characteristics.

Keywords Rockfall · Rock block characteristics · Bounce height · Velocity · Kinetic energy · Barrier

K. N. Manohara (✉) · A. Ujjwal · A. Dey · S. Sreedeep
Department of Civil Engineering, Indian Institute of Technology Guwahati, Assam, India
e-mail: n.manohara@iitg.ac.in

A. Ujjwal
e-mail: aujjwal@iitg.ac.in

A. Dey
e-mail: arindam.dey@iitg.ac.in

S. Sreedeep
e-mail: srees@iitg.ac.in

S. Mitra et al. (eds.), *Disaster Management and Risk Reduction: Multidisciplinary Perspectives and Approaches in the Indian Context*,
https://doi.org/10.1007/978-981-99-6395-9_7

1 Introduction

Rockfall is a vital problem which is frequently reported in North Eastern India [1]. Rockfall involves quick movements of rock blocks down the slope, from which they detach, under the effect of gravity [2]. The falling chunks of rock have massive energies and can cause enormous damages to life and property. Hence, it is important to understand the rockfall behavior so as to avoid such distress. Rockfalls are instigated by natural causes such as rain, seismic events, alternate freezing–thawing cycles and vegetative root penetration [1]. Anthropogenic activities like road cutting of slopes and blasting can also trigger rockfalls [3]. Slope geometry, properties of rock and contact features have been found to influence the behavior of rockfall [4]. Additionally, it has been observed that the characteristics of moving rock blocks can also affect their trajectories [4]. Therefore, it is wise to establish the rockfall behavior in relation to the characteristics of the falling chunks of rock. Simulation of rockfall is an effective way of studying the attributes of this hazard considering certain parameters like bounce height, maximum runout distance, velocity attained and kinetic energy involved. In this work, these parameters have been analyzed in terms of the characteristics such as shape, size and density of rock blocks. Furthermore, the effect of sharpness of the edges of rock blocks has been studied. The influence of the size and shape of falling rocks on the impact energy experienced by the barrier is also studied.

2 Methodology

2.1 Study Area

The studied rock slope is situated along NH 44A highway in Mizoram, India which connects the Aizwal city to Lengpui Airport [5]. It falls in a seismically active region and receives an annual rainfall of 2500 mm [5]. Hence, the considered slope is very prone to rockfall hazards. The slope geometry is shown in Fig. 1. The slope is composed of rock outcrop followed by some stretch of soil. The highway is represented by asphalt which is located very near to the slope.

2.2 Rockfall Simulation

Rockfall simulations have been done using Rocscience RocFall2 program [6] in which two-dimensional analysis of rockfall could be conducted. The slope section was represented as per the chosen geometry [5]. At a height of 12.8 m above the highway on the rock slope, a point seeder was assigned (Fig. 1). All the rock blocks

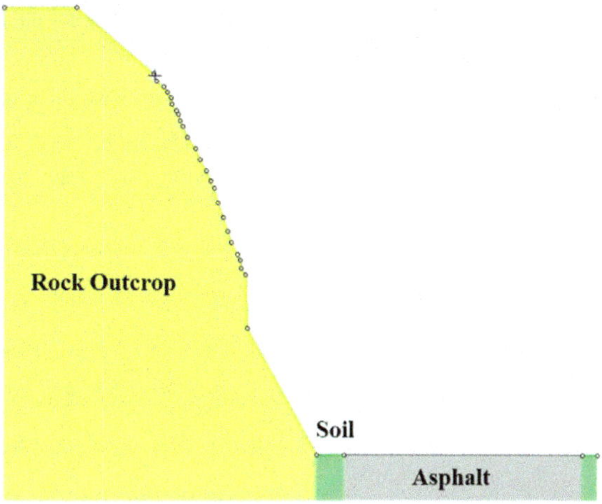

Fig. 1 The slope geometry adopted in this study [5]

were allowed to fall from this point only. The basic input parameters have been taken from the literature [5] as shown in Table 1.

In the present study, the rockfall parameters have been investigated from the perspective of rock block characteristics. Table 2 depicts the different shapes, sizes and densities of the rock blocks that have been considered in rockfall simulations. 500 rock blocks of each type have been allowed to fall from the seeder point. While studying the effects of size and density of the rock blocks, the shape was chosen

Table 1 Basic input parameters for the rockfall simulations [5]

Parameters	Values
Inclination of the rock slope	70–80°
Height of the point seeder	12.8 m
Number of the rock blocks	500
Rock block density	2700 kg/m^3
Mass of the rock blocks	500 kg
Slope roughness	2 (standard deviation)
Coefficient of Restitution (COR)	
1. Outcrop	
R_n (Normal)	0.45 (±0.04)
R_t (Tangential)	0.85 (±0.04)
2. Soil	
R_n (Normal)	0.3 (±0.1)
R_t (Tangential)	(±0.1)
3. Asphalt	
R_n (Normal)	0.4 (±0.04)
R_t (Tangential)	0.9 (±0.04)

Table 2 Shape, size and density of rock blocks taken for the rockfall simulations

Rock block characteristics	Values
Shape	Sphere, Triangle, Square, Pentagon, Octagon
Size (Radius of sphere, m)	0.25, 0.5, 0.75, 1, 1.25, 1.5
Density (kg/m^3)	2000, 2500, 3000, 3500

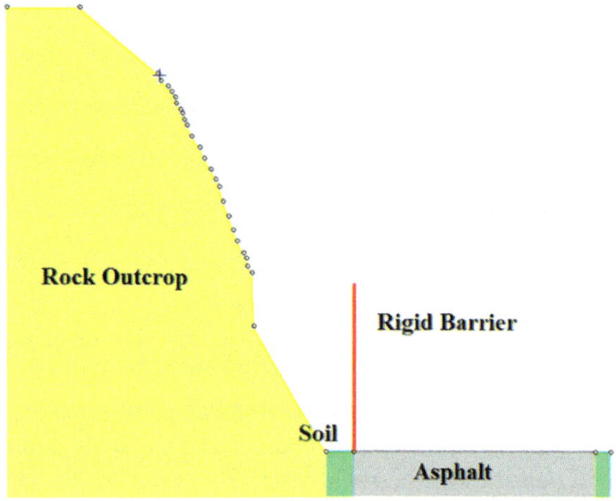

Fig. 2 Position of the rigid barrier in front of the rock slope

to be spherical. Furthermore, a rigid barrier was placed at the junction of highway (asphalt) and soil as shown in Fig. 2. The impact energy exerted on the barrier by falling rock blocks has been studied by considering different shapes and sizes of these blocks. Variation in the rockfall parameters due to sharpness of the edges of the rock blocks has also been analyzed for three different shapes.

3 Results and Discussions

3.1 Effects of the Shape of Rock Blocks

Different shapes of the rock blocks have been chosen as mentioned in Table 2. Mass and density of the rock blocks were kept constant as mentioned in Table 1. Figure 3 shows the variation of the total kinetic energy of rock blocks of different shapes. It could be seen that triangular rock blocks possess the highest kinetic energy for the chosen slope geometry followed by square, pentagonal and octagonal blocks which have similar variation in their kinetic energy along the trajectories. Spherical rocks have shown the least energy. Angular rock blocks have been found to possess a higher kinetic energy compared to the spherical blocks.

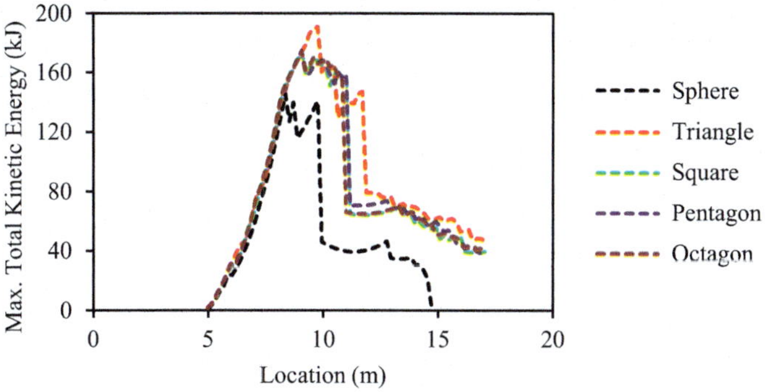

Fig. 3 Variation of maximum total kinetic energy for rock blocks of different shapes

Figure 4 shows the variation of bounce height of rock blocks based on their shapes. The triangular rock blocks have shown the highest bounce height followed by square, pentagonal, octagonal and finally, spherical rock blocks. This is in relation to the variation of kinetic energy as the shapes with higher kinetic energy have higher bounce height. Furthermore, maximum translational and maximum rotational velocity profiles have also been observed to follow similar trends in regards to the shape of the rock blocks where triangular shape showed highest values and spherical showed the least (Fig. 5a, b).

Figure 6 shows the distribution of number of rock blocks with runout distance for different shapes. From the graph, it is clear that the spherical rock blocks exhibit least runout distance of 10.01 m as compared to the other shapes which have a maximum runout distance of 12.12 m. Moreover, it can be seen that almost all the spherical rocks have traversed to a distance of 10.01 m (498 out of 500) and are concentrated there whereas the angular rock blocks are well distributed all along the runout area.

The observations are different from that reported in the literature [4] for a plane inclined slope where the spherical rock blocks have exhibited maximum values of

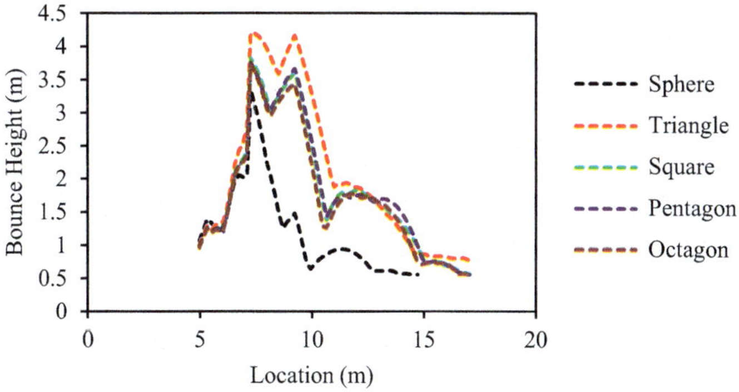

Fig. 4 Variation of bounce height for rock blocks of different shapes

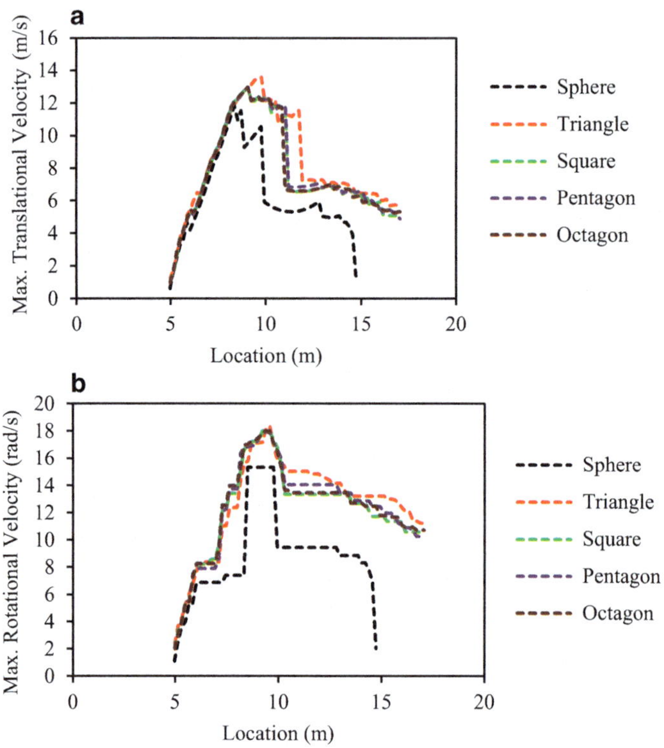

Fig. 5 **a** Variation of maximum translational velocity of rock blocks of different shapes. **b** Variation of maximum rotational velocity of rock blocks of different shapes

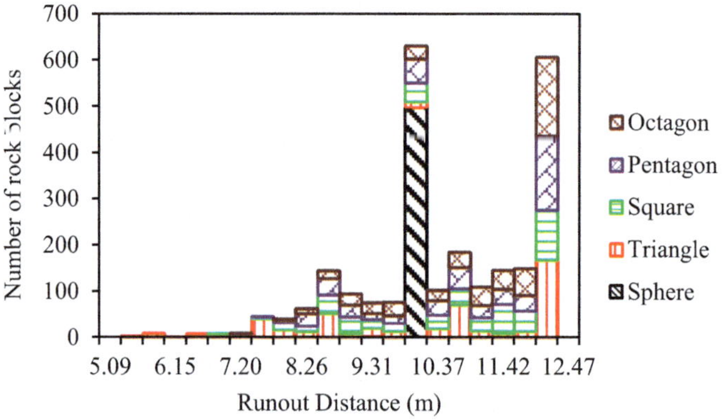

Fig. 6 Distribution of runout distance for rock blocks of different shapes

all the rockfall parameters. Hence, it could be inferred that slope geometry has a marked influence on the effects of the rock block characteristics. It is very important to consider the effect of slope geometry in comprehending the rockfall behavior. Further investigations would be required in this context.

3.2 *Effects of the Size of Rock Blocks*

Different radii of the spherical rock blocks were taken in order to study the effect of size on rockfall parameters as mentioned in Table 2. Figure 7 shows the variation of maximum total kinetic energy for different sizes of the rock blocks. Larger blocks clearly show a higher kinetic energy as compared to smaller blocks owing to their size.

Figure 8 shows the effect of the size of the rock blocks on the bounce height. Large sized rock blocks have bounced much higher than the smaller blocks. The blocks of radius 1.5 m have bounced to a maximum height of 5.66 m whereas those of 0.25 m radius could only reach up to a height of 2.64 m. Large sized blocks are characterized by greater kinetic energy and so they could bounce higher.

Figure 9a, b depict the variation of maximum translational and maximum rotational velocities of the rock blocks of different sizes. Translational velocities are similar for all the sizes of rock blocks but there is a marked difference in the values of rotational velocity. Rotational velocity of smaller rock blocks is much higher than that of larger blocks in a way that for blocks of radius 0.25 m, it is as high as 39.58 rad/s whereas for rock blocks of 1.5 m radius, it is only about 5.71 rad/s. As the rock blocks of different sizes translate with similar velocity, so, the smaller rocks must have a higher rotational velocity to keep up with the larger rock blocks.

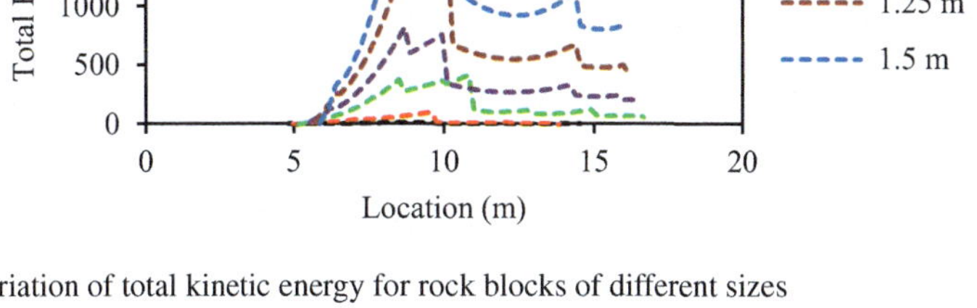

Fig. 7 Variation of total kinetic energy for rock blocks of different sizes

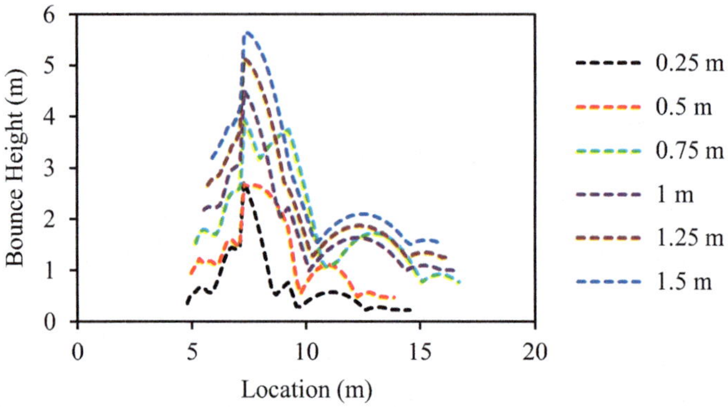

Fig. 8 Variation of bounce height for rock blocks of different sizes

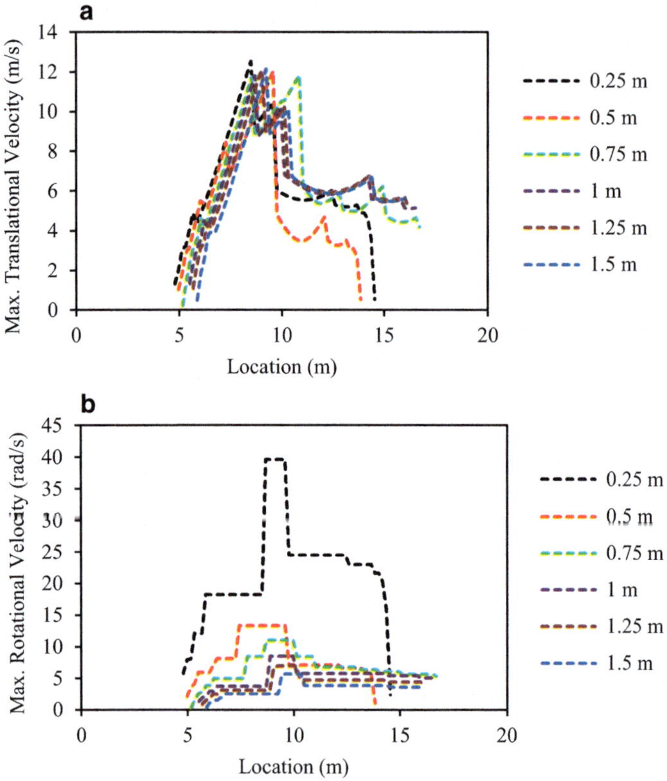

Fig. 9 **a** Variation of maximum translational velocity of rock blocks of different sizes. **b** Variation of maximum rotational velocity of rock blocks of different sizes

3.3 *Effects of the Density of Rock Blocks*

For this study, the density of rock blocks was varied keeping their shape as sphere and volume as 1 m^3. Different densities of the rock blocks adopted in this study are tabulated in Table 2. Figure 10 shows the variation of total kinetic energy with respect to the density of rock blocks. As the density increases, the total kinetic energy has also increased. Figure 11 shows the variation of the bounce heights for different densities. It can be observed that even with different densities, rock blocks have shown similar bounce heights and also follow a similar trajectory. Alongside, Fig. 12a, b show a similar trend for maximum translational and maximum rotational velocities. This analysis shows that there is no impact of density of rock blocks on the bounce height and velocities. The kinetic energy has increased only by the virtue of mass with the increase in the value of density. Hence, density of the rock blocks does not have an influence on all the rockfall parameters.

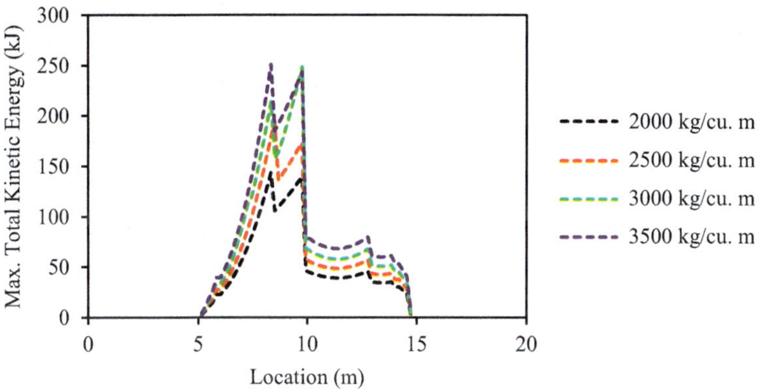

Fig. 10 Variation of maximum total kinetic energy for rock blocks of different densities

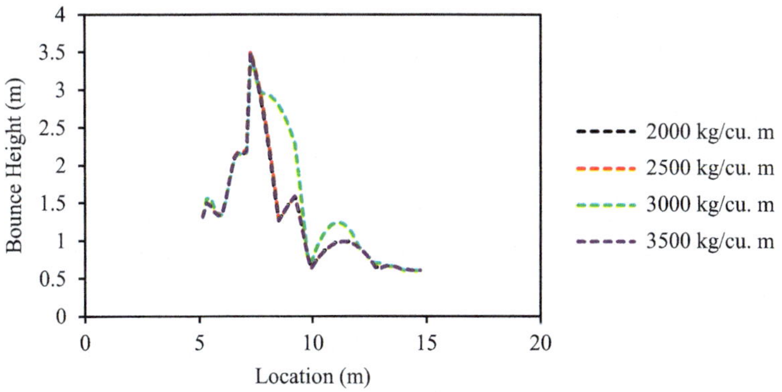

Fig. 11 Variation of bounce height for rock blocks of different densities

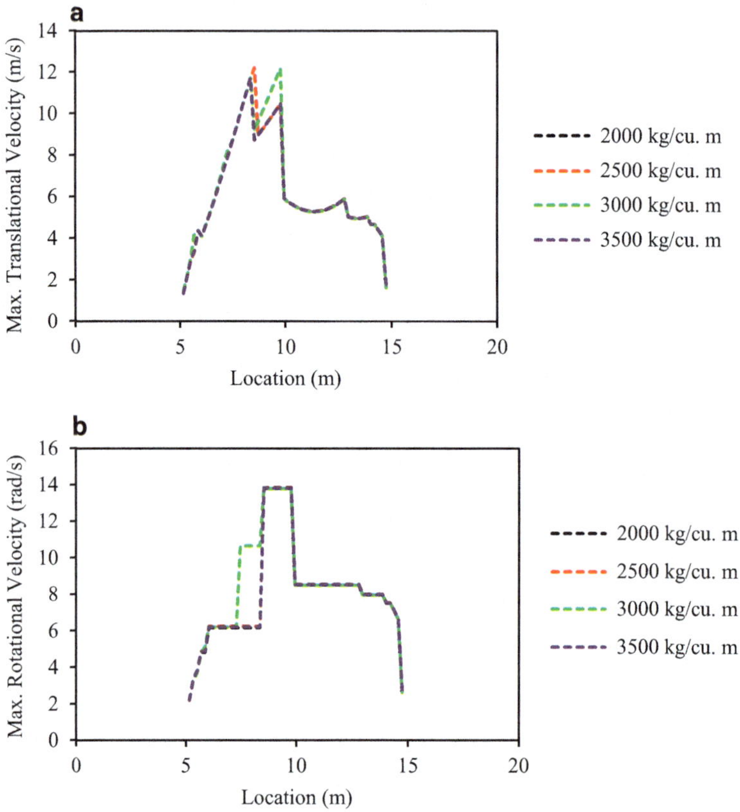

Fig. 12 **a** Variation of maximum translational velocity for rock blocks of different densities. **b** Variation of maximum rotational velocity for rock blocks of different densities

3.4 *Effects of the Sharpness of the Edges of Rock Blocks*

Rock blocks which detach from the rock mass could have sharp edges or smooth edges. The effect of this feature on rockfall parameters has been studied for triangular, square and hexagonal rock blocks. Figure 13 shows the shapes of rock blocks considered having smooth and sharp edges. Figure 14a, b, c show the variation of bounce heights of the triangular, square and hexagonal rock blocks having smooth and sharp edges. It could be observed that sharp-edged rocks exhibit a higher bounce height as compared to the smooth-edged rocks for all the considered shapes. It can also be seen that as the shape is varied from triangular to hexagonal, the difference in bounce height of smooth-edged and sharp-edged rock blocks has increased. Hence, as the edges of the rock blocks increase, the bounce height has increased as well.

Figure 15a, b, c show the variation of rotational velocity for triangular, square and hexagonal rock blocks having smooth and sharp edges. The rotational velocity

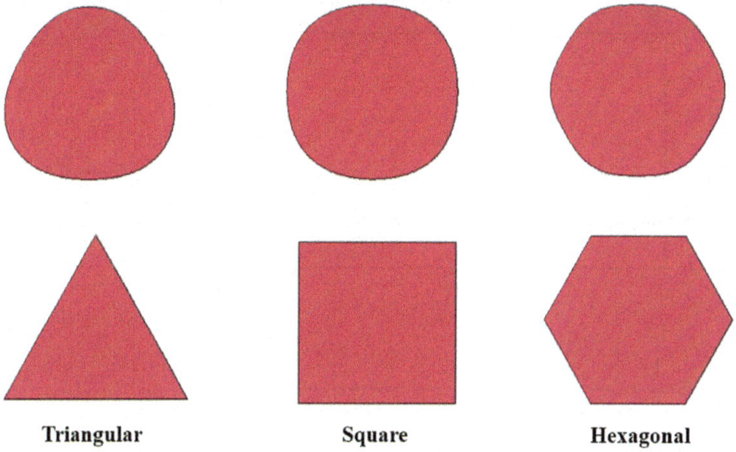

Fig. 13 Different shapes of the rock blocks are considered having smooth and sharp edges

showed an opposite trend when compared to that of bounce height. Smooth-edged rock blocks possess a higher rotational velocity than sharp-edged rocks for all the rock shapes. Rounded or smooth edges could enable higher velocity of rotation as compared to sharp edges. However, this difference in rotational velocity decreased as the shape is varied from triangular to hexagonal. Similar behavior could be seen for rotational kinetic energy as well as shown in Fig. 16a, b, c.

3.5 *Effect of the Rock Block Characteristics on the Impact Energy Experienced by the Barrier*

The effects of the shape and size of the rock blocks on the impact energy exerted on the rigid barrier were studied. The rigid barrier was positioned as shown in Fig. 2. The density of the rock blocks was kept constant as per Table 1. Table 3 shows the values of the impact energy exerted by different shapes of the rock blocks. It could be seen that spherical rock blocks impact the barrier with much lesser energy compared to other angular shapes. It is so because as per the chosen slope geometry, spherical rock blocks possessed the minimum kinetic energy. Hence, the impact energy exerted on the barrier is also dependent upon the slope geometry.

In order to understand the effect of size of rock blocks on the impact energy exerted upon the barrier, the spherical shape was chosen. Figure 17 shows the variation of impact energy with the size (diameter) of the rock blocks. It could be inferred that as the size of the rock block increases, the energy with which it impacts the barrier also increases. This could be related to the higher kinetic energy possessed by large sized rock blocks.

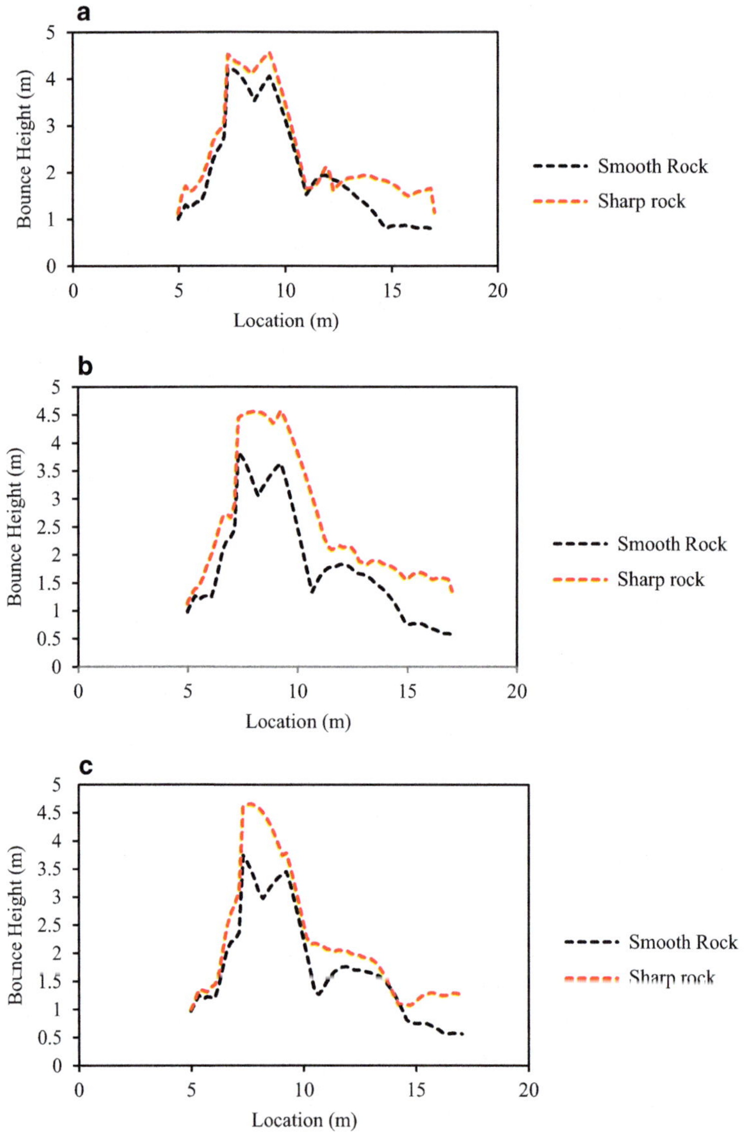

Fig. 14 **a** Variation of bounce height among smooth-edged and sharp-edged triangular rock blocks. **b** Variation of bounce height among smooth-edged and sharp-edged square rock blocks. **c** Variation of bounce height among smooth-edged and sharp-edged hexagonal rock blocks

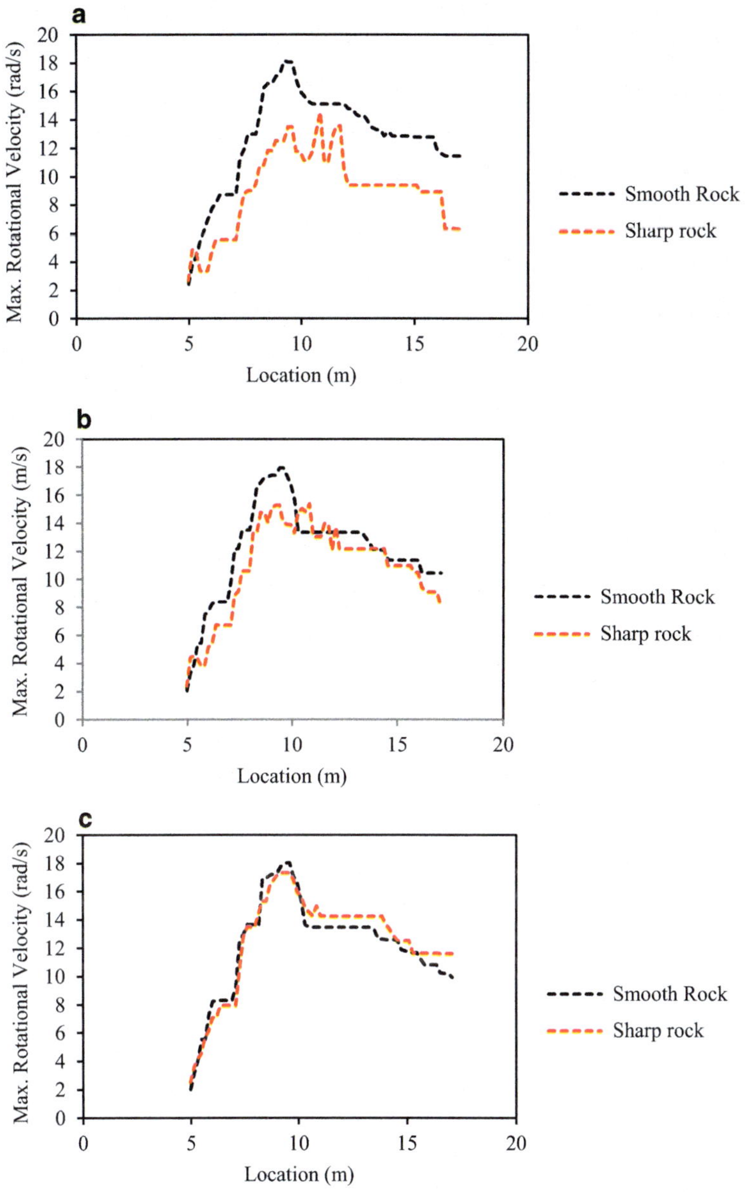

Fig. 15 **a** Variation of maximum rotational velocity among smooth-edged and sharp-edged triangular rock blocks. **b** Variation of maximum rotational velocity among smooth-edged and sharp-edged square rock blocks. **c** Variation of maximum rotational velocity of smooth-edged and sharp-edged hexagonal rock blocks

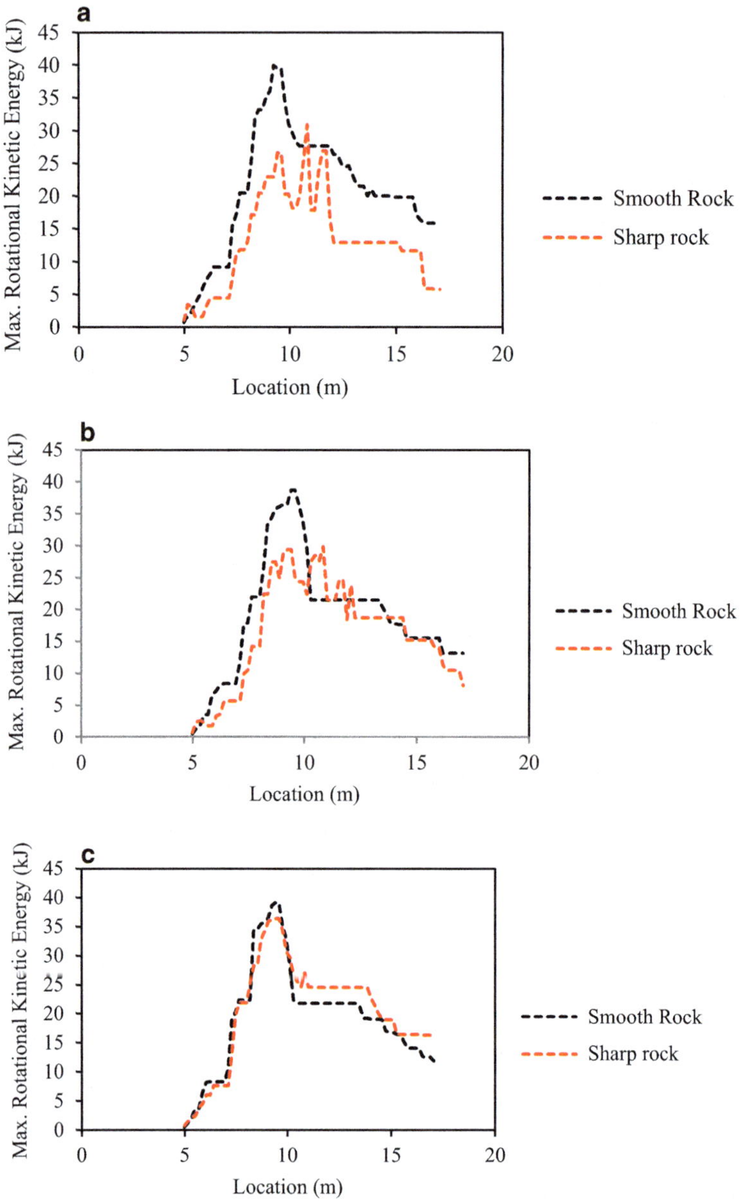

Fig. 16 **a** Vriation of maximum rotational kinetic energy of smooth-edged and sharp-edged triangular rock blocks. **b** Variation of maximum rotational kinetic energy of smooth-edged and sharp-edged square rock blocks. **c** Variation of maximum rotational kinetic energy of smooth-edged and sharp-edged hexagonal rock blocks

Table 3 Effect of shape of the rock block on the impact energy exerted on the barrier

Shape of the rock block	Impact energy exerted on the barrier (kJ)
Sphere	45.52
Triangle	151.47
Square	165.83
Pentagon	164.88
Octagon	169.61

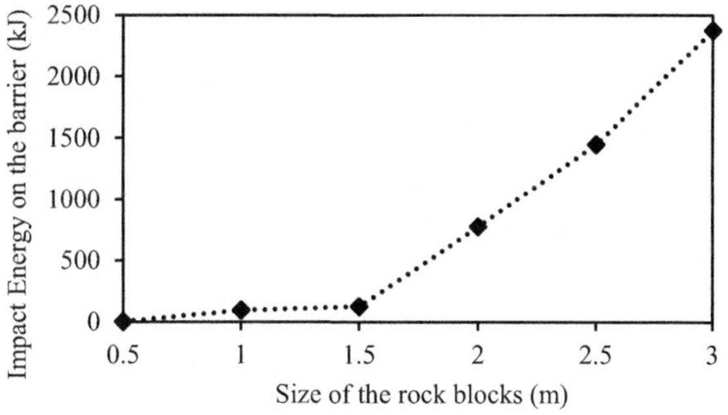

Fig. 17 Variation of impact energy on the barrier with the size of the rock blocks

4 Conclusion

In this work, an attempt has been made to establish the rockfall hazards from the perspective of rock block characteristics. Effects of the shape, size and density of rock blocks on the rockfall parameters were analyzed along with the effects of sharpness of the edges of rock blocks. Additionally, the energy with which the rock blocks impact the barrier was also studied in relation with their shape and size.

For the slope geometry chosen, triangular rock blocks were found to have the highest bounce height followed by square, pentagonal, octagonal and spherical rocks. Translational and rotational velocities also followed a similar trend. The runout distance was least for spherical rocks emphasising upon the fact that the slope geometry has a marked influence on the effect of the characteristics of the rock blocks. Further, as the size of the rock block increased, the bounce height and kinetic energy also increased. The density of the rocks did not affect the vital rockfall parameters and so, it does not influence the propensity of the rockfall hazards. Additionally, sharp-edged rocks experienced higher bounce height than the smooth-edged rocks. Smooth-edged rocks showed a higher rotational velocity and rotational kinetic energy than the ones having sharp edges. In terms of the impact energy exerted on the rigid barrier, spherical shape was least influencing as spherical rock blocks were found to exert much lesser impact energy compared to other angular shapes. The impact energy on the barrier was also found to increase with the increase in the size of the rock blocks. Hence, it could be concluded that the characteristics of moving rock blocks greatly influence the rockfall parameters and should be considered while understanding the behavior of rockfall hazards.

References

1. Verma AK, Sardana S, Sharma P, Dinpuia L, Singh TN (2019) Investigation of rockfall-prone road cut slope near Lengpui Airport, Mizoram, India. J Rock Mech Geotech Eng 11:146–158
2. Cruden DM, Varnes DJ (1996) Chapter 3-Landslide types and processes. In: Landslides: investigation and mitigation. Transportation Research Board, pp 36–75
3. Singh AK, Kundu J, Sarkar K, Verma HK, Singh PK (2021) Impact of rock block characteristics on rockfall hazard and its implications for rockfall protection strategies along Himalayan highways: a case study. Bull Eng Geol Environ 80:5347–5368
4. Nagendran SK, Ismail MAM (2019) Analysis of rockfall hazards based on the effect of rock size and shape. Int J Civ Eng 17:1919–1929
5. Verma AK, Sinha RK, Sardana S, Jaswal M, Singh TN (2021) Investigation into the Rockfall Hazard along Lengpui–Aizwal Highway, NH–44A, Mizoram, India. Indian Geotech J 51:732–745
6. Rocscience Rocfall2. https://www.rocscience.com/software/rocfall. Last accessed 15 May 2022

Tunnels, the Sustainable Means of Transportation Pathways in Natural Disaster Prone Darjeeling-Sikkim Himalaya (DSH)

Pratap Chandra Dhang

Abstract Sivok-Rangpo Rail Line Project (SRRP) is traversing along the Teesta River Valley through the seismically active Darjeeling and Kalimpong Hills in between MBT and MCT of Darjeeling-Sikkim Himalaya (DSH). Therefore, many environmentalists and geologists claimed that this project dominated by tunnels may invite natural disasters and the tunnels will not be safe under seismic events. From the studies like empirical correlations between measured peak ground accelerations (g) and observed damage in tunnels, the following tendencies have been revealed: (a) up to a peak acceleration of 0.2 g, slight damage, (b) from 0.2 g to 0.6 g, serious damage in unlined tunnels and in tunnels devoid modern lining, (c) from 0.6 g to 0.9 g, serious damage on tunnels having plain concrete lining (unreinforced). Though most severe earthquakes in Sikkim and adjoining areas had peak ground acceleration ranges from 0.15 g to 0.45 g only, which indicate that the modern lined tunnels constructed by NATM are very much safe under massive earthquakes, and NATM designs have all considerable factors for earthquake-induced ground movement in them. The transportation tunnels ensure less damage to the environments and biodiversity and are also used to avoid instable slopes which fail during earthquake-induced ground vibration. Therefore, tunnels are considered as safe and useful in sustainable infrastructural developments in this seismically active Himalayan region.

Keywords Himalaya · NATM · PGA · SRRP · Tunnels · Landslides

1 Introduction

The Himalayan orogenic system of about 2500 km has evolved in response to collision of the Indian and Eurasian plates, which caused sequential thrusting (Main Central Thrust or MCT, Main Boundary Thrust or MBT, Main Frontal Thrust or MFT, etc.), and riding of different thrust blocks on each other [2]. This long and

P. C. Dhang (✉)
9/1 Akshaytara Apartments, Siliguri, W.B 734 001, India
e-mail: pratap09.isi@gmail.com

S. Mitra et al. (eds.), *Disaster Management and Risk Reduction: Multidisciplinary Perspectives and Approaches in the Indian Context*,
https://doi.org/10.1007/978-981-99-6395-9_8

wide stretch of Himalayan region is devoid of proper infrastructural developments due to its inaccessibility and prone to natural disasters. However, recently Indian Railways and National Highway Authority of India have taken the initiative to extend the railways and roadways into the mighty Himalayan ranges. Therefore, numerous railway projects like Udhampur-Srinagar-Baramullah Rail Link (USBRL) Project in Jammu & Kashmir Himalaya, Rishikesh-Karnaprayag Railway Project in Uttarakhand, Sivok-Rangpo Rail Line Project (SRRP) in Darjeeling-Sikkim Himalaya, Bhalukpong-Tawang Railway Line Project in Arunachal Himalaya, etc., are undergoing. All these projects are characterized by 80–85% tunnels to ensure less harm to the environment and biodiversity and also to avoid unstable hill slopes. Frequent earthquakes and other natural hazards, such as flashfloods and landslides, etc., are inherent to the Himalayan regions (IS 1893 (Part 1); 2002), [5]. Himalaya is undergoing rapid uplift at rates of about 0.5 mm/year [44]. Present project is situated in between the Main Boundary Thrust (MBT) and Main Central Thrust (MCT); hence seismo-tectonically the region is very active, where earthquake of magnitude ≥ 6 may occur. Recent seismicity in Himalaya is restricted in two distinct zones, one is between MBT/MFT and MCT and the other is beyond Indus Tsangpo Suture (ITS) [31]. Therefore, the doubt regarding the safety of the tunnels comes into existence. Even local peoples and administrations have questions about the stability of the tunnels under the seismic activities.

Sivok-Rangpo Rail Line Project (SRRP) is traversing along the Teesta R. in Darjeeling Sikkim Himalaya (DSH). This region is considered as one of the most prone to natural disasters in terms of earthquakes and landslides. Therefore, many environmentalists and geologists questioned about the stability of the tunnels and also claimed that this project may invite further natural disasters. The tunnelling works in seismically active Himalayan region are new in India, however, the Alpine regions or the seismically very active Japan has prolific development in tunnelling. Recently the studies have been conducted in developed countries of Europe, America and Asia to analyze the effect of earthquakes on the tunnels [8, 13, 35, 38, 45, 47]. These researches show that underground structures are seismic resistant, however, there are geological conditions which may induce indirect earthquake damage to tunnels like fault zone activated by earthquake, sudden changes in condition due to contrasts in geomechanical properties of rocks, potentially liquefiable soils, pore fluids and marked anisotropy in local stress field [27]. These damages are often characterized by irreversible displacements, heavy water inflow, mechanical instability at tunnel portals, soil settlement and rupture due to liquefaction, etc.

The present work is based mostly on qualitative approach and on available literatures and case studies. Detailed analysis on seismic design for tunnels is beyond the scope of the present work. And in present study, Peak Ground Acceleration (PGA), the most important earthquake parameter has been used to depict the performance of tunnels in seismically active areas. The objectives of the present work are: (i) to get an overview of the seismo-tectonic set-up of the Sikkim-Darjeeling Himalaya (DSH) with respect to the Sivok-Rangpo Rail Line Project area, and (ii) to assess the probable impacts of seismic events or other natural disasters on the tunnels in the project area.

2 Seismotectonic Set-Up

Sivok-Rangpo Rail Line Project is traversing along the Teesta River through the Darjeeling and Kalimpong Hills of DSH. Several perennial and seasonal streams pass across the alignment of the project. The area is densely vegetated having high relative relief and high stream density. During monsoon (from May to October) these streams are used to be very active and debouches water directly or indirectly into the Teesta River. The area receives heavy to very heavy rainfall, i.e., more than 2600 mm per year. The slope of the hill is probably covered up by slope wash materials, colluvials and soil, which have thickness of about 0–30 m. The overburden material is featured by block in soil matrix. Block size ranges from 10 to 120 cm and are haphazardly embedded within the soil. Soil consists of variable grain size ranging from clay/silt, sand to gravel. Therefore, there are abundant voids and are moderately permeable with cohesion (c) and angle of internal friction (φ) are 0–10 kPa and 25–30° respectively. Thus, the hillslopes are fragile and prone to erosion and landslide (cf. [36]).

The region is tectono-stratigraphically sub-divided into four domains with characteristic stratigraphic and structural attributes, namely, (i) Foothill Belt, (ii) Inner Belt, (iii) Axial Belt, and (iv) Trans-axial Belt. The collision event is associated with multiple deformation phases and it is a part of an active fold-thrust belt. The major tectonic features are Main Frontal Thrust (MFT), Main Boundary Thrust (MBT), Main Central Thrust (MCT) NNW-SSE trending Tista and Gangtok lineaments, WNW-ESE trending Goalpara lineament and SW-NE trending Kanchanjangha Fault, Gish Transverse Fault and major structural feature, known as Rangit Window (Fig. 1) [10, 18, 26, 33]. The MBT and MCT are the main structural elements, along which the Proterozoic Daling rocks override the Gondwana rocks and the Lower Proterozoic high-grade metamorphic rocks of Darjeeling Gneiss override the Middle Proterozoic low-grade metamorphic rocks (schist and phyllite) of Daling Group (Table 1). Different lithological units are disposed in an arcuate regional fold pattern. The core is occupied by the Lesser Himalayan Belt of the Daling Group including the Lingtse-granitoid gneiss. The region also comprises medium- to high-grade crystalline rocks of the Higher Himalayan Belt. The MCT, a prominent ductile shear zone separates the two belts. Gondwana and molasse-type Siwalik sedimentary rocks of the Sub-Himalayan Zone occur in the southern part of the region. Several subsidiary thrusts are also present between MCT and MBT. Neotectonic evidences in frontal Darjeeling-Sikkim Himalaya are well exposed along Teesta R. (near Kalijhora village on NH 31) in the form of out-of-sequence thrusts and displacement of ~150 m eastward by the river while incising ~48 m vertically, giving rise to impaired disjointed strath terraces [32].

Frequent earthquakes and other natural hazards, such as flashfloods and landslides, etc., are inherent to the Himalayan regions [5, 9, 15, 28, 42]. The clustering of epicentres of earthquakes is recognised to occur between MCT and MBT [37] (Fig. 2). The stress continuously builds up along the major thrusts and sudden release of the stress causes frequent tremors. And strongest concentration of epicentres between 27

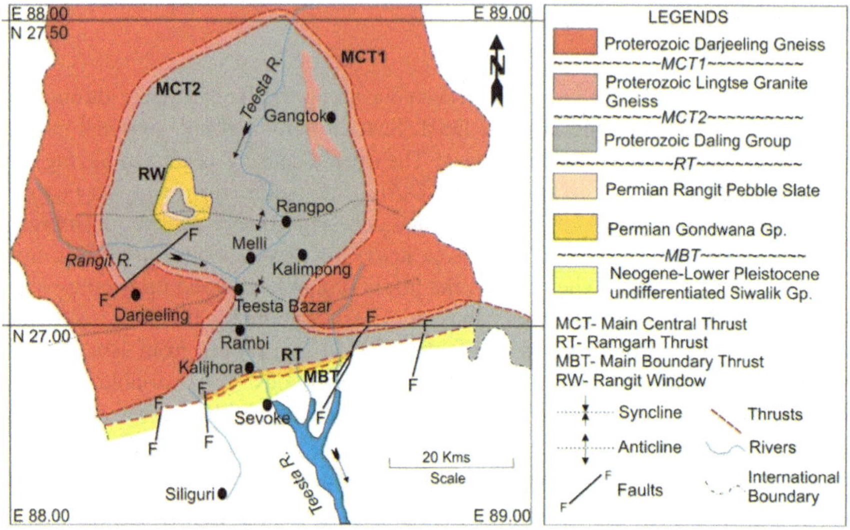

Fig. 1 Geological map of the Sikkim-Darjeeling Himalaya (modified and simplified after [18]). SRRP alignment traverses along the Teesta R. from Sevoke to Rangpo

and 27.5° N indicates that the Tista half-window recorded majority of seismic events [30]. It has been revealed from focal mechanism that the region experiences both thrust and strike-slip earthquakes occur in this area [20, 34]. The present project area lies in this active seismic zone, which is classified under Zone IV (IS 1893 (Part I) 2002). The seismic intensity in the region is VIII on Modified Mercalli intensity scale (MM scale) [43]. Most of the earthquakes occur at shallow depth. Seismic events here are associated with strain accumulation associated with the northward tectonic movement of Indian Plate [43]. The region is under compressive stress regime and $\sigma 1$ lies horizontally in N-S direction (cf. [21]). Recent studies revealed that northward motion of the Indian Plate exceeds 14 mm/year. GPS data indicate in the western most Himalaya near Kashmir, the convergence rate is ~12 mm/a, similarly, at the Central Himalaya, the rate attained 17 mm/a, whereas, around Assam and Sikkim area the rate is about 15 mm/a [5]. The seismic activity in this part is not uniform and Darjeeling-Sikkim Himalaya has not experienced major earthquakes, except a few. Nath et al. [33] presented a seismic hazard map of the Sikkim Himalaya, lying between Nepal and Bhutan Himalaya, in terms of horizontal peak ground accelerations with 10% exceedance probability over the next 50 years. They also showed that from the composite fault plane solution from the strong motion events about thrust faulting with strike slip component along MBT. Global Seismic Hazard Assessment Program (GSHAP) also characterized Sikkim region as surrounded by 8 seismic source zones. The impacts of the earthquakes and of consequent landslides in DSH can be seen through huge loss of life and properties.

Table 1 Tectonostratigraphy of Darjeeling-Sikkim Himalaya (DSH)

Group	Subgroup/Formation	Lithology	Age
Darjeeling Gneiss		Migmatitic gneiss	Proterozoic
MCT 1			
Lingste Granite-Gneiss		Granite-gneiss (Mylonite)	Proterozoic
MCT 2			
Daling Group	Buxa Fm	Massive thick dolomites, cherts and bands of phyllite	Proterozoic
	Reyang Fm	Slate, phyllite, quartzite	
	Gorubathan Fm	Phyllite, schist	
Ramgarh Thrust (RT)			
Rangit Pebble Slate (Talchir?)		Diamictite, rhythmite, quartzite, marl	Permian
Gondwana Group	Damuda Subgroup	Sandstone, carbo. shale and coal	
MBT			
Siwalik Group	Upper	Murti Boulder Bed	Neogene to L. Pleistocene
		Parbu Grit (pebbly-coarse-medium sandstone)	
	Middle	Geabdat Sandstone (medium to coarse sandstone, shale, pebble beds)	
	Lower	Chunabati (fine to medium sandstone, siltstone, mudstone, marl and conglomerate)	

3 Peak Ground Acceleration (PGA)

The Peak Ground Acceleration (PGA) is an important parameter for earthquake engineering. As per USGS Earthquake Glossary 'When the ground is shaking during an earthquake, it also experiences acceleration. The peak acceleration is the largest increase in velocity recorded by a particular station during an earthquake'. PGA is expressed as a fraction of gravitational acceleration. According to USGS, 'g' is the acceleration of gravity, i.e., 9.8 (m/s^2), when there is an earthquake, the forces caused by the shaking are measured as a percentage of gravity. On the other hand, seismic intensity scale is being used for measuring the intensity of shaking produced by earthquake, which is also known as Modified Mercalli Intensity Scale (MM scale). The correlation between PGA and MM scale has been provided in Table 2. The

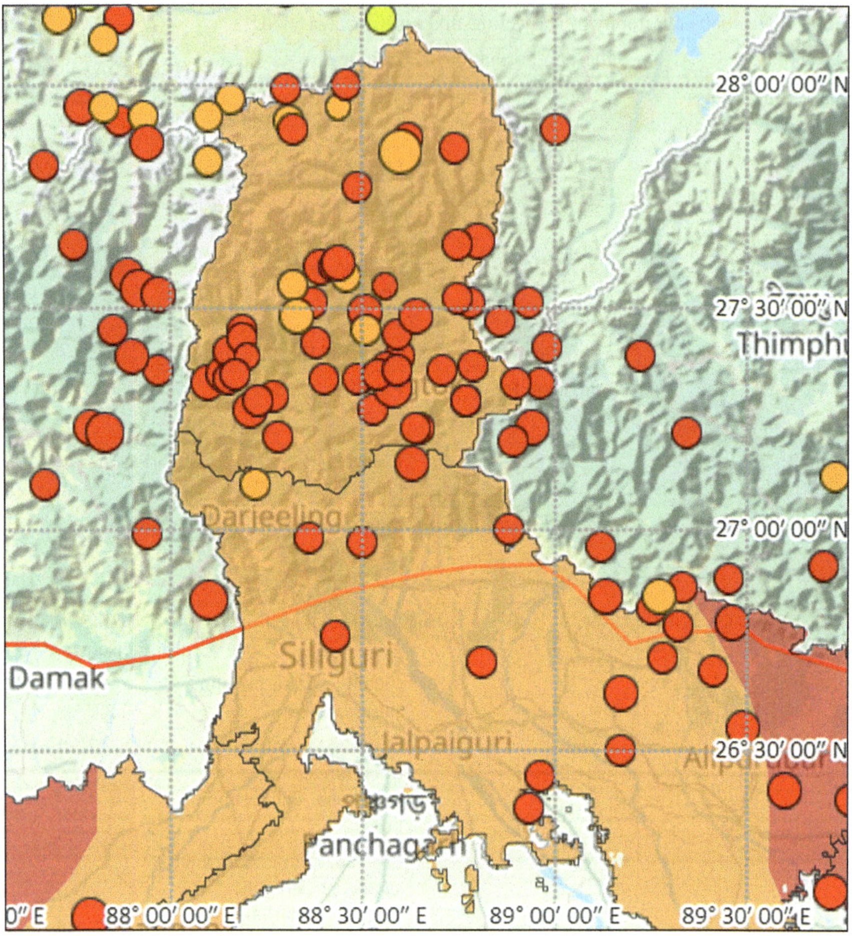

Fig. 2 Epicentres of earthquakes around Darjeeling-Sikkim Himalaya (DSH) and surrounding regions in last 70 years. *Source* National Centre for Seismology

seismic zones in India have been made on the basis of intensity of shaking [16, 23]. Raisinghani and Purohit [40] presented a correlation between the magnitude, PGA and intensity of some of the major earthquakes occurred in India between 1934 and 2015 (Table 3).

Extensive data regarding PGA in DSH is not available, however, recently some works have been undertaken. A study has been conducted by [33] at nine stations in Sikkim by installing strong motion accelerographs for continuous monitoring of the earthquake signals. They have collected data from 80 earthquakes (M = 3–5.6) recorded during 1998–2003 and analyzed the data to present a hazard scenario in the Sikkim Himalaya. The PGA has been estimated for the scenario earthquake of magnitude 8.3 at a depth of 26.25, and it has been found that there is spatial

Table 2 Correlation between MM scale, PGA and PGV (after USGS)

Mod. Mercalli intensity scale (MM)	Acceleration (g)	Velocity (cm/s)	Perceived shaking	Potential damage
I	<0.000464	<0.0215	Not felt	None
II–III	0.00297	0.135–1.41	Weak	None
IV	0.0276	1.41–4.65	Light	None
V	0.115	4.65–9.64	Moderate	Very light
VI	0.215	9.64–20	Strong	Light
VII	0.401	20–41.4	Very strong	Moderate
VIII	0.747	41.4–85.8	Severe	Mod. to heavy
IX	1.39	85.8–178	Violent	Heavy
X+	>1.39	>178	Extreme	Very heavy

Table 3 Details of major earthquakes occurred in India and seismic zones of India (modified after Raisinghani and Purohit 2016)

Sl. No	Event location	Year	Magnitude	PGA	MM scale
1	Bihar-Nepal	1934	8.2	0.3 g	IX
2	Assam	1950	8.7	–	X
3	India-Burma	1988	7.2	0.34 g	VIII
4	India-Bangladesh	1988	5.8	0.1 g	VI–VII
5	Garhwal	1991	7.1	0.3 g	VIII
6	Uttarkashi	1967	7.0	0.29 g	IX
7	Koyna	1967	6.5	0.4 g	VIII
8	Chamoli	1999	6.6	0.34 g	VIII
9	Bhuj	2001	7.7	0.38 g	VIII
10	Kashmir	2005	7.6	0.23 g	VIII
11	Sikkim	2011	6.9	0.35 g	VI
12	Nepal	2015	7.9	0.25 g	IX

variation with high at Singtam and adjoining areas and low at Lachen (Table 4). Research on recent high-intensity earthquakes in DSH and surrounding areas have been conducted, among them two major earthquakes are 18 September 2011 NW Sikkim Earthquake (M = 6.9) and 25 April 2015 Kathmandu Valley, Nepal Earthquake (M = 7.8) [15, 28]. In both the cases reported max. intensity was VII in MM scale and caused huge damages with economic losses and numerous casualties. But most of the damages were considered to be due to poor construction of houses and hilly landslide-prone terrain. Macroseismic field observation of Sikkim Earthquake was made by Mahajan et al. [29] and they found that there was an asymmetrical distribution and heterogeneous damage pattern in the region. The intensity distribution

Table 4 Spatial distribution of measured PGA values at nine strong motion stations in Sikkim by [33]

Sl. No	Stations	Lat/Long	PGA values	Remarks
1	Aritar	27.18N/88.67E	0.31	
2	Chungthang	27.60N/88.64E	0.22	
3	Gangtok	27.33N/88.62E	0.33	
4	Jorethang	27.13N/88.28E	0.31	
5	Lachen	27.71N/88.55E	0.20	Lowest
6	Mangan	27.52N/88.53E	0.28	
7	Melli	27.09N/88.45E	0.33	
8	Geyzing	27.28N/88.27E	0.30	
9	Singtam	27.23N/88.49E	0.38	Highest

map indicates max. PGA of 0.45 g at Chunthang, and it reduced toward north, 0.25 g at Theng, Bichhu and 0.15 g at Gangtok. Whereas, the max. PGA was 0.25 g (ranges 0.16–0.25 g) reported during Nepal Earthquake having shallow depth of 10–15 km [46].

4 Earthquakes and Tunnels

The general observations regarding seismic performance of underground structures [19] are:

(i) Underground structures suffer less damage than surface structures
(ii) Reported damage decreases with increasing overburden depth
(iii) Lined and grouted tunnels are safer than unlined tunnels
(iv) Tunnels are more stable under a symmetric load, which improves ground-lining interaction
(v) Damage may be related to PGA and velocity based on the magnitude and epicentral distance of the affected earthquake.

Further from the earlier studies like empirical correlations between measured peak ground accelerations or PGA (g) and observed damage in tunnels, the following tendencies have been revealed: (a) up to a peak acceleration of 0.2 g, slight damage, (b) from 0.2 g to 0.6 g, serious damage in unlined tunnels and in tunnels devoid modern lining, (c) from 0.6 g to 0.9 g, serious damage on tunnels having plain concrete lining (unreinforced). The underground structures are assumed to be seismic resistant since they are buried deeply in rock/soil layers, however, in case the tunnel experiences strong shaking and if it is traversing across faults or other adverse geological conditions, there will probability of damage. Zhang et al. [48] classified seismic damages in tunnels into five following patterns: (1) Cracks of tunnel lining, (2) Damage of construction joints, (3) Groundwater leakage, (4) Spalling and collapse of concrete lining, and (5) Damage of pavement. It has been observed that during earthquake, mostly the local cracks are developed on lining, in case there is any damage. The famous Bolu Tunnel, Turkey has been studied regarding seismic performance of tunnel [19]. The twin tunnels were constructed using NATM, where continuous monitoring of primary liner convergence was performed and modifications is support elements done accordingly. August 17, 1999 Koceali Earthquake had minimal impact, whereas, November 12, 1999 Duzce Earthquake caused a collapse of 300 m from the eastern portal. However, the collapse took place in clay gauge material in unfinished section and no tectonic displacement observed [27].

Recently the most severe earthquakes occurred at Kashmir (in 2005 with max. intensity of 7.6) and Sikkim Himalayas (in 2011 with max. intensity of 6.9). The peak ground acceleration of these earthquakes ranges from 0.15 g to 0.45 g only [4, 6, 41, 46]. Raisinghani and Purohit [40] presented a list of severe earthquakes in India with their magnitude, PGA and intensity (Table 3). From the list it has been clear that the most important parameter taken into consideration for the performance of structures in earthquake, i.e., PGA ranges from 0.1 g to 0.38 g, thus indicating that the tunnels may be affected by slight damage or serious damage, in case, the tunnels are devoid of lining. Therefore, the tunnels are very much safe under such massive earthquakes.

There are geological conditions which may induce earthquake damage to tunnels like fault zone activated by earthquake (fault displacement), sudden changes in condition due to contrasts in geomechanical properties of rocks, potentially liquefiable soils, pore fluids and marked anisotropy in local stress field [19]. These damages are often characterized by irreversible displacements, heavy water inflow, mechanical instability at tunnel portals, soil settlement and rupture due to liquefaction, etc. For these, modern geophysical and geotechnical investigation methods are being applied during the pre-design and design phase of these Himalayan tunnels as per guidelines of AFPS/AFTES. Multidisciplinary approach and incorporation of several factors (both static and dynamic conditions) into the designs of the underground structures to sustain the vibratory motion of the earthquakes have also been taken into consideration (cf. [1]).

Distortion of the cross section and increase in stress may occur due to compressive-tensile loads during earthquake-induced ground movement and stiffness of the lining, therefore, addition of earthquake joints design of flexible support systems is inherent here. The design of the tunnels in these regions is based on New Austrian Tunnelling Method (NATM), which put emphasis on ground behaviour reactions on creation of underground opening and to mobilize self-supporting capability of the ground by dual flexible lining without continuous deformation monitoring [25]. Structural verifications of the lining are used to be performed according to EN 1992 (Eurocode 2) for Ultimate Limit State (ULS) and Serviceability Limit State (SLS). Further to mention that tunnels can be subdivided into two parts, viz. tunnel portal areas (vertical cover up to 20 m) and deep structures (vertical cover beyond 20 m). Tunnel portal areas are considered as surface structures and heavy reinforced lining is used to done here. Whereas, the deep lying part of the tunnels is not significantly affected by earthquake damage because loads are attenuated at depth. However, special arrangements for active fault crossings zones, like enlargement of excavation section, stress controller, etc., are also supposed to be used.

5 New Austrian Tunnelling Method (NATM)

New Austrian Tunnelling Method (NATM) is an observational method, which is based on a concept whereby the ground surrounding an underground opening becomes a load bearing structural component through activation of a ring-like body of supporting ground [25, 39]. NATM is tunnelling concept starting from the initial design stages of an underground structure until the execution and construction. At present the designs of tunnels in Himalayan regions are mostly based on NATM, which is based on a concept whereby the ground surrounding an underground opening becomes a load bearing structural component through activation of a ring-like body of supporting ground. NATM also known as Sequential Excavation Method (SEM) works on understanding of the behavior of the ground as it reacts to the creation of an underground opening. The cross-section of the tunnels is a modified horse-shoe shape to promote smooth stress redistribution around newly created opening. Primary support (shotcrete, lattice girder and rockbolts) allowed a controlled ground deflection to mobilize inherent shear strength in the ground and to initiate load distribution. It also avoided development of wedge failure and generated a rock mass ring with significantly improved ground strength. Concavely rounded excavation surfaces initiate confinement forces and limit bending and tension forces (Fig. 3).

The primary stress in the surrounding ground before excavation depends upon the overburden pressure and the tectonic stresses. In case of the tunnels in SRRP, the overburden pressure is not very high but the tectonic stresses are high due to its nearness to the thrusts (MBT, MCT, etc.). After the excavation of tunnel, the tangential stresses increased. The induced tangential and radial stresses exceed strength of the surrounding ground, which yielded a plastic zone around the tunnel, which significantly controlled the tunnel behaviour (Fig. 3). Depending on the geological and geotechnical conditions, different failure modes are used to be defined, and depending on the potential failure modes, project specific requirements, boundary conditions and specific construction measures to ensure stability have been chosen. Considering the behavioural categories and in-situ stress depending on overburden, appropriate support measures have been used, which is made possible by considering precedent experiences in other similar tunnels. Supports were further verified by Confinement-Convergence Method (CCM) and numerical models. Based on the ground characteristics and the determined ground behaviour a feasible construction concept is chosen, consisting of excavation method, excavation sequence, support measures and auxiliary methods [7, 17, 22]. To evaluate the deformation, 3D monitoring and tunnel instrumentation are used to be done, for which bireflex targets as Deformation Monitoring Points (DMP) were installed in the tunnel roof and at

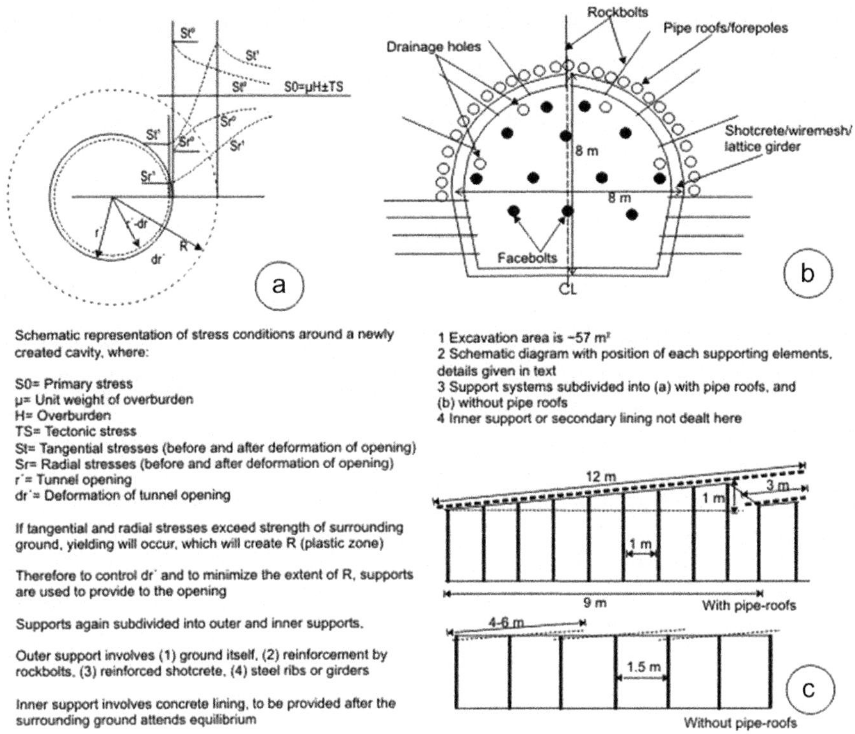

Fig. 3 Concept of NATM and general support systems in NATM

selected points along the tunnel walls (5 in a section). Vertical, horizontal and longitudinal (in tunnel direction) movements were measured from the targets (cf. [3, 27]). Other instruments like extensometer, strain gage, etc., were also used. When the surrounding ground of the tunnel attains its equilibrium state, after confirmation from deformation monitoring, final or secondary or permanent lining, that too, reinforced will be done.

In SEM-NATM, dynamic forces of earthquake have also been considered to design the permanent lining in the tunnels (cf. [24]). Similarly, the designs of the tunnels in DSH have also been done accordingly by following the relevant standards like [23] and guideline 'Seismic Design for Underground Structures' of ITA. Structural verifications of the lining are used to be performed according to EN 1992 (Eurocode 2) for Ultimate Limit State (ULS) and Serviceability Limit State (SLS). As the overall ground condition in this DSH is poor to very poor, the reinforced permanent lining has to be used. Tunnel portal areas have been considered as surface structures and heavy reinforced permanent lining has been designed. Whereas, the deep lying part of the tunnels is not significantly affected by earthquake damage because loads are attenuated at depth, the reinforced permanent lining has been designed as per the encountered groundmass condition, as numerous shear zones, local faults, etc., are encountered throughout. Special arrangements for active fault crossing zones, like enlargement of excavation section, stress controller, etc., can also be considered, if required. Even now-a-days advanced methods, like Tunnel Seismic Prediction (TSP) for predicting such zones in advance are also available, which will allow the tunnelling team to plan properly for these critical zones [11, 12].

6 Conclusion

Recent studies revealed that PGA in DSH is less than 0.4 g and also the estimated PGA for earthquake of magnitude 8.3 is also not high. It is to conclude here that the tunnels in seismically active areas are quite safe as peak ground acceleration during earthquakes never reaches 0.6 g to cause immense damage to the tunnel lining in Himalayan regions and the modern designing aspects of the tunnels have all considerable factors for earthquake induced ground movement in them. However, to increase the reliability of the underground structures and to accommodate larger events at closer distances, it is becoming necessary to consider the effect of dynamic loads generated by earthquake. Some geological features like faults and shear zones, etc., cross the tunnel alignment, therefore, reinforced permanent lining in NATM, use of stress controller, excavation of enlarged section, etc., have also been kept under consideration. Even modern technologies are available for the prediction of any critical zone (including behaviour of the zone), like faults, water bearing zones, etc., which may cause damage to underground structures during earthquakes. Most importantly, the tunnels ensure less damage to the environments and biodiversity and are also used to avoid instable slopes which fail during earthquake induced ground vibration. Tunnels are useful in sustainable infrastructural developments in the seismically active Himalayan regions. There are ample scopes for further study on this topic, as future researches on impact of earthquakes on tunnels quantitatively and estimation of site-specific PGA by instrumentation for considering that in seismic design are required

Acknowledgements The author is grateful to the Ircon International Limited and Indian Railways for providing the opportunity to work on the geological and geotechnical aspect before and during the construction of tunnels in SRRP. The author is very much thankful to the people who were engaged in the project. This research did not receive any specific grant from funding agencies in the public, commercial or not-for-profit sectors.

References

1. AFPS/AFTES (2001) Guidelines on earthquake design and protection of underground structures
2. Yin A (2006) Cenozoic tectonic evolution of the Himalayan orogeny as constrained by along-strike variation of structural geometry, exhumation history and foreland sedimentation. Earth Sci Rev 76:1–131
3. Barla G, Bonini M, Debernardi D (2008) Time dependent deformations in squeezing tunnels. In: International association for computer methods and advances in geomechanics (IACMAG), pp 1–6
4. Baruah S, Saikia S, Baruah S, Bora PK, Tatevossian R, Kayal JR (2016) The September 2011 Sikkim Himalaya earthquake Mw 6.9: is it a plane of detachment earthquake? GeomatS, Nat Hazards Risk 7:248–263. https://doi.org/10.1080/19475705.2014.895963
5. Bilham R (2019) Himalayan earthquakes: a review of historical seismicity and early 21st century slip potential. Geol Soc Lond, Spec Publ 483. https://doi.org/10.1144/SP483.16
6. Biswas SS, Pal R (2016) Causes of landslides in Darjeeling Himalayas during June to July, 2015. J Geogr Nat Disasters 6. https://doi.org/10.4172/2167-0587.1000173
7. Button E, Reidmuller G, Schubert W, Klima K, Medley E (2004) Tunnelling in tectonic melanges- accommodating the impacts of geomechanical complexities and anisotropic rock mass fabrics. Bull Eng Geol Env 63:109–117
8. Chen Z, Shi C, Li T, Yuan Y (2012) Damage characteristics and influence factors of mountain tunnels under strong earthquakes. Nat Hazards 61:387–401
9. Collins BD, Jibson RW (2015) Assessment of existing and potential landslide hazards resulting from the April 25, 2015; Gorkha Nepal earthquake sequence. U S Geol Surv, Open-File Rep 1142
10. De R, Kayal JR (2003) Seismotectonic model of the Sikkim Himalaya: Constraint from microearthquake surveys. Bull Seismol Soc Am 91:1395–1400
11. Dhang PC (2019) Report based on Tunnel Seismic Prediction (TSP) at a tunnel in lesser Himalaya, Jammu & Kashmir. J Geol Soc India 94:646. https://doi.org/10.1007/s12594-019-1372-9
12. Dickmann T (2014) 3D Tunnel Seismic Prediction: a next generation tool to characterize rock mass condition ahead of the tunnel face. J Rock Mech Tun Technol (JRMTT) 20:35–47
13. Dowding CH, Rozen A (1978) Damage to rock tunnels from earthquake shaking. J Geotech Eng Division ASCE 104:175–191
14. Eurocode 2 (1992) Design of concrete structures-Part 2-General Rules-structural fire design
15. Fan W, Shearer PM (2015) Detailed rupture imaging of the 25th April 2015, Nepal Earthquake using teleseismic P waves. Geophys Res Lett 42:5744–5752
16. Ghosh B, Pappin JW, So MML, Hicyilmaz KMO (2012) Seismic hazard assessment in India, 15 WCEE, LISBOA
17. Goricki A, Rachaniotis N, Hoek E, Marinos P, Tsotsos S, Schubert W (2006) Support decision criteria for tunnels in fault zones. In: Proceedings of the 55th Geomechanics Colloquium, vol 24. Salsberg, Felsbau, pp 1–14
18. GSI (2005) Geological map of Himalaya, the Director General, Geological Survey of India (2005)
19. Hashash YMA, Hook JJ, Schmidt B, Yao JIC (2001) Seismic design and analysis of underground structures. Tunn Undergr Space Technol 16:247–293
20. Hazarika P, Kumar MR, Srijayanthi G, Raju PS, Rao NP, Srinagesh D, Transverse tectonics in the Sikkim Himalaya: evidence from seismicity and focal- mechanism data. Bull Seismol Soc Am 100:1816–1822
21. Heidbach O, Tingay M, Barth A, Reinecker J, Kurfeß D, Müller B (2009) World stress map, II edition, Helmholtz Centre Potsdam–GFZ German Research Centre for Geosciences, Commission for the Geological Map of the World

22. Hoek, E (1999) Support for very weak rock associated with faults and shear zones. In: Distinguished lecture for the opening of the international symposium on rock support and reinforcement practice in mining. Kalgoorlie, Australia, pp1–20
23. IS 1893 Part I (2002) Criteria for earthquake resistant design of structures. Bereau of Indian Standards, p 45
24. Jaramilo CA (2017) Impact of seismic design on tunnels in rock-case histories. Undergr Space 2:106–114
25. Karakus M, Fowell RJ (2004) An insight into the New Austrian Tunnelling Method (NATM), ROCKMEC′2004-VIIth Regional Rock Mechanics Symposium, Sivas, Turkey
26. Kellet D, Grujic D, Mottram C, Mukul M (2004) Virtual field guide for the Darjeeling-Sikkim Himalaya, India, Jour. of the Virtual Explorer (Electronic version, ISSN 1441–8142, v. 47, paper 5. In: Montomoli C, Carosi R, Law R, Singh S, Rai SM (eds) Geological field trips in the Himalaya, Karakoram and Tibet
27. Kontogianni V, Stiros S (2003) Earthquakes and seismic faulting: effects on tunnels. Turkish J Earth Sc 12:153–156
28. Lay T, Ye L, Koper KD, Kanamori H (2017) Assessment of telesiesmically determined source parameters for the April 25, 2015 M_W 7.9 Gorkha, Nepal Earthquake and the May 12, 2015 M_W 7.2 aftershocks, Tectnophysics, 714 to 715, pp 4–20
29. Mahajan AK, Gupta V, Thakur VC (2012) Macroseismic field observations of 18th September 2011 Sikkim earthquake. Nat Hazards 63:589–603
30. Mishra OP, Chakrabortty GK, Singh OP, Ghosh DC, Mukherjee KK, Das (2010) A Report on Seismogenesis in the Sikkim-Darjiling Himalayas and Assimilation of Dynamic Snap Shots of the Region: Future Vulnerability: GSI Report
31. Mukhopadhyay B (2011) Clusters of moderate size earthquake along Main Central Thrust (MCT) in Himalaya. Int J Geosci 2:318–325
32. Mukul M, Jaiswal, M, Singhvi AK (2007) Timing of recent out-of-sequence active deformation in the frontal Himalayan wedge: Insights from the Darjiling sub-Himalaya, India. Geology 35:999–1003
33. Nath SK, Vyas M, Pal I, Sengupta P (2005) A seismic hazard scenario in the Sikkim Himalaya from seismotectonics, spectral amplification, source parameterization, and spectral attenuation laws using strong motion seismometry. J Geophys Res 110:1–24
34. Ni JF, Barazangi M (2007) Siesmotectonics of the Himalayan collision zone: geometry of the underthrusting Indian Plate beneath the Himalaya. J Geophys Res 89:1147–1163
35. Owen GN, Scholl RE (1981) Earthquake engineering of large underground structures, Final Report FHWA-RD-80–195. Washington DC, USA, Federal Highway Administration
36. Pal R, Biswas SS, Mondal B, Pramanik MK (2016) Landslides and floods in the Tista Basin (Darjeeling and Jalpaiguri districts) Historical Evidence, Causes and Consequences, pp 66–72
37. Pandey MR, Tandulkar RP, Avouac JP, Vergne J, Heritier T (1999) Seismotectonics of the Nepal Himala ya from a local seismic network. J Asian Earth Sci 17:703–712
38. Power MS, Rosidi D, Kaneshiro JY (1998) Seismic vulnerability of tunnels and underground structures revisited, North American Tunneling'98, Ozdemir (ed), p 243–250
39. Rabcewicz L (1964) The New Austrian Tunnelling Method, Part One, Water Power, p 453–457, Part Two, Water Power, p 511–515
40. Raisinghani B, Purohit SP (2016) Influence of parameters on performance evaluation of designed RC buildings: seismic hazards. Int J Struct Eng 7:378
41. Rossetto T, Peiris N (2009) Observations of damage due to the Kashmir earthquake of October 8, 2005 and study of current seismic provisions for buildings in Pakistan. Bull Earthq Eng 7:681–699
42. Sengupta A, Gupta S, Anbarasu K (2010) Landslides–investigations and mitigation in eastern Himalayan region. J Indian Roads Congr Paper No 560:133–142
43. Sharma ML, Sinvhal A, Singh Y, Maheshwari BK (2013) Damage survey report for Sikkim earthquake of 18 September 2011. Seismol Res Lett 84:49–56
44. Sharma RS (2009) Cratons and fold belts of India. Springer, p 304

45. Sharma S, Judd WR (1991) Underground opening damage from earthquakes. Eng Geol 30:263–276
46. Sunuwar SC (2018) Nepal Earthquake 25th April 2015: Hydro projects damaged, risks and lessons learned for design consideration. J Nepal Geol Soc 55:141–149
47. Zhang X, Jiang Y, Maegawa K (2020) Mountain tunnel under earthquake force: a review of possible causes of damages and restoration methods. J Rock Mech Geotech Eng 12:414–426
48. Zhang X, Jiang Y, Sugimoto S (2018) Seismic damage assessment of mountain tunnel: a case study on Tawarayama Tunnel due to the 2016 Kumamoto Earthquake. Tunn Undergr Space Technol 71:138–148

Subsurface Investigation Using Vertical Electrical Sounding Method to Evaluate Surface Instabilities in Kohima Town, Nagaland

Meripeni Ezung, C. Nokendangba Chang, Temsulemba Walling, and C. Chelladurai

Abstract Normal life is regularly disrupted in Kohima, the state capital of Nagaland, due to frequent incidences of land instabilities. Six unstable and two stable areas sharing similar geology in different parts of the town were selected for electrical resistivity study. Vertical Electrical Sounding (VES) was carried out to determine the subsurface conditions by analyzing variations in the resistivity values following the Schlumberger configuration. The objective of the present study is to evaluate the influence of subsurface conditions on surface instabilities. The VES data show three to five subsurface geo-electric layers, comprising of the topsoil followed by various rock layers that are found to be weathered and fractured. K type is the dominant curve which indicates higher saturation at the lower strata. Existing borehole data were used to validate the results of the VES. Thickness of soil overburden, types and condition of the subsurface rocks, the depth of water table have some influence on the type and incidences of slope instabilities.

Keywords Kohima · Land instabilities · Vertical Electrical Sounding (VES) · K type curve

1 Introduction

Kohima, the administrative capital of Nagaland, is located in the southern part of the state hosting several major offices and institutions besides facilitating as the main artery for the state of Manipur and several other towns, as the Trans-Asian Highway,

M. Ezung · C. Chelladurai
Department of Physics, St. Joseph University, Ikishe Model Village, Virgin Town, Dimapur, Nagaland 797115, India

C. N. Chang · T. Walling (✉)
Department of Geology, Nagaland University, Kohima Campus, Meriema, Nagaland 797004, India
e-mail: tem_wall@yahoo.com

S. Mitra et al. (eds.), *Disaster Management and Risk Reduction: Multidisciplinary Perspectives and Approaches in the Indian Context*,
https://doi.org/10.1007/978-981-99-6395-9_9

AH 1 passes through this town. Geologically, almost the entire Kohima town is made up of the Tertiary Disang Group of rocks consisting predominantly of shales with some siltstones and sandstone. These rocks are folded, faulted and highly crumpled, crushed and weathered [1–4]. The shale exposures are dark grey and splintery in nature and are weathered to clay in many areas and their interaction with water is responsible for several slope instabilities in the town [5–9]. The region experiences subtropical climatic condition and the copious amount of rainfall the town receives during the monsoon season contributes to weathering of the rock mass.

In hilly terrain, the occurrence of landslides is a frequent and natural process which is often triggered by external factors such as heavy rainfall, seismic and tectonic activities, change in water level, stream erosion, etc. [10–13]. In addition, developments on hill slopes for construction of roads, deforestations, buildings and other urban expansion activities, and also increase in various land use practices contribute to slope instability. The shear strength of rock mass is greatly decreased due to weathering and is considered an important factor for the occurrence of landslides. Shales that are often associated with landslides are weathered shale that has become clay-rich soil which has very low shear strength, especially when saturated with water. These materials easily move down slopes as landslides, causing much devastation. The maximum rainfall in the study area occurs during the monsoon season and is mostly confined between the month of May and August with average rainfall in last 5 years ranging between 1000 and 1800 mm (Source: Department of Soil and Water Conservation, Kohima, Nagaland). Surface waters and precipitation infiltrate through the soil and fractured rocks causing the shale to swell, and also leading to a build-up of porewater pressure in the subsurface bringing about a reduction in the slope stability [14]. Such infiltrations are considerably more prominent in barren, fresh cut surfaces and loose debris than in vegetated and covered areas.

Past workers have carried out case studies on several landslides in and around the township by employing various methods to determine the role of geology and geotechnical influence and generated significant information on causative factors. Slope classification maps of Kohima town and geotechnical reports were prepared by Sharda and Bhambay [15]. Some authors [16–18] have carried out slope stability investigations in the study area and the AH 1 to generate landslide hazard zonation maps and suggested remedial/mitigation measures. Various researchers have also added to landslide literature of the region with their studies on surface instability [9, 14, 16–25]. Other sub-surface factors relating to bedrock depth, fractures, voids, groundwater infiltration and interactions are important parameters for stability and require further investigation. Landslides are widespread and pose very high hazard in this hilly township, however, studies involving the use of electrical resistivity and corresponding information on the subsurface conditions are very limited. Hence, this study aims to focus on promising aspect of Vertical Electrical Sounding (VES) method to identify the inherent condition of the slope materials and possibly generate additional information on the causes of land instabilities in the area.

2 Objectives

To investigate the subsurface conditions at selected locations and evaluate their influence on surface instabilities.

3 Study Area

Eight (8) sites were selected for study in and around Kohima town. They are part of the Survey of India Topographic sheet number 83 K/2 and lies between 25°39′35″25°43′40″ North latitudes and 94°05′25″–94°07′50″ East longitudes (Fig. 1).

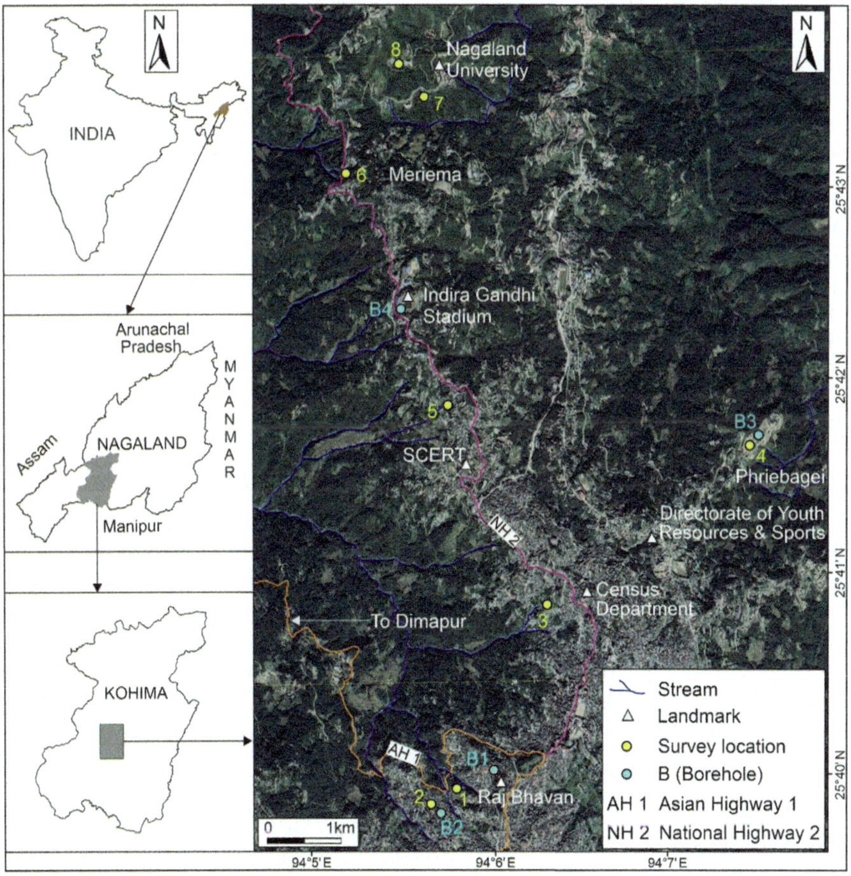

Fig. 1 Google Earth image showing location map of the study area

Table 1 Geographical co-ordinates for the study areas and bore hole locations

Sl. No	Location	Geographical coordinates	Bore hole location	Geographical coordinates
1	Officer's Hill	N 25°39′54.6″ E 94°05′47.8″	Officer's Hill (B1)	N 25°39′59.65″ E 94°06′02.17″
2	Merhulietsa	N 25°39′52.8″ E 94°05′41.0″	Merhülietsa (B2)	N 25°39′50.28″ E 94°05′41.20″
3	Pezielietsie	N 25°40′51.3″ E 94°06′21.5″	Bore hole data not available	
4	Phriebagei	N 25°41′36.57″ E 94°07′27.84″	Phriebagei (B3)	N 25°41′43″ E 94°07′36.72″
5	Perezie	N 25°41′51.31″ E 94°05′44.42″	Bore hole data not available	
6	Meriema	N 25°43′01.0″ E 94°05′08.4″	IG Stadium (B4)	N 25°42′20.58″ E 94°05′29.37″
7	Guest House, Meriema	N 25°43′23″ E 94°05′36″	Bore hole data not available	
8	English Department, Nagaland University	N 25°43′32″ E 94°05′25″		

Six unstable sites were identified for investigation along with two stable areas for comparative study. The unstable areas are located at Officer's Hill, Merhülietsa, Pezielietsie, Phriebagei, Perezie, Meriema and the two stable sites are Guest House, Meriema and English Department, Nagaland University (Table 1).

4 Materials and Methods

Electrical resistivity surveys are achieved by applying two procedures; Vertical Electrical Sounding (VES), which is based on the resistivity variations with depth, and the horizontal profiling, based on the resistivity variations with lateral extent.

One dimensional (1D) VES method for measuring the electrical resistivity is based on estimation of the electrical conductivity or resistivity of the medium. In this type of sounding, centre of the electrode system is kept fixed and measurements are taken at different locations for various values of current electrode separations starting from small initial values to several meters. Here, four electrodes are used, two outer electrodes for the current (source) electrode and two inner electrodes for the potential (receiver) electrode. The current (I) is introduced between one pair of the current electrodes, say, A and B and the potential difference (δV) produced as a result of the current flow is measured with the help of the other pair of electrodes which is the potential electrodes, say, M and N. All the four electrodes should be in a line, with the outer electrode spacing kept large as compared to the inner electrode spacing

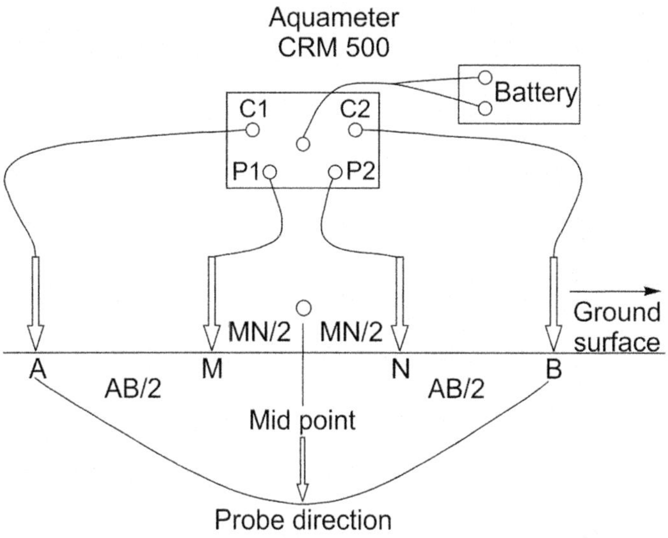

Fig. 2 Schematic diagram for the Schlumberger array

(Fig. 2). Thus, variation in the electrical characteristics with depth is obtained from the apparent resistivity variation with the current electrode spacing. The apparent resistivity (ρ_a) of geologic formation is then computed by using the formula [26].

$$\rho_a = K\frac{\delta V}{I}, \text{ where K} = \frac{\pi l}{2}\left[(L/l)^2 - 1\right]$$

is the geometrical spacing factor. (1)

The electric topology of the geological bedding model of the earth, studied by means of sounding curve characteristics is used. In order to classify the sounding curves, the four basic categories for subsurface layers with resistivity ρ1, ρ2 and ρ3, with ρ1 at the top followed by ρ2 and ρ3 are given as:

$\rho 1 < \rho 2 < \rho 3$: A-type.
$\rho 1 < \rho 2 > \rho 3$: K-type.
$\rho 1 > \rho 2 < \rho 3$: H-type.
$\rho 1 > \rho 2 > \rho 3$: Q-type.

The present study utilized the VES method employing the Schlumberger configuration [27]. In this type, current electrode spacing is kept greater than the potential electrode spacing by an order of 5, i.e., AB/2 ≥ 5 MN/2. This configuration gives a good resistivity contrast between the saturated and unsaturated layers and also the software available for interpretation is convenient.

A resistivity meter, Aquameter CRM 500 was used to acquire the resistivity data sets. Each VES location has varied profile lengths due to surface restrictions (Table 2).

Table 2 Profile length for each study location

Sl. no	Location	Profile length (m)
1	Officer's Hill	100
2	Merhülietsa	80
3	Pezielietsie	200
4	Phriebagei	50
5	Perezie	120
6	Meriema	200
7	Guest House, Meriema	40
8	English Department, Nagaland University	80

Each apparent resistivity value was computed using Eq. (1). The field data were interpreted using the iterative resistivity sounding interpretation software IPI2win [28, 29].

5 Results and Discussion

VES data demarcate the subsurface of the study area in terms of the soil cover followed by various rock layers that are weathered and fractured (Table 3) [30]. Sounding curves of the types K, KH and HKH delineate to show three to five layers (Fig. 3).

The curves presented at Officer's Hill, Merhülietsa, Phriebagei and Perezie areas show increasing resistivity to a maximum and then decreasing to lower resistivity values ($\rho 1 < \rho 2 > \rho 3$) and are thus designated as K type curves. Fresh shales or shale with sandstone intercalation is indicated between the top soil and the lower saturated weathered shales or clay. The resistivity of the first layer comprising of the top soil or shale exposures ranges between 50.3 and 195 Ω m with a thickness of 0.9–4.26 m. In the second layer, shale with sandstone intercalations with low saturation show higher resistivity values ranging from 106 Ω m to 827 Ω m with thickness of 1.26–7.814 m. The third layer with lower resistivity values ranging between 11.8 and 97.25 Ω m at greater depths infers a zone of saturation of the rocks which may have led to the weathering of the shales. These areas are considered suitable for groundwater extraction [31, 32].

The KH type with four-layer sounding model of ($\rho 1 < \rho 2 > \rho 3 > \rho 4$), is shown at the Guest house, Meriema, which is one of the stable areas in the present study. This area has top soil thickness of 0.627 m, followed by more compact rock layer comprising of fresh shale or shale with sandstone intercalations at a depth of 1.712 m. The third layer indicates the presence of the fractured rocks with thickness of 10.42 m at 12.13 m depth, followed by the region with highly weathered shale, which may be considered as the suitable zone for groundwater reservoir [33–38].

Table 3 Layer parameters for the study areas

VES station	Resistivity (Ω m)	Thickness (m)	Depth (m)	Inferred lithology	Curve type
Officer's Hill	$\rho1 = 182.7$	h1 = 1.816	d1 = 1.816	Top soil/shale exposure	K
	$\rho2 = 442.2$	h2 = 7.814	d2 = 9.63	Shale/shale with sandstone intercalation	
	$\rho3 = 97.25$	–	–	Weathered shale/ clay	
Merhülietsa	$\rho1 = 50.3$	h1 = 4.26	d1 = 4.26	Top soil/shale exposure	K
	$\rho2 = 106$	h2 = 4.97	d2 = 9.23	Shale/shale with sandstone intercalation	
	$\rho3 = 11.8$	–	–	Highly weathered shale/clay	
Pezielietsie	$\rho1 = 136$	h1 = 1.94	d1 = 1.94	Top soil/ shale exposures	HKH
	$\rho2 = 9.98$	h2 = 2.36	d2 = 4.3	Highly weathered shale/clay	
	$\rho3 = 100$	h3 = 5.37	d3 = 9.67	Shale/shale with sandstone intercalation	
	$\rho4 = 4.3$	h4 = 12.8	d4 = 22.5	Groundwater/ clayish material	
	$\rho5 = 2990$	–	–	Sandstone	
Phriebagei	$\rho1 = 195$	h1 = 1.99	d1 = 1.99	Top soil/Shale exposure	K
	$\rho2 = 827$	h2 = 2.88	d2 = 4.87	Fresh shale layer/ shale with sandstone intercalation	
	$\rho3 = 34.1$	–	–	Weathered shale/ clay	
Perezie	$\rho1 = 122$	h1 = 0.9	d1 = 0.9	Top soil/ shale exposures	K
	$\rho2 = 405$	h2 = 1.26	d2 = 2.16	Shale/shale with sandstone intercalation	
	$\rho3 = 89.5$	–	–	Weathered shale/ clay	
Meriema	$\rho1 = 137$	h1 = 0.827	d1 = 0.827	Top soil/shale exposure	HKH

(continued)

Table 3 (continued)

VES station	Resistivity (Ω m)	Thickness (m)	Depth (m)	Inferred lithology	Curve type
	$\rho 2 = 24.26$	h2 = 0.364	d2 = 1.19	Weathered shale layer/clay	
	$\rho 3 = 67.08$	h3 = 13.09	d3 = 14.28	Weathered shale layer	
	$\rho 4 = 18.15$	h4 = 32.66	d4 = 46.94	Highly weathered shale/clay	
	$\rho 5 = 6954$	–	–	Sandstone	
Guest House, Meriema	$\rho 1 = 516.9$	h1 = 0.627	d1 = 0.627	Top soil/sandstone exposure	KH
	$\rho 2 = 992.8$	h2 = 1.085	d2 = 1.712	Shale/shale with sandstone intercalation	
	$\rho 3 = 215$	h3 = 10.42	d3 = 12.13	Fractured shale	
	$\rho 4 = 39.93$	–	–	Highly weathered shale/clay	
English Department, Nagaland University	$\rho 1 = 753$	h1 = 2.41	d1 = 2.41	Top soil/sandstone exposure	HKH
	$\rho 2 = 179$	h2 = 1.29	d2 = 3.7	Fractured shale	
	$\rho 3 = 571$	h3 = 4.56	d3 = 8.26	Shale/shale with sandstone intercalation	
	$\rho 4 = 10.6$	h4 = 6.76	d4 = 15	Highly weathered shale/clay	
	$\rho 5 = 4272$	–	–	Sandstone	

2 (two) unstable areas at Pezielietsie and Meriema and 1 (one) stable area located at English Department, Nagaland University show the five-layer sounding HKH curves type with resistivity values given as $\rho 1 > \rho 2 < \rho 3 > \rho 4 < \rho 5$. The first layer has top soil coverage with thickness between 0.827 and 2.41 m and resistivity values ranging between 136 and 753 Ω m. Highly weathered/fractured rocks are found at the second and the fourth layers with data showing thickness from 0.364 m to 32.66 m and resistivity values ranging between 4.3 and 179 Ω m indicating two consecutive groundwater levels or clay material. These layers are therefore characterised by weathered rocks. The third layer with resistivity values between 67.08 and 571 Ω m indicates layers containing shale with less moisture content. In the fifth layer, presence of sandstone is found with higher resistivity values at greater depth.

The results obtained above are in consonance with local geology derived from available borehole data of four areas near the case studies (Fig. 4) which also confirms the presence of fractured/weathered shale layer as the near-surface materials.

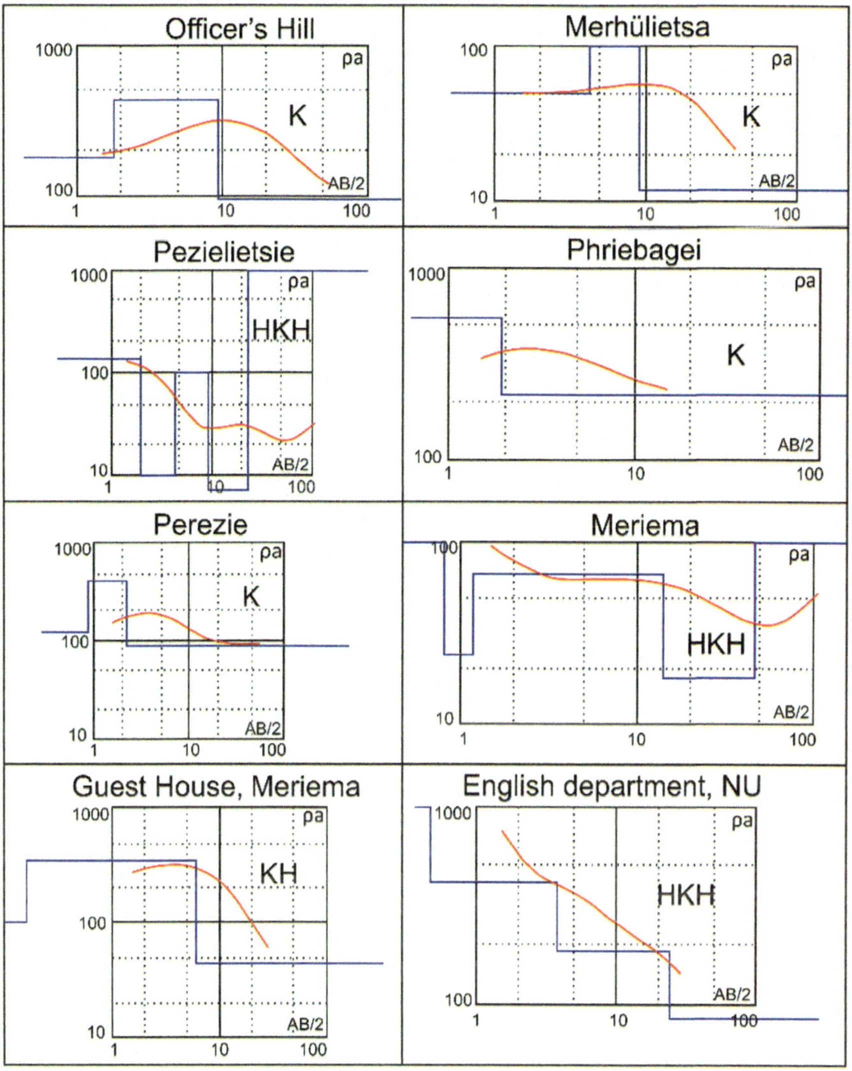

Fig. 3 Resistivity sounding curves of K, KH and KHK types

6 Conclusion

Resistivity analyses of some selected sites have been determined through geoelectrical survey in Kohima town. Analysis of the sounding data delineates the subsurface layers based on the resistivity values. The subsurface profiles from available borehole data show the bedrock containing weathered/fractured layers which agrees with the resistivity sounding analysis. The first layer, occupying the surficial areas shows moderate—high resistivity values of 50.3–753 Ω m, indicating top soil at depths of

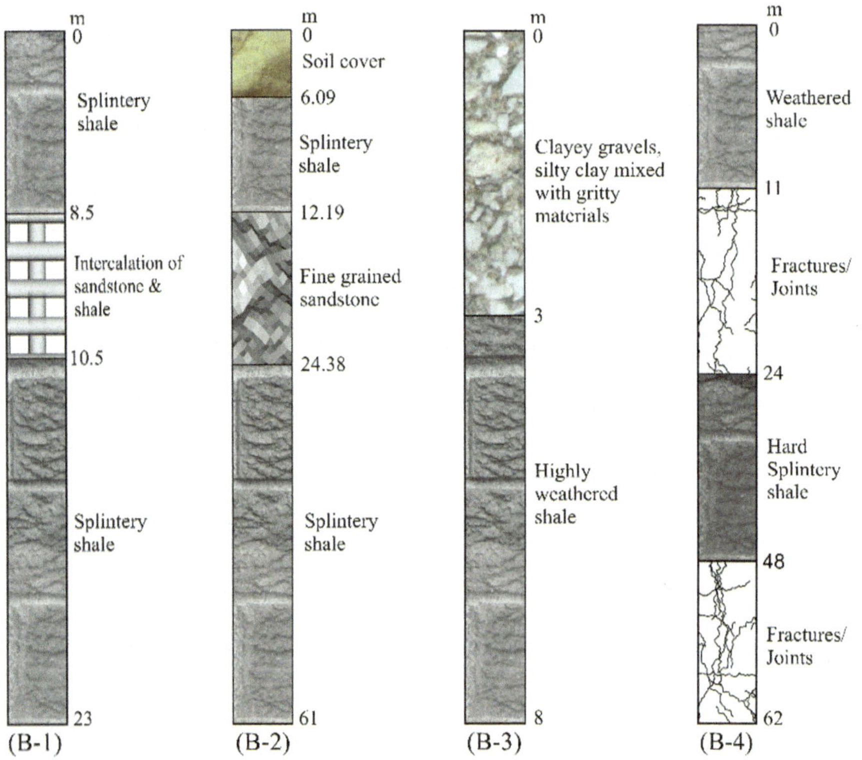

Fig. 4 Soil profiles for bore hole B-1, B-2, B-3 and B-4. *Source* Directorate of Geology and Mining, Dimapur, Nagaland and J. R. Engineers & construction, Guwahati

0.627–4.26 m. Six unstable areas and one stable area show soil thickness of $\approx$ or >1 m and one of the stable areas with <1 m. This indicates that the thickness of soil cover has less influence on the surface instabilities. Layers showing lower resistivity values indicate that weathered rocks with clay layers are encountered at various depths in the study areas. All the unstable areas have their zone of weathering within 10 m depth from the surface. The stable sites have groundwater depth more than 10 m. It is therefore concluded that the depth of water table influences incidences of slope instabilities as groundwater can be correlated with weathering of the country rocks which corresponds to the depth of the weathered rocks in the study area. Moreover, the presence of weathered rocks at shallow depths below the surface is more prone to surface instabilities, irrespective of the type of rock, fractures and thickness of the soil cover.

Acknowledgements The authors wish to acknowledge the Department of Geology, Nagaland University for providing the necessary logistics and equipment to carry out the investigation. The support given by Mr. Mehilo Apon, Mr. Notoka and Ms. Olivia Richa, Research scholars, Department of Geology, during the field works is also gratefully acknowledged.

References

1. Directorate of Geology and Mining, Nagaland Miscellaneous Publication, No 1 (1978)
2. Nandy DR (2000) Geodynamics of Northeastern India and the adjoining region. ACB Publications, Dehradun
3. Geological Survey of India (2011) Geology and Mineral resources of Manipur, Mizoram, Nagaland and Tripura. Miscellaneous Publication, No. 30, Part IV, vol 1(Part-2)
4. Gogoi M, Mathur N, Kumar TS, Walling T, Pukan S (2021) Geochemical characterization of shales of the Eocene Disang Group Kohima Syncline India: inferences to hydrocarbon potential and depositional environment. Pet Res 6(1):42–56. https://doi.org/10.1016/j.ptlrs.2020.09.001
5. Lotha KA (1994) A note on the geotechnical investigation on landslide at Cheipfutsiepfe, Lower AG colony, Kohima Town. Nagaland. Unpubl. Rep. Geoenvir. Cell, DGM, Nagaland
6. Aier I, Supongtemjen Khalo M, Thong GT (2009) Geotechnical assessment of the Mehrülietsa slide in Kohima, Nagaland. In: Kumar A, Kushwaha RAS, Thakur B (eds) Journal of Earth System Science, vol 1. Concept Publishing Company, New Delhi, pp 81–88
7. Aier I, Thong GT (2011) Supongtemjen: Geological evaluation of surface instability along NH 39 (180 km), west of Raj Bhavan, Kohima, Nagaland. In: Singh TN, Sharma YC (eds) Slope stability—natural and man made slope. Vayu Education of India, New Delhi, pp 192–201
8. Aier I, Singh MP, Thong GT, Ibotombi S (2012) Instability analyses of Merhülietsa slide, Kohima. Nagaland Nat Haz 60:1347–1363. https://doi.org/10.1007/s11069-011-9913-6
9. Chang CN (2022) Slope stability studies of some weak zones between hospital colony and Jotsoma bypass, Kohima, Nagaland. Unpublished PhD thesis, Nagaland University, Kohima
10. Kumar CR, Kesiezie N, Singh N, Maiti S (2016) Seismic site response studies for microzonation and hazard assessment of Kohima, Nagaland, North Eastern Region of India. Indian J Geosci 71:501–518
11. Jamir I, Gupta V, Kumar V, Thong GT (2017) Evaluation of potential surface instability using finite element method in Kharsali Village, Yamuna Valley. Northwest Himalaya J Mount Sci 14:1666–1676. https://doi.org/10.1007/s11629-017-4410-3
12. Jamir I, Gupta V, Thong GT, Kumar V (2019) Litho-tectonic and precipitation implications on landslides, Yamuna Valley, NW Himalaya. Phys Geog 41(4). https://doi.org/10.1080/02723646.2019.1672024
13. Roccati A, Faccini F, Luino F, Ciampalini A, Turconi L (2019) Heavy rainfall triggering shallow landslides: a susceptibility assessment by a GIS-approach in a Ligurian Apennine Catchment (Italy). Water 11(3):605. https://doi.org/10.3390/w11030605
14. Chang CN, Ezung M, Apon M, Supongtemjen WT, Thong GT (2021) Assessment of landslides along NH 29 in the Kevüza Area, Kohima, Nagaland. Indian Geotech J 51:841–860. https://doi.org/10.1007/s40098-021-00566-z
15. Sharda YP, Bhambay GC (1980) Kohima town, Nagaland: a decade of environmental geoscientific studies. GSI Sp. Publ. Series, p 9
16. Walling T (2005) Geological investigation of land instability in Kohima Town, Nagaland. Unpublished PhD thesis, Nagaland University, Kohima
17. Supongtemjen (2013) Geological investigation of land instability between Kohima and Zhadima. Unpublished PhD thesis, Nagaland University, Kohima
18. Khalo M (2016) Slope stability investigations between Kohima and Mao with special reference to NH-39. Unpublished PhD thesis, Nagaland University, Kohima
19. Kemas K, Thong GT, Walling T (2004) Chokidzü debris slide—a case study. Nagaland Univ Res J 2:89–94
20. Walling T, Lotha KA, Thong GT, Aier I (2005) Chiepfütsiepfe slide, Kohima, Nagaland. Causes and mitigation mcasures. Proc NRDMS (DST) Sem Landslide Haz Miti NE Ind 48–54
21. Aier I, Supongtemjen, Thong GT (2009) Slope mass rating and kinematic analyses along part of NH 61, Nagaland, NE India. Int J Earth Sci Eng 2:520–526
22. Sothu HN, Aier I, Thong GT (2009) Evaluation of two unstable sites along NH 150, SE of Kohima Town. Nagaland India Landslides 2:27–30

23. Supongtemjen, Thong GT (2014) Risk analyses along part of NH 2, north of Kohima town, Nagaland. Indian Landslides 7:35–44
24. Supongtemjen, Walling T, Tep S, Thong GT (2015) Stability analyses of two fresh cut slopes along NH 2, Meriema, Nagaland. Proceedings of the National Seminar on Landslides: Management and Mitigation Strategies. J Eng Geol XL:158–170
25. Walling T (2016) Supongtemjen, Chang, C.N., Thong, G.T.: Geotechnical analyses of debris slide near Secretariat junction, Kohima, Nagaland. In: Srivastava SK (ed) Recent trends in earth science research with special reference to NE India. Today and Tomorrow Publishers, New Delhi, pp 277–289
26. Telford WM, Geldart LP, Sherrif RE (1990) Applied geophysics, 2nd edn. Cambridge University Press
27. Ganesh R, Gowtham B, Manive T, Senthilkumar S, Sundrajan M (2017) Application of Resistivity methods in landslide investigations along Mettupalayam to Coonoor highway, NilgirisDistrict, Tamilnadu, India. Sch J Eng Tech 5(11):661–667
28. Vender Velpen BPA (1998) A computer processing package for DC Resistivity interpretation for IBM compatibles. ITC J 4:1–4
29. Gopinath VST, Vinodh K, Gowtham B, Arulprakasam V (2015) Geolectrical characterization of substrata by using Geoelectrical Imaging Technique in Gagilam river sub basin, Tamilnadu, India. Int J Sci Eng Appl Sci 1(6):451–457
30. Loke MH, Barker RD (1996) Practical techniques for 3D resistivity surveys and data inversion. Geophys Prospect 44:499–523
31. Sultan AS, Santos FAM (2008) 1D and 3D resistivity inversions for geotechnical investigation. J Geophys Eng 5:1–11
32. Gouet DH, Meying A, Nkoungou HLE, Assembe SP, Nouck PN, Mbarga TN (2020) Typology of sounding curves and lithological 1D models of mineral prospecting and groundwater survey within crystalline basement rocks in the east of Cameroon (Central Africa) using electrical resistivity method and Koefoed computation method. Int J Geophys. https://doi.org/10.1155/2020/8630406
33. Alile OM, Amadasun CVO, Evbuomwan AI (2008) Application of vertical electrical sounding method to decipher the existing subsurface stratification and groundwater occurrence status in a location in Edo north of Nigeria. Int J Phys Sci 3(10):245–249
34. Bairu A, G/her Y, G/her G (2013) Application of vertical electrical sounding and horizontal profiling methods to decipher the existing subsurface in river Segan dam site, Tigray, Northern Ethiopia. J Environ Earth Sci 3(10). ISSN 2224–3216
35. Kana JD, Djongyang N, Dadjé A, Raïdandi D (20150 Vertical electrical soundings for subsurface layers and groundwater investigations in the Mayo Kani area in Cameroon. Int J Sci Res ISSN (Online) 2319–7064
36. Hassan E, Rai JK, Anekwe UO (2017) Geoelectrical survey of groundwater in some parts of Kebbi state, Nigeria, a case study of Federal Polytechnic bye-pass Birnin Kebbi and Magoro Primary health centre Fakai local Government. Geosci J 7(5):141–149
37. Manimegalai MK, Gowtham B, Vinodh K (2017) Investigation of subsurface and groundwater state at Gadilam river sub-basin, Tamil Nadu, India. Int J Modn Res Revs 5(10):1632–1638
38. Umar DU, Igwe O (2019) Geo-electric method applied to groundwater protection of a granular sandstone aquifer. Appl Water Sci 9–112

The 2021 Chamoli Disaster: Is It GLOF or LLOF?

Vishal Singh, Shubham Maurya, and Arindam Dey

Abstract On February 7 2021, a debris flow in Dhauliganga and Rishiganga Valley in Chamoli (30.2937° N, 79.5603° E) wreaked havoc on the downslope. This debris flow caused the death of at least 80 people and the missing of 124 people. It nearly swept away the Rishiganga hydropower plant and severely damaged the Tapovan Vishnu gad Hydropower. The initial claim was that the cause of this disaster was 'Glacial lakes outburst flood' (GLOF). However, geologists and glaciologists scrapped this claim after studying the images before and after the event. There was no detailed experimental research or modelling to understand the origin of the flash flood in this paper. In this review, an attempt has been made to understand the cause of the flood event and challenges to the infrastructural project. Such an attempt is significant for the North-eastern region of India. The mountains in this region have a similar composition as those in the Chamoli district. Studies have shown that rapid infrastructure in this region of the country harms its mountains. Hence, the lesson learnt from the Chamoli disaster will help to shape the disaster risk reduction programs and future mitigative measures in the northeast region.

Keywords Debris dammed lake · Infrastructural project · Remote sensing · Chamoli Disaster · Rishiganga flash flood

---

V. Singh (✉) · S. Maurya · A. Dey
Center for Disaster Management and Research, Indian Institute of Technology, Guwahati, India
e-mail: vishalakhilesh@iitg.ac.in

S. Maurya
e-mail: smaurya@iitg.ac.in

A. Dey
e-mail: arindam.dey@iitg.ac.in

S. Mitra et al. (eds.), *Disaster Management and Risk Reduction: Multidisciplinary Perspectives and Approaches in the Indian Context*,
https://doi.org/10.1007/978-981-99-6395-9_10

1 Introduction

The Himalayas is the youngest mountain range which is a 'fold' type mountain formed when the Indian plate collided with the Eurasian plate. It is also known as the 'Third pole' of the Earth. Although Himalayan glaciers are near to the subtropical region, they have the highest concentration of frozen freshwater outside the polar area. In the hilly terrain, the loss of glaciers, snow, and permafrost alters the frequency, size, and location of natural hazards. This leads to increasing hazard exposure to people and infrastructure that is primarily a consequence of population expansion, tourism, and improvement in socioeconomic conditions [1]. The Himalayan area has seen a significant rise in temperature over the previous several decades, much above the world average [2]. Increased melting in the Himalayan cryosphere has resulted in the formation/expansion of glacial lakes, permafrost thaw, and snow/ice/debris avalanches [3].

The presence of fragmented, sheared, and crushed rocks [4] and steep slopes [5], along with neotectonics and structural activities alongside the main boundary thrust (MBT) [6, 7], made the western Himalayas a hotspot of mass movement [8] and slope instability [9]. Since, the late nineteenth century, the Uttarakhand Himalayan region has seen more than fifteen life-threatening events, including earthquakes, landslides, avalanches, and lake outburst flash floods [8], such as 'GLOF' in Chorabari lake that has caused the death of at least 6000 people.

Due to unfavourable terrain and extreme weather conditions, field inquiry is limited in understanding the natural hazards produced in high-altitude Himalayan sites. Satellite-based remote sensing data with numerical modelling has been more helpful in recent years in understanding these hazards better and quantifying the natural hazard process in high mountain environments [10–12]. The analytical hierarchy technique and remote sensing assisted in successfully delineating possible avalanche-prone zones in numerous sites throughout the western Himalayas [13, 14]. For classifying and vulnerability assessment of the glacial lakes in the Himalayas [15], remote sensing data is used. The risk of the Gya lake is better understood by using remote sensing data [16]. The use of remote sensing data and numerical modelling to determine the significant elements involved in the Kedarnath event [17] and the Balia Nala landslide in Nainital [4], Uttarakhand, has been proven successful. Based on satellite remote sensing, numerical modelling, and field investigations [11] reported the collapse of two adjacent glaciers in western Tibet in July and September 2016. It was discovered that the collapse caused by climate-driven and weather-driven external forces, which massively reduced the basal friction connected to subglacial water and fine-grained bed lithology. Using the Landsat archive, Sentinel-2 pictures, and the ASTER imageries, [18] found multiple Glacier detachments and rock-ice avalanches that occurred in the Petra Pervogo range, Tajikistan, between 1973 and 2019.

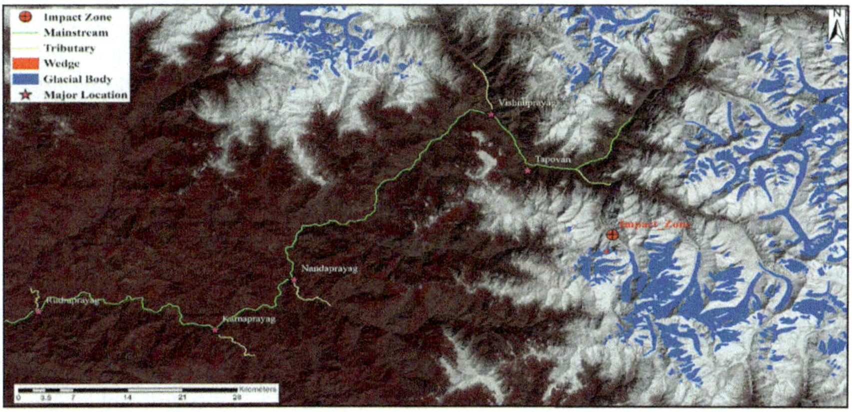

Fig. 1 Map of the study area [26]

2 Methodology

An insightful review of existing literature is the current study's research strategy because it has been done previously in many research works. A systematic review is a fundamental scientific study approach which is used to analyze, synthesize, and convey the findings and implications of a wide range of research publications on a particular problem. The approach for the review was broken down into three steps, as follows. The first stage is to determine the research problem and then formulate it. The comprehensiveness of the collected datasets was a factor in the selection of relevant literature. It is critical to identify a list of keywords and other relevant terms connected to research problems before beginning a database search. Following the definition of keywords in the preceding phase, the following step is to search for and choose the associated attempts to study this hazard. The next stage has been analyzing and reporting the findings of the literature. Figure 1 provides the map of the study area.

3 Study Area

The Rishiganga catchment covers 690 square kilometres in the western Himalayas, with numerous big valleys and hanging glaciers such as Changband, Hanuman, Trishul, Bethartoli, Ronti, and others [19]. The region has a rough topography that ranges from 1200 to 7800 m above sea level. The catchment is encircled by eight peaks. The catchment also houses the UNESCO World heritage site, namely, Nanda Devi National park and Valley of flowers. Due to hanging glaciers with steep slopes and ice calving from the glacier's terminus, the region is prone to avalanches [20, 21]. Weathered crushed and broken rock layers characterize the terrain along the

Rishiganga and Dhauliganga valleys, which sit between the Vaikrita thrust (VT) and the South Tibetan Detachment (STD) [19]. Muscovite-schist, Kyanite-biotite, leucogranite, and gneiss make up most of the area's formations [22]. The geology of the region is composed of Vaikrita group high-grade metamorphic rocks. The Ronti Glacier is a hanging glacier bordered to the east by Nanda Gunti (6390 m a.s.l) and to the west by Trishul (7120 m a.s.l.). On the northeastern slope of Trishul mountain, beside the Valley of the Ronti glacier, the research area also lies within the Alaknanda catchment. The Indian Summer Monsoon (ISM) has the most influence on this area. However, the mid-latitude westerlies also contribute to the snowfall that feeds the region's numerous glaciers in the winter. The cryosphere, terrain dynamics, ecosystems, and socioeconomic fabric of the people in this region are all controlled by these weather systems.

A cascading hazard happened on February 7, 2021, when a significant avalanche occurred during the separation of a rock-ice composite wedge formation from Trishul's northeastern slope, causing debris flow.

4 Critical Insight into the 2021 Chamoli Disaster

ICIMOD [23] begins the study by comparing the images before and after the event and found that a crack was formed before the event. The mass movement was calculated using available data and imagery. This study concluded that glacial lake outburst flooding (GLOF) did not cause the flood previously claimed in the news because no glacial lake was near the study area. A massive rockslide just below the Ronti peak caused the debris flash flood. Heavy precipitation contributed to high flows downstream. Lack of ice cover destabilized the rock and increased the freeze–thaw cycles, substantial snow falls before the event and rapid meltwater generation contributes to ideal fracturing conditions. The study also suggests that finding the role of climate change in a single event like the Chamoli disaster is difficult. Shugar et al. [24] conducted the research using various data sources, including remote sensing, eyewitness recordings, numerical modelling, seismic data, and reconnaissance field observations. The study identified the three-key geology of the rock, i.e., multiple directions of planar weaknesses fractured the rocks and the failing mass which detached was created by four of these fractures. The rock mass is located near a significant thrust fault and contained numerous local shear fractures, which combined with other discontinuities, would have aided aqueous chemical weathering. Rocks disintegrated were found to be very fine when they collided because of their type. For the numerical modelling of the cascading hazards r.ava flow was used. The research concludes that the failure of the wedge shape rock caused avalanche. The combination of the critical rock and ice ratio (80% rock and 20% ice) along with large avalanche height increased the debris flowability. The location of the hydropower plant increased the severity of the Chamoli disaster.

Pandey et al. [25] used satellite pictures, a digital elevation model from the Indian Cartosat-1 satellite (10 m spatial resolution), ALOS DEM (12 m spatial resolution), and TabDEM-X (30 m spatial resolution) for hydrological simulations and 3D view generation of the topography. The MIKE-11 software was used for its appropriateness in modelling the narrow valley basins of the Himalayas. Global precipitation measurement (GPM) calculated the amount of precipitation before the event. A seismogram of nearby observatories was also analysed to find the seismic origin of the rockfall. This research concluded that abrupt weather changes and climatic variation influenced bedrock stability. The idea of the occurrence of the rock fall triggered by the earthquake was discarded. The sediments from the lesser-known flood of the low magnitude also added to the debris flow. Microwave remote sensing, numerical modelling, and rockfall simulation were used by Mondal and Bharti [26] to study the failure mechanism of the overhang wedge. For topographic visualization, Google pro Earth imagery, Advance land observing satellite (ALOS), and Phased array type -L band synthetic aperture radar (PALSAR) DEM were used. Rocscience software was used for kinematic as well as numerical analysis. This research concluded that glacial mobility has not caused the failure of the wedge. From the kinematic slope analysis, the slope was not susceptible to structurally controlled failure. From the probabilistic back analysis, the slope was susceptible to stress-controlled failure.

Srivastava et al. [27] analyzed the climatic conditions during the Chamoli disaster which was predicted by a high-resolution model. The study revealed that a combination of abrupt, unexpected cooling and heavy rainfall for a few days before the occurrence and persistently strong north-westerly winds may have had a role in the Chamoli rock-ice avalanche. Meena et al. [28] used remote sensing data to determine the impact of the Chamoli tragedy on the water quality of the Rishi Ganga River upstream near Tapovan and the Ganga River downstream near Haridwar. Five locations were used to analyze water quality changes across the two study areas. Three different indices were used, including the Normalized difference water index (NDWI), Normalized difference turbidity Index (NDTI), and Normalized difference chlorophyll index (NDCI). This research found qualitative changes in the parameter related to the water qualities. Data from satellite imagery, scientific case studies, and reports were used by Siddque et al. [29] to discover the geological and meteorological interaction during the Chamoli disaster. Changes in spectral fingerprints of planet scope imageries and backscattering coefficient from sentinel-1 synthetic aperture radar (SAR) data at six different sites were examined by Meena et al. [30]. This study concludes that pronounced changes in the backscattering coefficient were associated with the deposition of the snow dust in the Valley. Hence, the noisy signal recorded by seismogram from a nearby observatory was related to the fall of the glacier block. The snow dust increased the melting rate of the snow, which increased the water level in the Rishiganga river. To study the characteristic and dynamic processes of the Chamoli disaster, Zhang et al. [31] used multi-temporal satellite imagery and divided the cascading hazard into five zones. The study discussed the possible cause of the Chamoli disaster and the relevance of the satellite images and seismic network in the early warning system.

Qi et al. [32] tried to detect the precursors of the Chamoli landslide using remote sensing. It was found that the slope moved a distance of more than 20 m till 2020. Comparing multitemporal DEM and geomorphic analyses, Zhou et al. [33] concluded that the combined effects of climatic and weather changes acting on a steeply sloped and fracture-prone geological condition were primarily responsible for the rock-ice avalanche. Verma et al. [34] concluded that heavy snowfall over the Nanda Gunti peak and intense wind speed might lead to the initiation of the avalanche or landslide. Van Wyk de Vries et al. [35] concluded that Snow-loading in a deep headwall fracture and permafrost degradation in the extensively jointed bedrock caused the collapse. Pandey et al. [36] concluded that temperature variations in the vicinity of the surface resulted in permafrost freeze–thaw-freeze processes, which may have accelerated ice/ snow melting and caused the rock-ice avalanche. Singh and Kansal [37] attempted the flood mapping of the Chamoli flash flood using sentinel-1 imagery. Bhardwaj and Sam [38] concluded that while making the hazard assessment in the mountainous region, it is necessary to take note of the previous conditions prevalent in that region. Kropáček et al. [39] discussed how remote sensing may provide fast and unbiased information on the Chamoli disaster mechanisms, scope, and course. Fan et al. [40] concluded that advances in Earth observation and seismic monitoring systems can provide information on the location and time of approaching catastrophic disasters in high mountain areas.

5 Processes Involved in the Cascading Hazards

Figure 2 provides a step-by-step schematic of the cascading hazard experienced in this 2015 Chamoli disaster event. Figure 3 provides a pictorial representation of the entire region. Due to glacial buttressing and stress release fracturing [41], a crack originated which in turn, was detected in 2016 through feature tracking. Rock block displacement continued in consecutive years since then, with increased displacement in the summers of 2017 and 2018. This occurrence produced an 80 m wide gash in the glacial ice and rock mass under the uppermost section of the wedge, proving the instability and rockfall vulnerability of Trishul mountain's north-eastern side. On February 7, 2021, a prismatic wedge (approximately 71 m deep) with a triangular face (850 m × 550 m) accounting for approximately 60 million metric tonnes of mass (80% rock and 20% metamorphosed ice) from the northern slope of Trishul mountain crashed over the Ronti gad valley floor [42] with the velocity of 127–134 miles per hour. Initially, the rock and ice mass was airborne (free fall) because they travelled 1742 m before colliding with the valley floor. The rock and ice avalanche's high potential energy quickly transformed into kinetic and thermal energies (frictional ice heating). The large block of ice cooled from −8 °C to 0 °C as it travelled, then melted due to frictional heating, thereby, releasing a massive amount of water. The ice heating in the avalanche generated liquid water which permitted the change in flow characteristics, thus, becoming more fluid down the Valley. A tremendous air blow caused by the debris flow levelled roughly 20 hectares of forest. Boulders, ice blocks, meltwater, and sediments contributed to debris formation. These boulders contributed significantly in forming a debris-dammed lake about 40 m thick where Ronti gad meets the Rishiganga river. With time, a free-flowing river transformed into a 700 m-long lake upstream of the Rishiganga valley. The debris flow from the Ronti Nala caused damage to two hydropower plants. This indeed caused significant loss of lives.

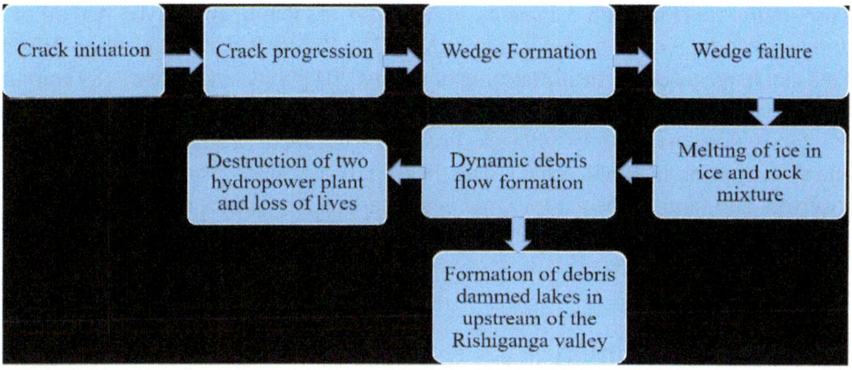

Fig. 2 Process of cascading hazard experienced in 2021 Chamoli Disaster

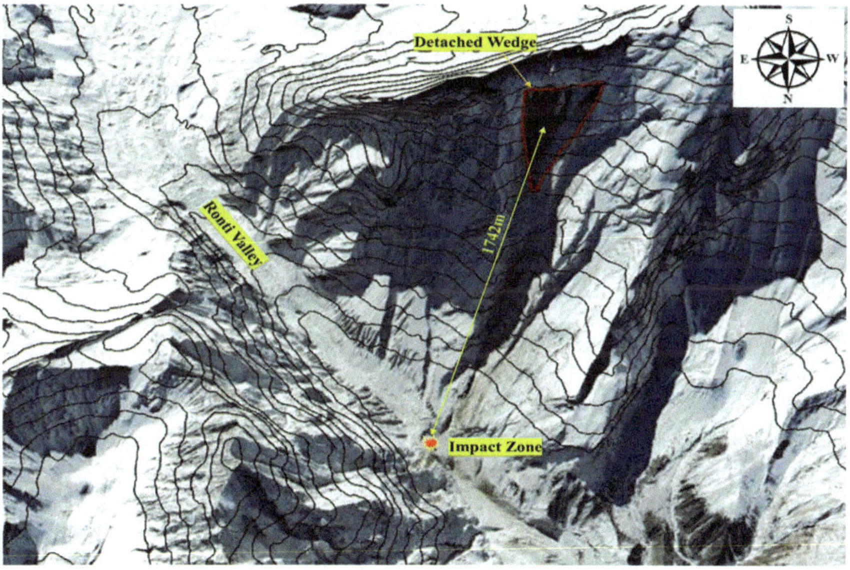

Fig. 3 Image depicts slope and impact location (Google Earth Pro image of 11-Feb-2021) [26]

6 Aftermath of the Chamoli Disaster

The debris flow nearly swept away the Rishiganga hydropower plant with a capacity of 13.2 MW and Tapovan Vishnu gad hydropower project of National Thermal Power Corporation (NTPC) with a capacity of 520 MW at Dhauliganga river. A total of 6 bridges including motor bridge was constructed by Border Road Organisation (BRO). This incident caused the death/missing of at least 204 people and at least 189 animals and livestock lost their lives. One temple and one pucca house were destroyed by the debris flow. At least 13 villages directly or indirectly affected by this disaster due to disruption in supply chain. The disaster also impacts psychologically to the survivors and families of the missing/ death people. The debris lake formed by this disaster poses threat even today [43].

7 Possible Mitigation

More authority and access to resources should be given to National Disaster Management Authority (NDMA) and National Disaster Response force (NDRF) for effective response to the disaster. Role of the NDMA should not limit to only in disaster management. It should play a significant role in planning and site selection of the project in the country. Climate change certainly increases the meteorological and cryospheric hazards, so data used for modelling and prediction of such type of disaster should be enhanced and a single repository for such data should be maintained. Governments should allocate adequate budget for financing the cost of maintaining the repositories. First responder of any disaster was the community so role of the community in disaster preparedness and response should not be underestimated. Disaster preparedness plan must be made from discussion with community. Country must end the burning of the fossil fuel sooner than later and focused on more sustainable energy to meet the demand. Hydroelectric energy is indeed a clean energy and has a roll to reduce carbon emission but it has repercussion on the flora and fauna as well as nearby villages of the hilly area. So, site for hydroelectric plant must be selected keeping the balance between protection of the environment and socio-economic benefits. Indian government must collaborate with the nations that are leaders in green energy to reduce the emissions of the greenhouse gases. Indian government must increase the finance for disaster resilience activity. Indian government should focus on the long-term impact while taking any decision. Awareness in people regarding climate change should be increased using non-structural measures. At individual level, we must keep the distance from disinformation and also alert the others.

8 Conclusion

From the literature review it was evident that Chamoli disaster does not cause by GLOF/LLOF. Several geological and meteorological conditions, such as wedge failure, rock and ice avalanches, and heavy precipitation that occurred before the event, were part of the Chamoli disaster. Seepage of the water in bedrock due to thawing- freezing- thawing cycle of the snow destabilized the rock. Global warming, avalanche in 2016, and increase in the temperature before the event increased the seepage rate of water in the bedrock. Finding the impact of the climate change on the whole cascading process was difficult but the single event such as the thawing of the snow might be affected by the climate change. The Himalayan region is susceptible to cascading hazards such as GLOF in Chorabari lake as well as in Gya lake. Due to the human interferences in the hilly region, impact of the disaster could aggravate. From this literature review, it was clear that continuous monitoring and an early warning system combined with seismological observations could have decreased the impact of the Chamoli disaster. The use of remote sensing proved instrumental for doing the research in disaster management. To avoid the destruction of future infrastructure and loss of lives in hilly region, therefore it is, necessary that the construction plan should also incorporate the disaster response plan. The construction activity must use the principle of sustainability. There is a need for detailed investigation of the dynamics between mountains and glaciers in order to have a better understanding of the cryospheric disaster. Increased research and ground survey with satellite imagery can be used to mitigate the effect of the disaster. Governments in the Himalayan state and government of India come with the policy that specially caters for the Himalayas. A balance must be maintained between the development and the protection of ecosystems in the Himalayan region. The Himalayas directly or indirectly affect the food security, water security, and financial security of the country, if the Himalayas is protected then nation security will be protected.

References

1. Hock R, Rasul G, Adler C, Cáceres B, Gruber S, Hirabayashi Y, Jackson M, Kääb A, Kang S, Kutuzov S, Milner A, Molau U, Morin S, Orlove B, Steltzer (2019) High mountain areas. In: IPCC special report on the ocean and cryosphere in a changing climate. Cambridge University Press, pp. 131–202
2. Shrestha AB, Wake CP, Mayewski PA, Dibb JE (1999) Maximum temperature trends in the Himalaya and its vicinity: an analysis based on temperature records from Nepal for the period 1971–94. J Clim 12(9):2775–2786
3. Pandey P, Ali SN, Champati Ray PK (2021) Glacier-glacial lake interactions and glacial lake development in the central Himalaya, India (1994–2017). J Earth Sci 32(6):1563–1574
4. Kumar M, Rana S, Pant PD, Patel RC (2017) Slope stability analysis of balia Nala landslide, kumaun lesser Himalaya, Nainital, Uttarakhand, India. J Rock Mech Geotech Eng 9(1):150–158
5. Paul SK, Mahajan AK (1999) Malpa rockfall disaster, Kali valley, Kamaun Himalaya. Curr Sci 76(4):485–487

6. Valdiya KS, Bartarya SK (1989) Problem of mass-movements in a part of Kumaun Himalaya. Curr Sci (Bangalore) 58(9):486–491
7. Valdiya KS (2003) Reactivation of Himalayan frontal fault: implications. Curr Sci 1031–1040
8. Kargel JS, Leonard GJ, Shugar DH, et al. (2016) Geomorphic and geologic controls of geohazards induced by Nepal's 2015 Gorkha earthquake. Science 351(6269), aac8353(2016)
9. Panikkar SV, Subramanyan V (1997) Landslide hazard analysis of the area around Dehra Dun and Mussoorie, Uttar Pradesh. Curr Sci 73(12):1117–1123
10. Ambrosi C, Strozzi T, Scapozza C, Wegmüller U (2018) Landslide hazard assessment in the Himalayas (Nepal and Bhutan) based on Earth-Observation data. Eng Geol 237:217–228
11. Kääb A, Leinss S, Gilbert A et al (2018) Massive collapse of two glaciers in western Tibet in 2016 after surge-like instability. Nat Geosci 11(2):114–120
12. Kirschbaum D, Watson CS, Rounce DR, et al. (2019) The state of remote sensing capabilities of cascading hazards over High Mountain Asia. Front Earth Sci 197
13. Snehmani, Bhardwaj A, Pandit A, Ganju A (2014) Demarcation of potential avalanche sites using remote sensing and ground observations: a case study of Gangotri glacier. Geocarto Int 29(5):520–535
14. Sethya KK, Pandey P, Chattoraj S, Manickam S (2018) Mapping, modeling and simulation of snow avalanche in Alaknanda Valley, Central Himalaya: Hazard Assessment. In IGARSS 2018–2018 IEEE international geoscience and remote sensing symposium, pp. 5150–5153. Valencia, Spain
15. Worni R, Huggel C, Stoffel M (2013) Glacial lakes in the Indian Himalayas—From an area-wide glacial lake inventory to on-site and modeling-based risk assessment of critical glacial lakes. Sci Total Environ 468:71–84
16. Schmidt S, Nüsser M, Baghel R, Dame J (2020) Cryosphere hazards in Ladakh: the 2014 Gya glacial lake outburst flood and its implications for risk assessment. Nat Hazards 104(3):2071–2095
17. Champati Ray PK, Chattoraj SL, Bisht MPS, Kannaujiya S, Pandey K, Goswami A (2016) Kedarnath disaster 2013: causes and consequences using remote sensing inputs. Nat Hazards 81(1):227–243
18. Leinss S, Bernardini E, Jacquemart M, Dokukin M (2021) Glacier detachments and rock-ice avalanches in the Petra Pervogo range, Tajikistan (1973–2019). Nat Hazard 21(5):1409–1429
19. Rana N, Sharma S, Sundriyal Y, Kaushik S, Pradhan S, Tiwari G, Juyal NA (2021) Preliminary assessment of the February 7 2021 flashflood in lower Dhauli Ganga valley, Central Himalaya, India. J Earth Syst Sci 130(2):1–10
20. Pralong A, Funk M (2006) On the instability of avalanching glaciers. J Glaciol 52(176):31–48
21. Margreth S, Funk M, Tobler D, Dalban P, Meier L, Lauper J (2017) Analysis of the hazard caused by ice avalanches from the hanging glacier on the Eiger west face. Cold Reg Sci Technol 144:63–72
22. Mukherjee PK, Jain AK, Singhal S et al (2019) U-Pb zircon ages and Sm-Nd isotopic characteristics of the Lesser and Great Himalayan sequences, Uttarakhand Himalaya, and their regional tectonic implications. Gondwana Res 75:282–297
23. ICIMOD,https://www.icimod.org/article/understanding-the-chamoli-flood-cause-process-impacts-and-context-of-rapid-infrastructure-development/, last accessed 2022/06/07.
24. Shugar DH, Jacquemart M, Shean D et al (2021) A massive rock and ice avalanche caused the 2021 disaster at Chamoli. Indian Himalaya. Science 373(6552):300–306
25. Pandey P, Chauhan P, Bhatt CM et al (2021) Cause and process mechanism of rockslide triggered flood event in Rishiganga and Dhauliganga River Valleys, Chamoli, Uttarakhand, India using satellite remote sensing and in situ observations. J Indian Soc Remote Sensing 49(5):1011–1024
26. Mondal SK, Bharti R (2022) Glacial burst triggered by triangular wedge collapse: a study from Trisul Mountain near Ronti glacier valley. Geomat Nat Haz Risk 13(1):830–853
27. Srivastava P, Namdev P, Singh PK (2022) February 7 chamoli (Uttarakhand, India) rock-ice avalanche disaster: model-simulated prevailing meteorological conditions. Atmosphere 13(2):267

28. Meena SR, Chauhan A, Bhuyan K, Singh RP (2021) Chamoli disaster: pronounced changes in water quality and flood plains using Sentinel data. Environ Earth Sci 80(17):1–13
29. Siddque T, Haris PM, Pradhan SP (2022) Unraveling the geological and meteorological interplay during the 2021 Chamoli disaster. Natural Hazards Research, India
30. Meena SR, Bhuyan K, Chauhan A, Singh RP (2021) Snow-covered with dust after Chamoli rockslide: inference based on high-resolution satellite data. Remote Sens Lett 12(7):704–714
31. Zhang T, Yin Y, Li B, et al. (2022) Characteristics and dynamic analysis of the February 2021 long-runout disaster chain triggered by massive rock and ice avalanche at Chamoli, Indian Himalaya. J Rock Mech Geotech Eng
32. Qi W, Yang W, He X, Xu C (2021) Detecting Chamoli landslide precursors in the southern Himalayas using remote sensing data. Landslides 18(10):3449–3456
33. Zhou Y, Li X, Zheng D et al (2021) The joint driving effects of climate and weather changes caused the Chamoli glacier-rock avalanche in the high altitudes of the India Himalaya. Sci China Earth Sci 64(11):1909–1921
34. Verma S, Sharma A, Yadava PK, Gupta P, Singh J, Payra S (2022) Rapid flash flood calamity in Chamoli, Uttarakhand region during Feb 2021: an analysis based on satellite data. Nat Hazards 1–15
35. Van Wyk de Vries M, Bhushan S, Jacquemart M, et al. (2021) Pre-collapse motion of the February 2021 Chamoli rock-ice avalanche, Indian Himalaya. Nat Hazards Earth System Sci Discuss 1–29
36. Pandey VK, Kumar R, Singh R et al (2022) Catastrophic ice-debris flow in the Rishiganga River, Chamoli, Uttarakhand (India). Geomat Nat Haz Risk 13(1):289–309
37. Singh S, Kansal ML (2022) Chamoli flash-flood mapping and evaluation with a supervised classifier and NDWI thresholding using Sentinel-2 optical data in Google earth engine. Earth Sci Inform 1–14 (2022)
38. Bhardwaj A, Sam L (2022) Reconstruction and characterization of past and the most recent slope failure events at the 2021 rock-ice avalanche site in Chamoli, Indian Himalaya. Remote Sensing 14(4):949
39. Kropáček J, Vilímek V, Mehrishi P (2021) A preliminary assessment of the Chamoli rock and ice avalanche in the Indian Himalayas. Remote Sensing 18:3489–3497
40. Fan X, Yunus AP, Yang YH et al (2022) Imminent threat of rock-ice avalanches in High Mountain Asia. Sci Total Environ 836:155380
41. Mao W, Wu L, Singh RP et al (2022) Progressive destabilization and triggering mechanism analysis using multiple data for Chamoli rockslide in February 7, 2021. Geomat Nat Haz Risk 13(1):35–53
42. GAPHAZ. 2021. First insights into the Chamoli disaster, Report https://www.gaphaz.org/files/210220Update_Chamolidisaster_20210207_GAPHAZ.pdf , last accessed 2022/05/09
43. NDMA Report, https://ndma.gov.in/sites/default/files/PDF/Reports/Detalied_report_UK_Disaster.pdf last accessed, 2022/05/09

Critical Zone Mapping for Hazard Susceptibility Analysis

Alok Kumar, Arnab Kumar Pal, Adnan Ahmad, Dhritilekha Deka, and Archana M. Nair

Abstract The Critical Zone thickness (CZT) is defined as the depth from the top of the canopy to the topmost zones of the water table. Critical Zone monitoring is important, as it is the most dynamic part of our atmosphere and it is very vulnerable and suffers the most due to human influence. In this study, we have derived the CZT of Assam. CZT was derived by adding the vegetation height and water table depth. The analysis shows that the CZT for Assam varies from 5.228 m to 78.638 m. The vegetation height varies from 1.632 m to 74.529 m. We have classified the region as having low, medium, and high CZT. Based on the comparison with the Landslide Hazard Zonation map derived from the previous literature, it was found that the CZT correlates with the Landslide Hazard Zonation map. Hence, this study could prove substantial for Disaster Management authorities to take necessary steps for mitigation purposes.

Keywords Critical zone thickness · Vegetation height · Landslide hazard zonation

A. Kumar · A. K. Pal · A. Ahmad · D. Deka · A. M. Nair (✉)
Department of Civil Engineering, Indian Institute of Technology Guwahati, Assam, India
e-mail: nair.archana@iitg.ac.in

A. Kumar
e-mail: k.alok@iitg.ac.in

A. K. Pal
e-mail: arnab.pal@iitg.ac.in

A. Ahmad
e-mail: adnan176104005@iitg.ac.in

D. Deka
e-mail: dhrit174104139@iitg.ac.in

S. Mitra et al. (eds.), *Disaster Management and Risk Reduction: Multidisciplinary Perspectives and Approaches in the Indian Context*,
https://doi.org/10.1007/978-981-99-6395-9_11

1 Introduction

The Critical Zone (CZ) is the zone where rock meets life. It is considered as the skin of the Earth and is made up of numerous biotic and abiotic components [1]. The CZ provides all the essential resources needed for the survival and existence of life. Simultaneously, it is the most vulnerable layer of the Earth's atmosphere and is most influenced by human activities. CZ has numerous ongoing processes, which may be physical, chemical, and biological. These processes are generally coupled with each other in a highly complex system. The CZ comprises various energy cycles, which govern the weather and climate of the Earth. The processes in the energy cycles are generally irreversible and can't be traced back [2]. These processes are archaic and slow going, but anthropogenic human activities have accelerated these processes, creating pressure on the available resources. The most affected part is the near-surface layer of the Earth. For instance, when soil formation occurs through the disintegration of rocks, it undergoes mechanical and chemical weathering, then transportation, and then accumulation occurs. However, intensive land use and climate change have taken place due to increasing economic demand. Some of the reasons may be the increasing human and livestock population, uncontrolled consumption of natural resources, and increased land use [3]. To track the interdependencies of different cycles with each other, track the usage of the resources, and create an action plan for adapting to this climate change, it is of high importance to study the critical zone. Critical Zone Science is the study of analyzing the impacts of accelerated anthropogenic and natural activities on the CZ. Many Critical Zone Observatories (CZO) have been set up worldwide to advance the research in CZ Science. These CZOs work in a network, integrate ecological and geological sciences and perform computational analysis to predict the Earth's past and future. CZ science can provide insight into long-term adaptability to human disturbances.

In 2001, US National Research Council (NRC) defined the CZ as the depth from the top of the canopy to the topmost zones of the water table. Technically the critical zone at a location is the sum of canopy height, water table depth, and groundwater thickness [4]. There has been indecision as to which definition of CZ should be considered the best of the bunch. The skepticism lies with the water table boundary because the lower boundary of the active water cycle is yet to be defined [5]. In this study, we have tried to map the critical zone for the year 2020 for Assam state as per the definition given by NRC, 2001 and study the contribution of different components in the critical zone.

2 Materials and Methods

2.1 Study Site

The study was conducted for the state of Assam. Assam is situated on the foothills of the eastern Himalayas with 89°42′–96°30′ E longitude and 22°19′–28°16′ N latitude, as shown in Fig. 1. It has quite uneven topography, hills, plains, and rivers. An apprising report by the Department of Environment and Forest Government of Assam states that annual rainfall has declined by 2.96 mm/year, whereas the mean average temperature has increased by 1.7–2.2 °C from 1971 to 2000. Extreme rainfall events have increased by 5 to 38%. Cases of cloudbursts have increased in recent years due to extreme rainfall, leading to flash floods [6]. If we observe the water table, freshwater scarcity is already a problem in Assam in the summer seasons. The situation is predicted to worsen due to rapid climatic changes. Additionally, extreme rainfall events followed by temperature variation make Assam susceptible to soil weathering and subsequently make it a landslide-prone area.

2.2 Data Sets and Methodology

As per the definition given by NRC, 2001 the near-surface critical zone thickness constitutes the vegetation height and the water table height. The aboveground part consists of the vegetation height, and the underground part is extended up to the water table depth. The critical zone thickness was estimated by following the methodology as shown in Fig. 2. In September 2018, NASA's Earth Observing System launched the Ice, Cloud, and land Elevation Satellite-2 (ICESat-2) satellite mission, a follow-up mission to ICESat-1. ICESat-2 has an improved laser system called ATLAS (Advanced Topographic Laser Altimetry System). The ICESat-2 mission was focused on the cryosphere study. However, taking advantage of the viewpoint of space, global surfaces like land, oceans, and vegetation were also measured [7]. ATL08 data product of ICESat-2 was used to perceive the vegetation height for this study [8]. The data was retrieved and processed (filtered) for the satellite path from Open Altimetry [9]. The data were interpolated by following the inverse distance weighted (IDW) algorithm and the final Canopy Height Map was obtained in the GIS platform as shown in Fig. 3a. The Water Table Depth was extracted from India Water Resources Information System (India-WRIS portal) [10] and the corresponding map was obtained after interpolation by following a similar algorithm as shown in Fig. 3b. These mapped components were then overlaid on each other to estimate the critical zone thickness as presented in Fig. 3c.

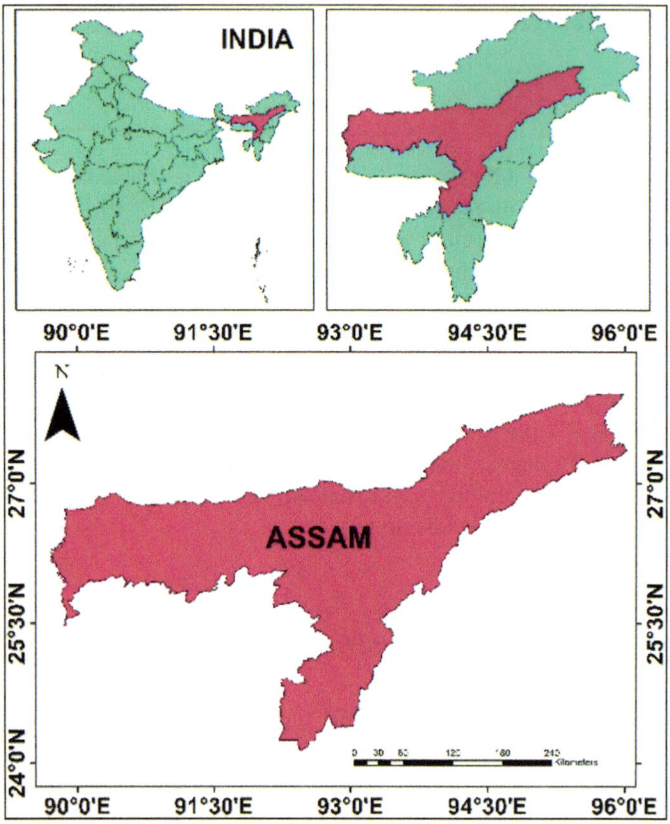

Fig. 1 Geographical location of the study area: Assam

3 Results and Discussion

The critical zone thickness (CZT) map was obtained using this methodology presented in the flowchart (Fig. 2). The vegetation height, presented in Fig. 3a, has a proportion of 80.93 $\pm$ 10.43% (mean $\pm$ one standard deviation). The water table depth was found to have maximum value of 19.508 m as shown in Fig. 3a,b. The CZT for Assam varies from 5.228 m to 78.638 m as shown in Fig. 3c. The proportion of underground components was found to be 19.07 $\pm$ 10.43%.

We have classified the study region as having low, medium and high CZT as shown in Table 1. We have also found that the Critical Zone thickness is more in the middle Assam region (Karbi Anglong West, Hojai, Karbi Anglong East, Dima Hassao, Cachar). Based on the comparison with the Landslide Hazard Zonation map extracted from the Building Material and Technology Promotion Council website [11] as shown in Fig. 4, the zone having high CZT is prone to Landslides.

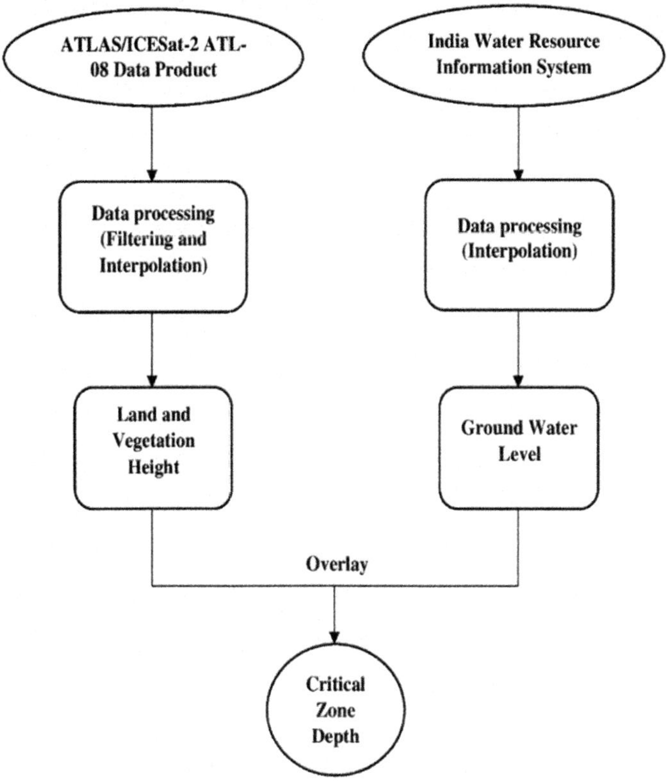

Fig. 2 A brief overview of the methodology followed in this study

4 Conclusions

The study shows that the critical zone thickness is mainly composed of Canopy Height. Compared with the landslide hazard zonation (LHZ) map with the critical zone thickness map, more focus needs to be laid on the portion with high critical zone thickness in Assam. Hence, this study could prove substantial for Disaster Management authorities to take necessary steps for mitigation purposes.

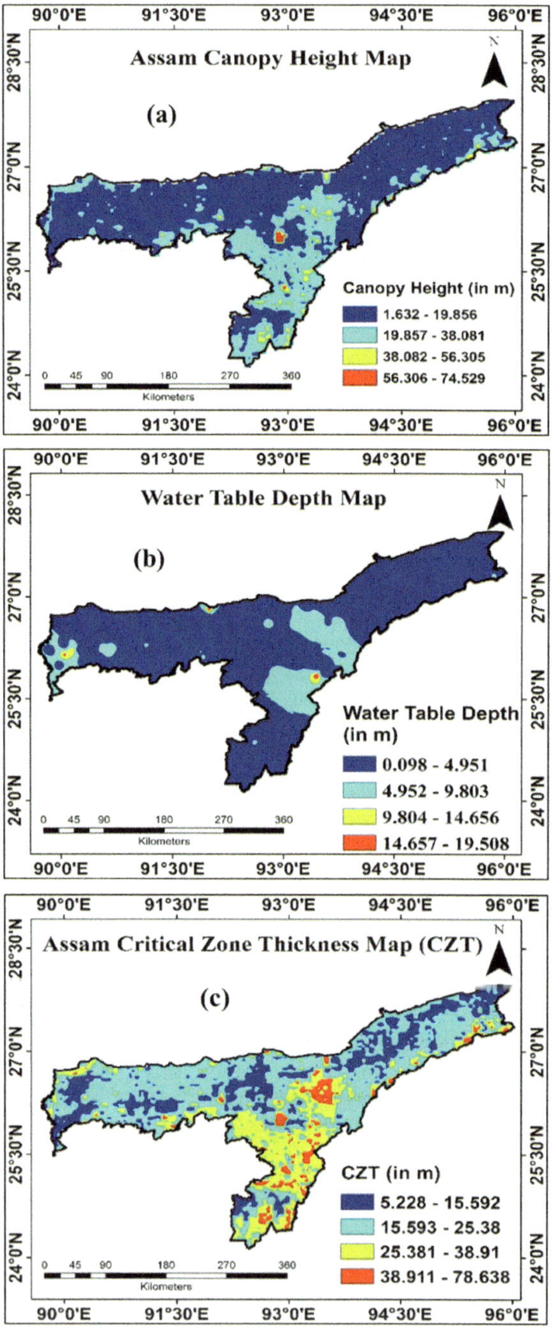

Fig. 3 **a** Spatial distribution of Canopy Height across Assam; **b** Spatial distribution of groundwater table depth across Assam; **c** Spatial distribution of Critical Zone thickness across Assam

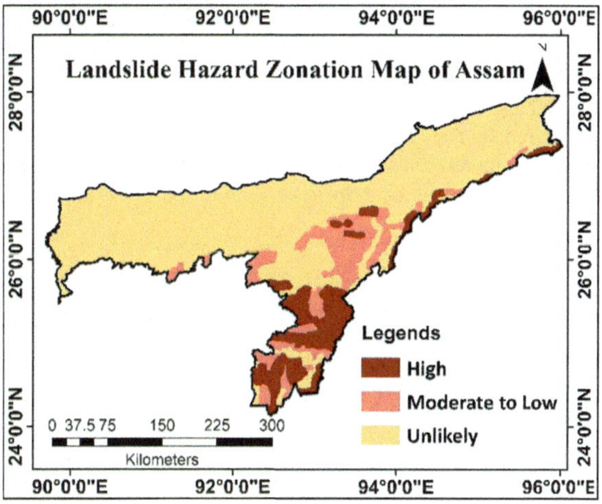

Fig. 4 Landslide hazard zonation (LHZ) map of North East India modified after [11]

Table 1 Critical zone thickness classification followed in this study

Critical zone thickness (CZT) (meter)	Thickness class
5.228–15.592	Low
15.593–38.91	Medium
38.911–78.638	High

Acknowledgements The author acknowledges the contribution of Earth System Science and Engineering laboratory of Department of Civil Engineering, Indian Institute of Technology Guwahati.

References

1. Lewis JS, Farnsworth ML, Burdett CL, Theobald DM, Gray M, Miller RS (2017) Biotic and abiotic factors predicting the global distribution and population density of an invasive large mammal. Sci Rep 7(1):1–12
2. Brantley SL, White TS, White AF, Sparks D, Richter D, Pregitzer K, Amundson R (2005) Frontiers in exploration of the Critical Zone: Report of a workshop sponsored by the National Science Foundation (NSF), October 24–26, 2005. Newark, DE, 30. (2006)
3. Hooke RL, Martín Duque JF, Pedraza Gilsanz JD (2013) Land transformation by humans: a review. Ene 12:43
4. Xu X, Liu W (2017) The global distribution of Earth's critical zone and its controlling factors. Geophys Res Lett 44:3201–3208
5. Lin H (2010) Earth's Critical Zone and hydropedology: concepts, characteristics, and advances. Hydrol Earth Syst Sci 14(1):25–45
6. Dimri AP, Chevuturi A, Niyogi D, Thayyen RJ, Ray K, Tripathi SN, Mohanty UC (2017) Cloudbursts in Indian Himalayas: a review. Earth-Sci Rev 168:1–23
7. Neuenschwander AL, Magruder LA (2019) Canopy and terrain height retrievals with ICESat-2: A first look. Remote Sens 11(14):1721

8. Neuenschwander AL, Pitts KL, Jelley BP, Robbins J, Klotz B, Popescu SC, Nelson RF, Harding D, Pederson D, Sheridan R (2020) ATLAS/ICESat-2 L3A Land and Vegetation Height, Version 3
9. Khalsa SJS, Borsa A, Nandigam V, Phan M, Lin K, Crosby C, Lopez L (2020) OpenAltimetry-rapid analysis and visualization of Spaceborne altimeter data. Earth Sci Inform 1–10
10. Website reference: http://www.india-wris.nrsc.gov.in/, last accessed 2022/05/06
11. Website reference: https://www.bmtpc.org/, last accessed 2022/05/03

Pushover Analysis for Seismic Requalification of Pile Foundation

Bidisha Borthakur and Arup Bhattacharjee

Abstract Civil engineers can play a crucial role by contributing to the field of disaster management in two important ways. First is by setting of design and safety standards, and second by actually designing and constructing infrastructure such as to prevent damage and losses caused by hazards. Past earthquake case histories show that structures supported by different types of foundations are affected in a different manner depending upon the type of earthquake and the type of foundation. Thus proper seismic investigation of already existing structures along with conducting seismic studies while designing new structures has become very important. In this research work, static pushover analysis is conducted on different piles embedded in stratified soil having different combinations of surrounding soil layers. An attempt is made to incorporate static pushover analysis, which is simple and less time consuming into the seismic requalification procedure provided by Krishna et al. (2014), to simplify the entire process of estimation of the damage level of the foundation and determine if seismic requalification is required. The seismic analyses are conducted in OpenSees PL which has the ability to reproduce SSI effects due to earthquake loading computationally and thus can be used to conduct FE computations to obtain realistic results of seismic analyses.

Keywords OpenSees PL · Pushover analysis · SSI

B. Borthakur (✉)
Department of Civil Engineering, Golaghat Polytechnic, Furkating, Assam, India
e-mail: bidishaborthakur17@gmail.com

A. Bhattacharjee
Department of Civil Engineering, Jorhat Engineering College, Jorhat, Assam, India
e-mail: bhatta_arup@yahoo.com

S. Mitra et al. (eds.), *Disaster Management and Risk Reduction: Multidisciplinary Perspectives and Approaches in the Indian Context*,
https://doi.org/10.1007/978-981-99-6395-9_12

1 Introduction

1.1 General

Natural disasters are unpredictable and can cause serious threat to communities nationwide. Earthquake is one such natural disaster whose neither magnitude nor duration is foreseeable. Thus it becomes very challenging for civil engineers to construct in earthquake prone areas. However, the effects of earthquake can be minimized by taking certain preventative measures as per standard scientific guidelines. Any construction in such earthquake prone area requires proper designing with consideration of proper standards in order to make the structures resistant to earthquake. From past earthquake case histories, it can be seen that structures supported by different types of foundations are affected in a different manner depending upon the type of earthquake and the type of foundation.

Pile foundations have always been a solution to civil engineers when load has to be transferred from the superstructure through weaker soil strata onto less compressible soil or rock. It is seen that failure patterns of piles vary depending upon the peak ground acceleration, geometric configuration of pile foundation and surrounding soil layers. However, it has been found from various seismic investigations that piles are the most vulnerable components of the entire structure and failure of pile has resulted in the failure of the entire structure. The main reason behind this could be the unaccountability of surrounding soil condition while designing the pile foundation. Thus pile foundation should be designed by taking all the necessary factors into consideration such that there is no failure of foundation due to seismic loading and proper seismic analyses should be conducted for that. Even though dynamic analysis has been the conventional seismic analysis, from recent researches it has been seen that static nonlinear pushover analysis can be used in seismic design as it has the ability to simulate equivalent peak load occurring on the structure during earthquake. Pushover analysis can imitate the actual peak load acting on the structure during earthquake, which helps to realistically predict the response of the structure. The ability of pushover analysis to simulate the peak dynamic response of structure decides the accuracy of this method [5]. The pushover analysis helps to improve understanding of post-yield structural behavior and results in more accurate prediction of global displacement along with realistic prediction of earthquake demand in individual structural elements [2]. This can further help in evaluating the yield moment and the yielding zone of pile, thus helping to evaluate the failure pattern of pile. Thus in this research work, an attempt is made to incorporate static pushover analysis into the seismic requalification procedure to simplify the entire process of estimation of the damage level of the foundation and determine if seismic requalification is required.

1.2 Soil-Structure Interaction Using OpenSees PL

A user-friendly interface of OpenSees, known as OpenSees PL was created using the using the pre- and post-processing efforts of OpenSees. OpenSees PL was created for 3D foundation-ground analyses such that the complicated soil-structure interaction mechanism could be incorporated while analyzing the foundation under seismic loading in order to represent the actual geometric configuration that is involved due to soil-structure interaction. It is a FE graphical user-interface for 3D ground-structure interaction response which allows conducting pushover analysis as well as seismic simulations [6]. This Finite Element Analysis software utilizes object-oriented design principles and programming approach and can incorporate element formulation, material relations, analysis algorithms and solution strategies.

1.3 Validation of Seismic Analysis Using OpenSees PL

The seismic analysis of pile foundation using OpenSees PL is validated with the results obtained by Lu et al. [4] and analytical results obtained by Abedzedah et al. [1]. A circular free-head pile of 10.15 m length and radius 203.20 mm, fully embedded in a 20.12 m soil domain of submerged unit weight 9.87 kN/m^3 is modeled in OpenSees PL [4]. The pile is modeled using linear beam-column elements so that bending moment, axial loads and shear force could be viewed easily with rigid beam-column elements representing the diameter and interface with the surrounding soil elements. The soil is modeled using 8-node brick elements with *MultiYield* material to capture seismic events accurately. Lateral incremental pushover loading is applied monotonically at the pile head up to a total load of 140.12 kN. Figure 1. shows the pile deflection and the bending moment experienced by the pile throughout its length due to pushover loading. From the results, it is seen that pile response in terms of deflection and bending moment obtained from pushover analysis in OpenSees PL is similar to the analytical results of pushover analysis obtained by Abedzedah et al. [1]. Thus for the seismic analysis of pile foundation in OpenSees PL for this study, modeling is done as per Lu et al. [4].

2 Pushover Analyses of Pile Embedded in Stratified Soil

In this research work, pushover analysis is conducted on single piles of 10 m length each having different diameters surrounded by different combinations of multilayered soil with water table upto the pile head. Force-based pushover analysis is conducted on each pile of 0.8 m and 1.2 m diameter by applying step-wise incremental horizontal loading of 10 kN at the pile head. The pushover loading is applied till pile head displacement of 0.2 m, i.e., 2% of pile length is obtained. However, pile

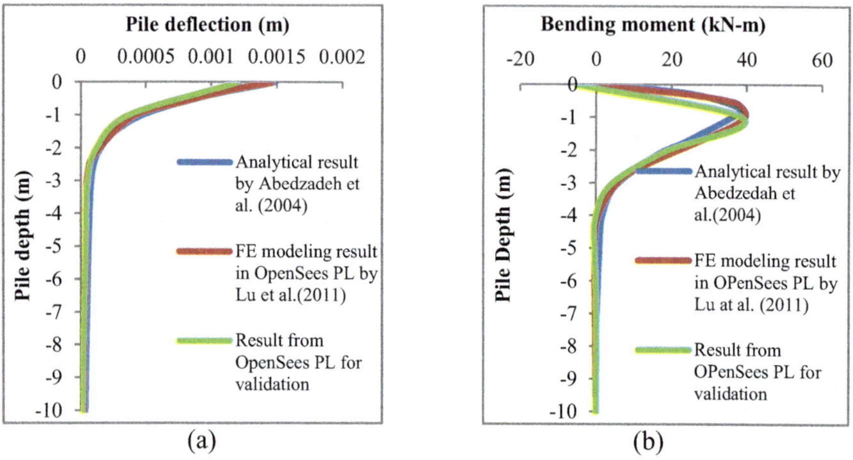

Fig. 1 Comparison of analytical results and FE modeling results in terms of **a** Pile deflection and **b** bending moment of pile for seismic analysis of pile-soil system for validation

embedded in different types of soil is seen to require different number of loading steps to reach 0.2 m displacement of pile head. Different combination of cohesionless soils and cohesive soil layers having different thickness is considered to be surrounding the piles.

2.1 Numerical Modeling of Pile-Soil System

Pile and soil elements used by Lu et al. [4] have been used for modeling the pile-soil system for this research work. The circular pile of 10 m length is modeled using linear beam-column elements. Rigid beam-column elements, which are 10^4 times stiffer than the pile elements axially and flexurally, are used for representing the cross-sectional diameter and the interface with the soil elements surrounding the pile. The pile is considered to be fixed at the top and bottom and the water table is considered to be up to the pile head. The mass density of the considered pile is taken as 2400 kg/m^3. The Young's modulus and the shear modulus are 3×10^7 kPa and 1.154×10^7 kPa respectively.

The 10 m soil domain consists of layers of cohesionless dense sand and cohesive medium soil in three different combinations. The nonlinear soil modeling is done by modeling the soil domain with 8-node brick elements using *MultiYield* material. Sand is modeled using *PressureDependMultiYield* soil model and cohesive medium soil is modeled using *PressureIndependMultiYield* model. The details about the soil elastic properties, soil nonlinear properties, fluid properties, dilatancy properties and liquefaction properties for the saturated cohesionless soil and cohesive soil are given in Table 1. The water table is considered up to the pile head so as to consider liquefaction analysis while conducting dynamic or static analyses. The soil combinations are:

1. Saturated cohesionless dense sand (5m) + Cohesive medium soil (5m)
2. Cohesive medium soil (5m) + Saturated cohesionless dense sand (5m)
3. Cohesive medium soil (2m) + Saturated cohesionless dense sand (6m) + Cohesive medium soil (2m)

The boundary condition is rigid box type and is considered to be fixed at the bottom in all directions. The plane of symmetry for half mesh configuration is fixed in Y direction while keeping it free in Z and X direction to model 3D full mesh scenario (Fig. 2).

Table 1 Soil properties of cohesionless dense sand and cohesive medium soil

	Cohesionless dense sand	Cohesive medium soil
Soil elastic properties		
Saturated mass density (Mg/m^3)	2.1	1.5
Reference pressure (kPa)	80	100
Reference Shear modulus (kPa)	130,000	60,000
Reference Bulk modulus (kPa)	390,000	300,000
Soil Nonlinear Properties		
Friction (deg)	40	0
Cohesion(c) multiplied by ((sqrt(3))/2)	0.3	37
Fluid Properties		
Fluid mass density (Mg/m^3)	1	1
Horizontal permeability (m/s)	6.60E−05	1.00E−09
Vertical permeability (m/s)	6.60E−05	1.00E−09
Dilatancy/liquefaction Properties		
Phase transformation angle (deg)	27	–
Contraction parameter	0.03	–
Dilation parameter 1	0.8	–
Dilation parameter 2	5	–
Liquefaction parameter 1	0	–
Liquefaction parameter 2	0	–
Liquefaction parameter 3	0	–

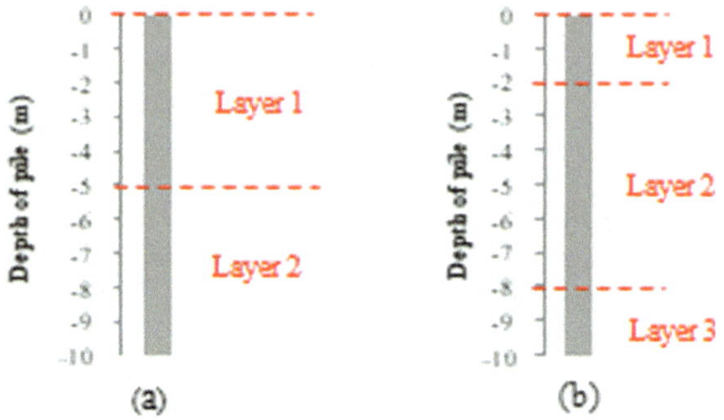

Fig. 2 Soil layering

2.2 *Pushover Analysis Results of Embedded in Stratified Soil*

Force-based pushover analysis is conducted on each pile of 0.8 m and 1.2 m diameter by applying step wise incremental horizontal loading of 10 kN at the pile head until pile head displacement of 0.2 m, i.e., 2% of pile length is obtained. The response of pile for the different cases of pile diameter and surrounding soil condition were obtained in terms of bending moment. The bending moment profile for 0.8 m diameter and 1.2 m diameter pile surrounded by the different combinations of soil layers is shown below in Fig. 3.

From the results, it is seen that the bending moment witnessed by the pile is highest at the pile head for all the diameters of pile. It is also observed that change in position of soil layers lead to different bending moments on the pile. It is observed that the

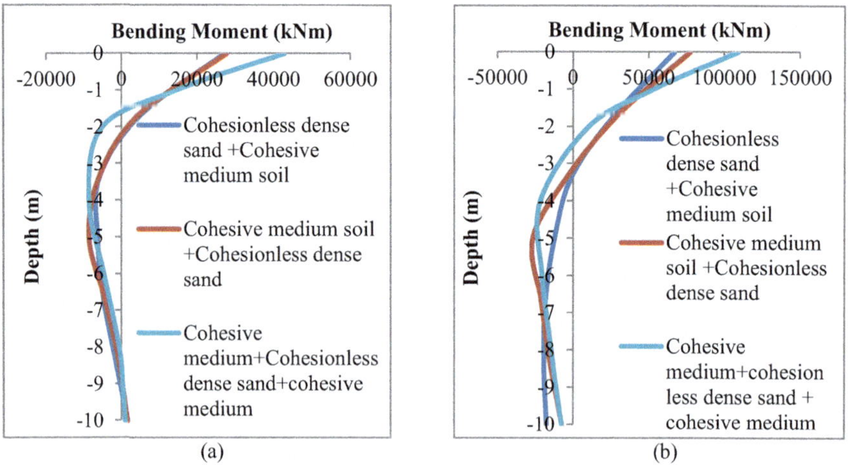

Fig. 3 Bending moment profile of **a** 0.8 m and **b** 1.2 m diameter piles embedded in multi layered soil

Table 2 Maximum bending moment witnessed by each pile for different cases of multilayered soil

Soil combination	0.8 m diameter Pile	1.2 m diameter pile
Saturated cohesionless dense sand + Cohesive medium soil	27209kN m	67,247.6kN m
Cohesive medium soil (5 m) + Saturated cohesionless dense sand (5 m)	27,712.6 kN m	77,523.6 kN m
Cohesive medium soil (2m) + Saturated cohesionless dense sand (6m) + Cohesive medium soil (2m)	42800 kN m	109,332.4kN m

number of location of change in bending behavior of pile is mainly influenced by the pile diameter. The maximum magnitude of bending moment witnessed by each case of pile is given in Table 2.

2.3 Determination of Yield Moment of Pile

Yield moment of pile for various conditions of surrounding soil is obtained by drawing double tangents on moment curvature graphs for the different diameters of pile throughout the depth of pile at various depths. From Fig. 4, it can be seen that the yield moment obtained from moment curvatures of 0.8 m diameter pile embedded in Cohesionless dense sand placed above Cohesive medium soil is 1750 kN m. On reversal of soil layers, the yield moment increases to 2000 kN m. However. When cohesionless dense sand is sandwiched between layers of cohesive medium soil, the yield moment is 1500 kN m.

From the moment curvatures graphs plotted for 1.2 m diameter pile (Fig. 5), it is seen that when cohesionless dense sand is placed above cohesive medium soil, the yield moment is as high as 5000 kN m. However, on reversal of soil layers, the yield moment is 2000 kN m. For cohesionless dense sand placed between layers of cohesive medium soil, the yield moment obtained is 3000 kN m.

The yield zone of the pile was seen to vary depending upon the pile diameter as well as the soil surrounding the pile. From Fig. 6 it is also seen that the position of the soil layer also influences the yielding zone of the pile.

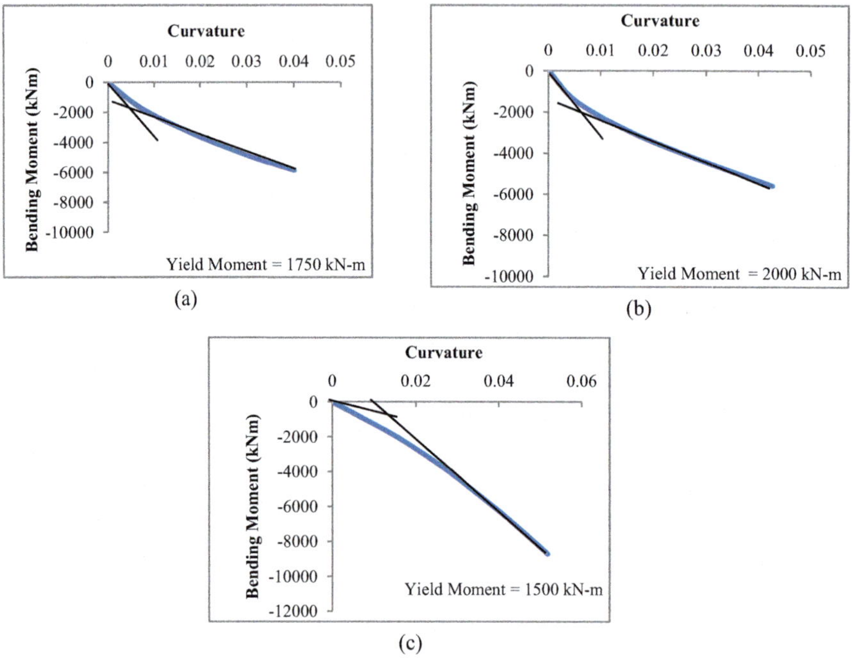

Fig. 4 Moment–curvature of 0.8 m diameter pile embedded in double layered soil consisting of **a** Cohesionless dense sand + Cohesive medium soil **b** Cohesive medium soil + Cohesionless dense sand and **c** Cohesive medium soil + Cohesionless dense sand + Cohesive medium soil

2.4 Application of Pushover Analysis for Seismic Requalification

Incorporation of new and improved seismic analysis methods such as pushover analysis for seismic requalification helps to provide a simple and accurate solution to earthquake assessments of previously built as well as new structures. Since geotechnical aspects play an important role in seismic response of a structure, it is mandatory to take into account the soil conditions surrounding a foundation for pushover analysis of the foundation. Also, pushover analysis can imitate the actual peak load acting on the structure during earthquake, which helps to realistically predict the response of the structure. Proper prediction of response and failure pattern of structure can help to reduce impact on the structure as well as reduce cost of seismic retrofitting. Krishna et al. [3] provided various steps that are required to be followed for seismic requalification of geotechnical structures. As per the procedure of seismic requalification provided, selection of earthquake level, ground motion simulation, site-specific ground response and liquefaction analysis is the first step for seismic hazard identification and analysis. Since pushover analysis of foundation involves consideration

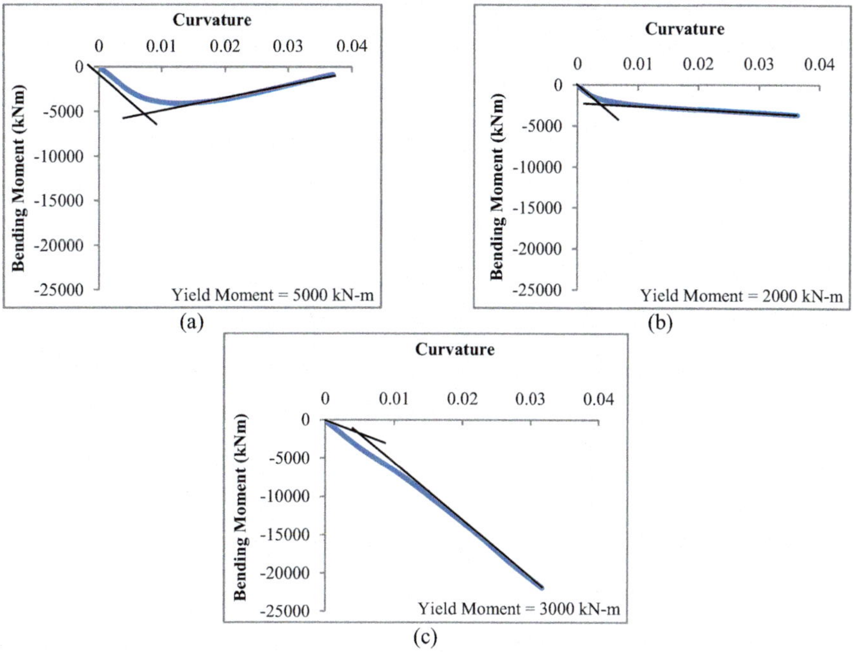

Fig. 5 Moment–curvature of 1.2 m diameter pile embedded in double layered soil consisting of **a** Cohesionless dense sand + Cohesive medium soil **b** Cohesive medium soil + Cohesionless dense sand and **c** Cohesive medium soil + Cohesionless dense sand + Cohesive medium soil

of geotechnical aspects, potential liquefaction of soil along with simulation of actual peak load on structure due to earthquake, it can be used for seismic analysis of foundation for seismic requalification. The second step involves evaluation of the results obtained from seismic analysis in terms of deformations and stresses acting on the structure. The results obtained from pushover analysis help to evaluate the yield moment and the yielding zone of pile, thus helping to evaluate the failure pattern of pile. Prediction of accurate failure pattern can help in proper estimation of the damage level of the foundation and determine if seismic requalification is required. Since pushover analysis provides the desired conditions and response for seismic requalification of foundation, it can be included in the seismic requalification procedure provided by Krishna et al. [3]. Figure 7 shows the seismic requalification procedure chart with inclusion of pushover analysis.

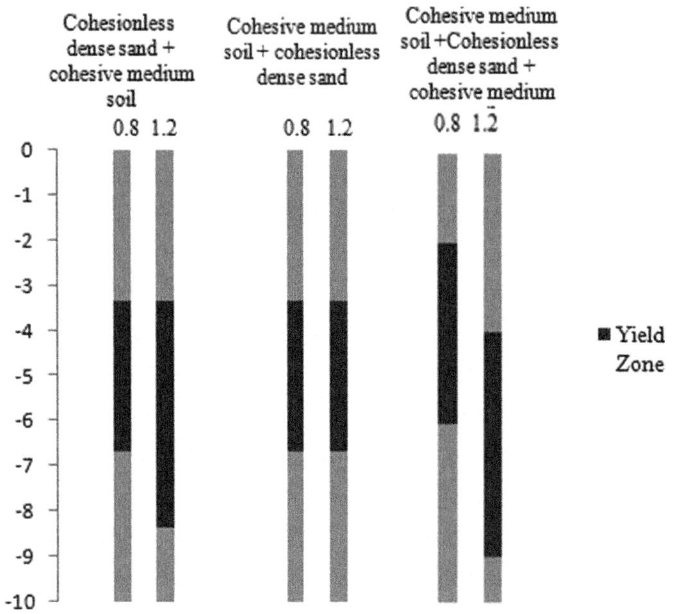

Fig. 6 Yield zone of 0.8 m and 1.2 m diameter pile surrounded by different surrounding soil condition

2.5 Conclusion

From the yield moment and yield zone of pile obtained after conducting pushover analysis of single piles surrounded by different cases of multilayered soil, it can be concluded that:

- The yield moment of each diameter pile depends on the surrounding soil type along with positioning of soil layers in multilayered soil.
- The zone of yielding is also seen to depend on the number of soil layers along with the positioning of soil layer and pile diameter.
- Since pushover analysis can estimate the yield moment of the pile along with the yield zone, it can thus be used for conducting seismic analysis while doing seismic requalification studies. Thus it can be incorporated into the seismic requalification procedure chart as given by Krishna et al. [3].

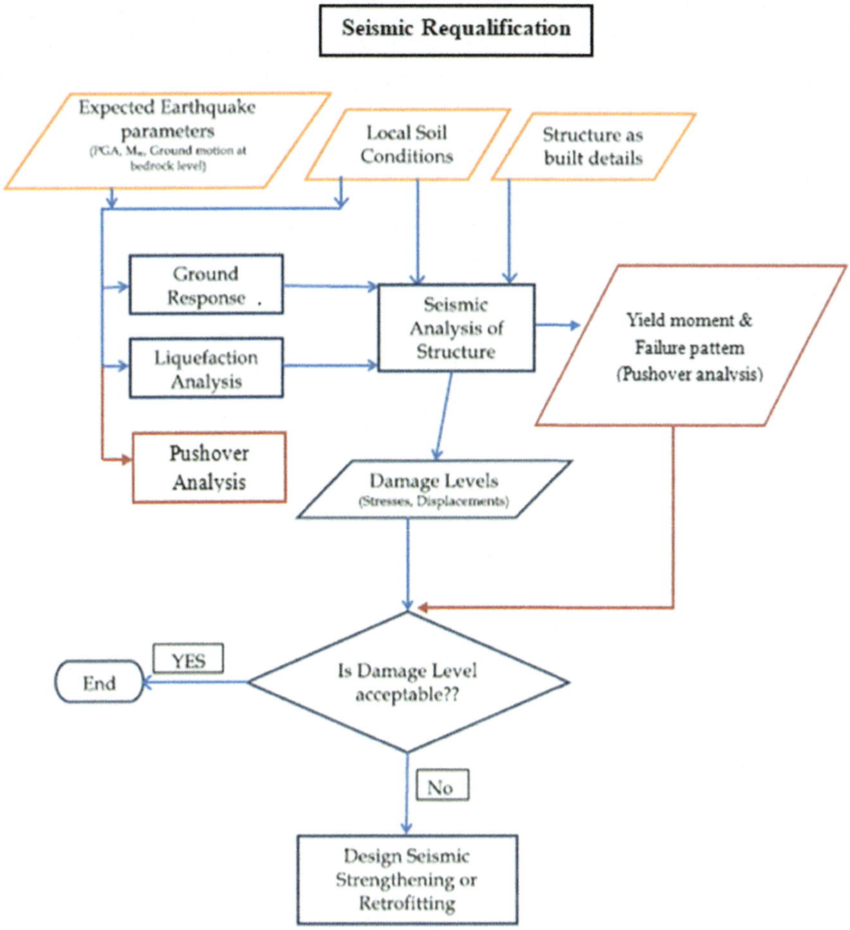

Fig. 7 Modified Seismic requalification chart

References

1. Abedzadeh F, Pak RYS (2004) Continuum mechanics of lateral soil-pile interaction. J Eng Mech 130(11):1309–1318
2. Kircher AC (1997) Overview of earthquake analysis methods with emphasis on nonlinear static methods. EERI Technical Seminar Series, Earthquake Engg Research Inst., Mountain View, California
3. Krishna AM, Bhattacharya S, Choudhury D (2014) Seismic requalification of geotechnical structures. Indian Geotech J 44(2):113–118. https://doi.org/10.1007/s40098-014-0115-5
4. Lu J, Elgamal A, Yang Z (2011) OpenSees PL: 3D lateral pile-ground interaction. User Manual. University of California, San Diego

5. Mukhopadhyay M, Choudhury D, Phanikanth VS, Reddy GR (2008) Pushover analysis of piles in stratified soil, The 14Th World Conference on Earthquake Engineering. Beijing, China
6. Wang N (2015) Three-dimensional modeling of ground-pile systems and bridge foundations. Phd Thesis. University of California, San Diego

Damage Detection Using Experimentally Obtained Frequency Response Data with an Application of Artificial Neural Network-Based Improved Mode Shape Curvature

Sonu Kumar Gupta and Surajit Das

Abstract Artificial neural networks and the mode shape curvature technique are used to assess the severity and locations of damages using experimentally recorded displacement mode shapes as input data. The Bruel and Kjaer instrument with an impact hammer is used to obtain frequency responses such as displacement mode forms with varied damage levels. Three-mode forms are first considered. It is shown that without further investigation, the recorded frequency response (displacement mode shapes) is insufficient to localize the damage. Artificial neural network training processes are utilized to reduce measurement error from the measured frequency response data set. Using central difference approximation, the training data sets are then used to generate the mode forms curvatures.

Keywords Frequency response function · Artificial neural network · Damage location · Improved curvature damage index

Nomenclature

FR	Frequency-response
ANN	Artificial neural network
FFT	Fast flourier transform
M	Mass
D	Damping
K	Stiffness
$X(\omega)$	Output response
$F(\omega)$	Input force
$H(\omega)$	Frequency response for displacement

S. K. Gupta (✉) · S. Das
Department of Civil Engineering, National Institute of Technology Agartala, Agartala, India
e-mail: sngupta77@gmail.com

S. Mitra et al. (eds.), *Disaster Management and Risk Reduction: Multidisciplinary Perspectives and Approaches in the Indian Context*,
https://doi.org/10.1007/978-981-99-6395-9_13

v	Number of input/output
y''	Model curvature
h	Element size
S_{mm}	Power spectrum
e_k	Experimentally obtained data
p_k	ANN-based predicted data
N_k	Normalized value
Z_k	Input/output data
$Z_{k,\, min}$	Minimum value
$Z_{k,\, max}$	Maximum value
MSE	Mean square error
MAPE	Mean absolute percentage error
R	Regression coefficient
MCDI	Modified curvature damage index
PCA	Principal component analysis
$\gamma^2(\omega)$	The coherence

1 Introduction

In reality, every civil and mechanical structure deteriorates with time as a result of environmental and other factors. Structural health monitoring guarantees that the structure remains safe and sound even after repeated stresses. Damage detection techniques and instruments that are effective, dependable, and cost-efficient are required to assess the state of structures during their service life. External loads, accidents, blasting, military activities, and natural disasters including earthquakes, floods, storms, and landslides all have the potential to inflict structural damage. The goal of structural damage assessment is to evaluate whether a structure has damage or a reduction in stiffness, and if so, where the damage is located and how severe it is. One of the most likely ways for accurately depicting the state of structures is modal analysis with frequency response function. To differentiate between healthy and damaged structures, the observed frequency response function was combined with an artificial neural network and principal component analysis [1]. The frequency response function [2] was used to propose a method for evaluating a free-free beam structure. Using frequency response data and the interpolations damage detection technique, the sensitivity of a bridge structure was evaluated at various temperatures [3]. The pattern of nonlinear damages in a building model was predicted using a frequency response function and artificial neural networks [4]. Mode shapes and operational deflection forms were employed to diagnose flaws in a cantilever beam using the frequency response function [5].

Natural frequency can also be used to examine the overall health of existing structures. An aluminum plate and a cross-ply carbon-fiber-reinforced plate were successfully detected and calculated using inadequate sets of natural frequency data of a plate construction [6]. Natural frequencies were used in a technical experiment to assess the sensitivity of the eigenvalues owing to stiffness loss in a steel-framed structure [7]. A review published [8] discovered several strategies based on natural frequencies, mode shapes, and curvature-mode-shapes, as well as comparative benefits and cons for each method. A recent study [9] showed that using observed natural frequencies in combination with noise response rate can improve the sensitivity of damage detection systems.

Because mode form is a function of structural stiffness, even minor changes in stiffness can result in significant changes in mode shapes, it is critical for structural status monitoring. The mode shape and curvature mode shape technique, which has proven to be an excellent tool for detecting and localizing structural faults, was used in the majority of previous investigations. References [10, 11] published a well-researched article using second derivatives of mode shape to locate structural flaws. The difference in mode form curvatures between healthy and injured structures, according to this study, can efficiently indicate the damage site. A composite plate is analyzed to investigate size and position of de-lamination [12]. Damage areas in steel tubes and composite laminates were found and quantified using vibration response data such as natural frequencies and mode shape data [13, 14]. Many faults in a cantilever beam were discovered using an enhanced type of curvature mode forms termed TEO-WT curvature-mode-shapes [15]. The damage in a cantilever beam was identified using a damage detection technique based on operational deflection shapes (ODS) [16]. Damage to wind turbine blades was detected using differences in mode form curvature [17].

An artificial neural network is a novel method for identifying and localizing deterioration in unexplained structural systems. Unknown structural changes and the existence of flaws in the structures were characterized using artificial neural network techniques [18–20]. Using lamb-wave testing data trained using ANN [21], a fatigue fracture developing in a steel structure was found.

End-users care about the safety of a serviceable building, which appears to be a common reason for the SHM. To evaluate the performance of structural systems, SHM, and damage detection are required in all constructions. For more than 40 years, a variety of methodologies have been used to assess the health of a structure and to predict structural behavior in real-time. For structural damage detection, the majority of previous studies used simulated FR data. The main goal of this study is to use ANN to reduce measurement error in experimentally obtained FR data sets, then use the trained FR data sets to compute the mode shape curvatures corresponding to undamaged and damaged structures, and finally compare the mode shape curvatures corresponding to undamaged and damaged structures to estimate structural damages.

1.1 The Objective of the Present Work

The key objectives listed for the present investigation are as follows:

- Determine the displacement mode shape using the experimental FRF data point.
- Evaluate the resilience of FR data sets using the Coherence function.
- To look for structural degradation by observing natural frequency shifts and abrupt changes in coherence function.
- To train a suitable ANN to extract error-free FR data and in that way construct error-free displacement mode shapes from experimentally obtained FR data sets.
- Create mode shape curvature using error-free mode shapes.
- The curvature damage index is calculated using the mode shape curvature, which aids in the detection of damage in the model structure.

2 Theoretical Background

2.1 Frequency Response Function

Because the frequency response function fluctuates owing to deficiency in the structures, the dynamic attributes of structures can be derived from the frequency response data. A system with multiple degrees of freedom (MDOF) has a basic dynamic equation of equilibrium:

$$[M]\{\ddot{x}(t)\} + [D]\{\dot{x}(t)\} + [K]\{x(t)\} = F(t) \tag{1}$$

$[M]$, $[D]$, and $[K]$ represent the n × n mass, damping, and stiffness matrices respectively. By applying the Fast Fourier Transform (FFT) to the frequency response, the following formulas are obtained [22].

The external force and displacement are also expressed as:

$$f(t) = \{F(\omega)\}e^{j\omega t}$$

$$x(t) = \{X(\omega)\}e^{j\omega t}$$

Equation (1) is arranged in the form

$$\left\{-\omega^2[M] + j\omega[D] + [K]\{x(t)\}\right\}e^{i\omega t} = \left\{F(\omega)e^{i\omega t}\right\} \tag{2}$$

Defining $H(\omega)$ as:

$$[H(\omega)] = \left(-\omega^2[M] + j\omega[D] + [K]\right)^{-1} \tag{3}$$

Equation (2) is expressed in the form:

$$H(\omega) = \frac{X(\omega)}{F(\omega)} \tag{4}$$

In general, $H(\omega)$ is expressed in the matrix form:

$$H(\omega) = \begin{bmatrix} H_{11} & H_{12} & \dots\dots & \dots\dots & H_{1m} \\ H_{21} & H_{22} & \dots\dots & \dots\dots & H_{2m} \\ . & . & . & . & . \\ . & . & . & . & . \\ H_{n1} & H_{n2} & \dots\dots & \dots\dots & H_{nm} \end{bmatrix}$$

H(ω) is expressed in the form of column matrix:

$$H(\omega)_{i1} = \left[H_{11}\ H_{21}\ ..\ H_{n1} \right]^T$$

and,

$$\alpha(\omega)_{i1} = \left[\alpha_{11}\ \alpha_{21}\ ..\ \alpha_{n1} \right]^T \tag{5}$$

2.2 *Absolute Mode Shape Curvature Difference Method (AMSCD)*

The following terms, such as flexural rigidity *EI* and the bending moment *M*, are related by the structural curvature y'' at a beam cross-section:

$$y'' = \frac{M}{EI} \tag{6}$$

This clearly shows that if a specific area of the beam is damaged, the curvature will increase in proportion. A single damage point in a cantilever beam is discovered using mode shape curvature derived from observed frequency response function data. This work uses the experimentally obtained modal data to characterize the healthy and damaged beams. The normalized displacement mode shape is trained to reduce inaccuracy using the Matlab ANN tool. According to [10], the modal parameters are used to calculate the curvature mode shape using the central difference approach.

$$H(\omega)''_{i,j} = \left[H(\omega)_{i+1} - 2H(\omega)_i + H(\omega)_{i-1} \right] / h^2 \tag{7}$$

$$\left| \Delta H(\omega)''_i \right| = \left| \left(H(\omega)''_i \right)^d - \left(H(\omega)''_i \right)^u \right| \tag{8}$$

Here, $H(\omega)_{i,j}$ denotes the FRF measurement for accelerometer at location *i* and force location at *j*. *h* is the distance between nodal points *i* and $i + 1$.

2.3 Coherence Function

The coherence function is a signal spectrum which is affected by dynamic excitation. Checking the frequency response function for quality and non-linearity in frequency bands or resonance regions is a good indicator. It is also a standard spectrum analyzer with software that provides precise information on the quality of FRFs. The coherence function for linear systems may be divided due to measurement noise in the output. For this function, the time-domain equation of motion is [23, 24].

$$x(t) = S[f(t)] + m(t) \tag{9}$$

The operator 'S' is a function of the structure's elastic, inertial, and energy dissipation properties, and *m*(*t*) represents the operational noise. In the frequency domain, the coherence function $\gamma^2(\omega)$ is written as:

$$\gamma^2(\omega) = \frac{|H(\omega)|^2 S_{ff}(\omega)}{S_{xx}(\omega)} = 1 - \frac{S_{mm}(\omega)}{S_{xx}(\omega)} \tag{10}$$

In the given formulation, $S_{ff}(\omega)$, $S_{xx}(\omega)$, and $S_{mm}(\omega)$ are the power spectrum of input, response and noise respectively. The fraction of output power will be linearly correlated with the input. $\gamma^2(\omega)$ always lies between 0 and 1 ($0 \leq \gamma^2 \leq 1$). If $S_{mm}(\omega) = 0$, then $\gamma^2 = 1$, which indicates the absence of any measurement noise in the measured output. $\gamma^2 < 1$ indicates the presence of measurement noise in the output. The coherence map in Fig. 4 shows the existence of noise in the experimentally collected FRF data in this study (Figs. 1, 2 and 3).

3 Experimental Setup and Procedure

To test the dynamic properties of a cantilever beam, a specimen with a length of 520 mm was chosen. The beam's width and thickness are 20 and 6 mm, respectively. For the experiment, mild steel beam material with parameters such as density (7850 kg/m^3), modulus of elasticity (200 GPa), and poison ratio (0.29) is used. A steel platform is permanently fastened to one end of the beam and secured tightly to prevent movement in any direction. A bolted connection secures the steel platform to a concrete surface.

To extract the dynamic modal properties of the model structure, a test setup using B& K equipment is developed. Using a B&K Fast Fourier Transform (FFT) analyzer, this experiment attempts to extract the frequency response function and, as a result, acquire the natural frequency, mode shapes, and modal acceleration. Figure 1a depicts two beam specimens, while Fig. 1b depicts specimens with various cut depths.

Three channels are employed in the hardware module, one for an impact hammer and the other two for accelerometer. The accelerometers are adhered to the bottom

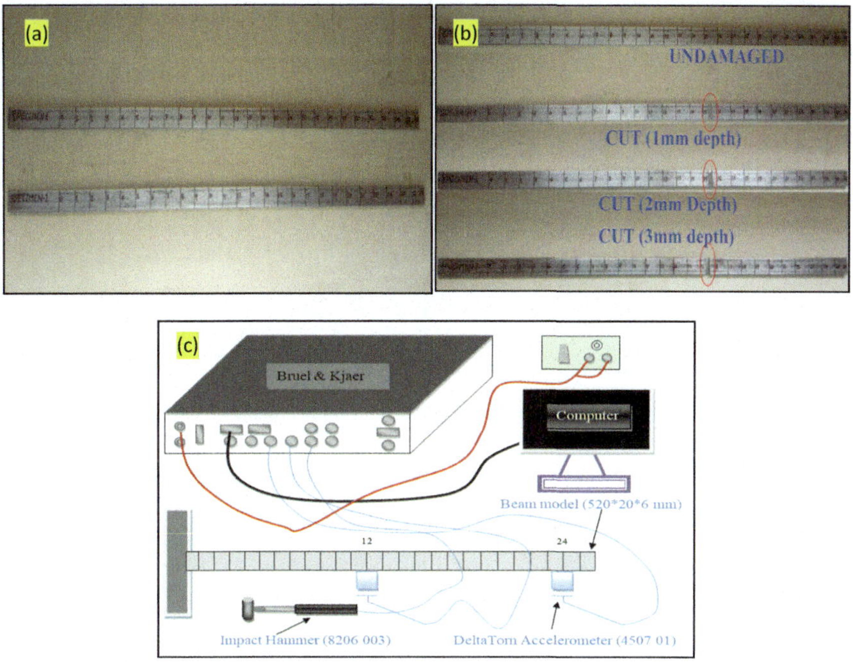

Fig. 1 **a** Beam specimens 1 and 2, **b** beam with saw cut and, **c** experimental setup with line diagram

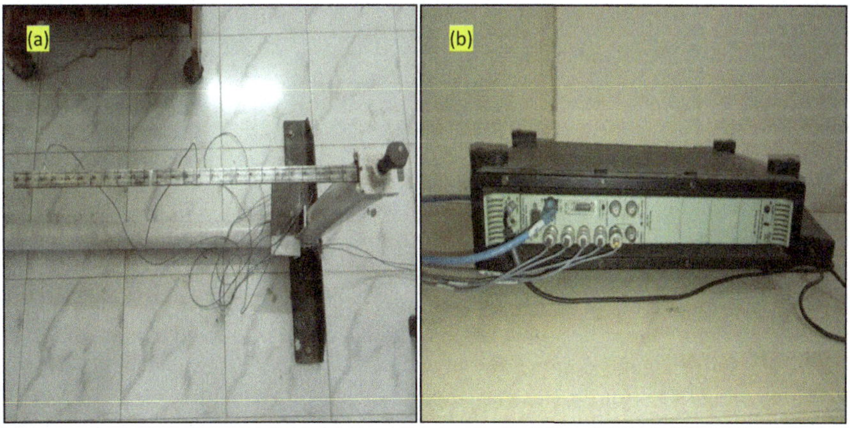

Fig. 2 **a** Beam model and, **b** Bruel and Kjaer module

surface of the beam specimens with adhesive at node locations 12 and 24, as shown in Fig. 2a.

This type of impact hammer works best for medium- to small-scale structures. The hammer has a strong plastic head that provides impact pressure on the structures

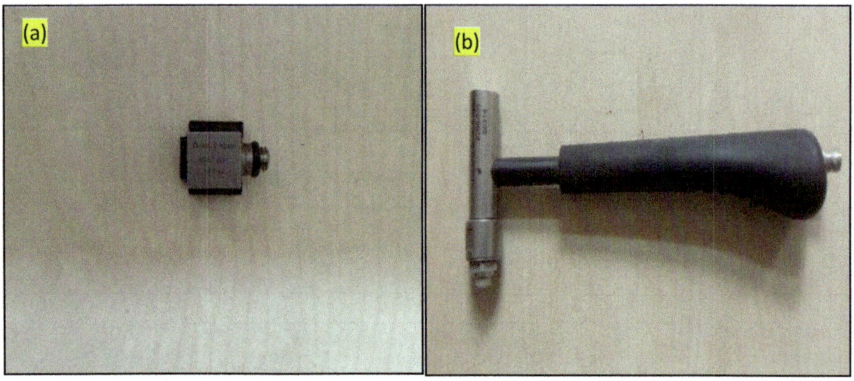

Fig. 3 **a** DeltaTorn accelerometer type (4507 001) and, **b** impact hammer (8206 002)

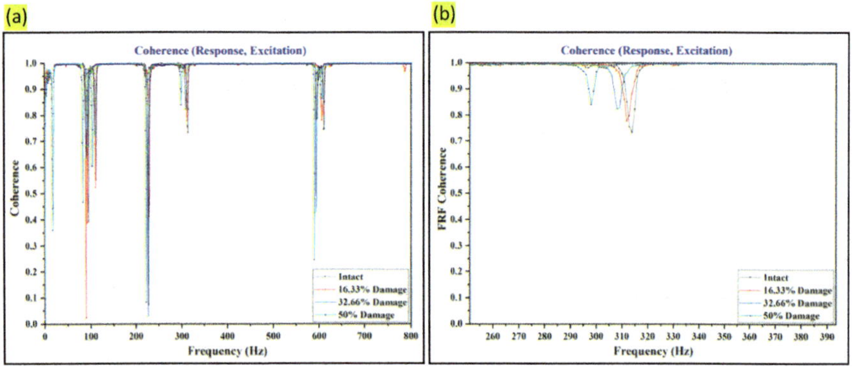

Fig. 4 **a** Coherence variations with different damage cases, **b** enlarged plot of coherence

for easy excitation. Highly sensitive accelerometers are utilized to obtain the modal parameters. The size and weight of the accelerometers are designed so that they have no detrimental impact on the measured data. 800 Hz and 800 lines, respectively, are the fixed frequency range and frequency line. The beam specimen is attached to a robust base before the experiment begins, and its length is marked at equal intervals to divide it into elements connected by nodes. DeltaTorn accelerometers are adhered to the specimen's bottom surface at precise nodes using adhesive material. As seen in Fig. 1a, these specimens had 27 nodes and 26 sub-elements. The reaction should be monitored multiple times at each position to generate a noise-free response. The coherence plot [24] illustrated in Fig. 4a, in which the coherence values are confined within the interval of 0 and 1, is used to assess the quality of FRF data. A response with a value of 1 has no measurement noise, while a response with a value of 0 has the most noise. The imaginary part's peaks are used to form the modes of structure, as shown in Fig. 7. The same method is followed in the event of a damaged beam specimen as it does in the case of an undamaged beam structure. Saw cuts cause damage in three

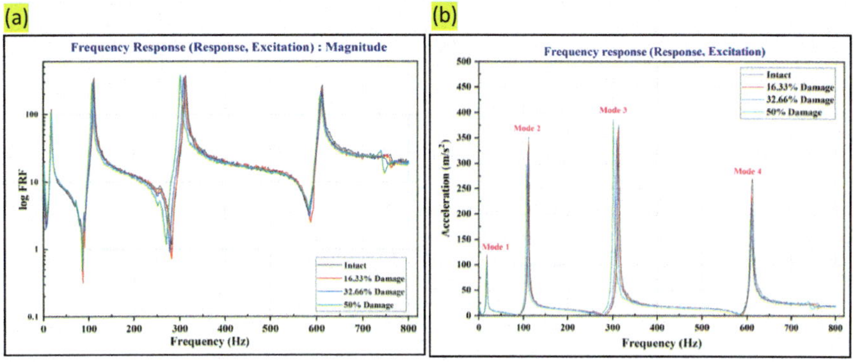

Fig. 5 **a** Log FRF plot, **b** imaginary FRF plot

phases, with 16.33%, 32.66%, and 50% of the thickness, respectively, resulting in cutting depths of 1 mm, 2 mm, and 3 mm, as illustrated in Fig. 1b. The data from the frequency response function at each position is crucial in constructing the structure's mode shapes (Figs. 5 and 6).

4 Results and Discussion

4.1 *Use of Coherence Function in Frequency Response Extraction*

The existence of damage and loss of linearity in structures can be successfully identified by monitoring sudden changes in the coherence function of recorded FR^s, as shown in Fig. 4 for the current investigation. Figure 4a depicts the total fluctuation over a frequency range (0–800 Hz).

The coherence shifts toward lower frequency in between (290–320 Hz) with increasing damage level in the structure, as shown in Fig. 4b, which has also been demonstrated in [4, 24, 25]. This pattern indicates that the structure has lost stiffness as a result of damage.

4.2 *Damage Prediction Using Frequency Response Function*

The quality of frequency response depends on coherence function. The presence of structural faults or modifications is indicated by deterioration in frequency response quality or noisy frequency response data. In software configuration, the frequency range and frequency lines are set to 800 Hz and 800 lines, respectively. The measured

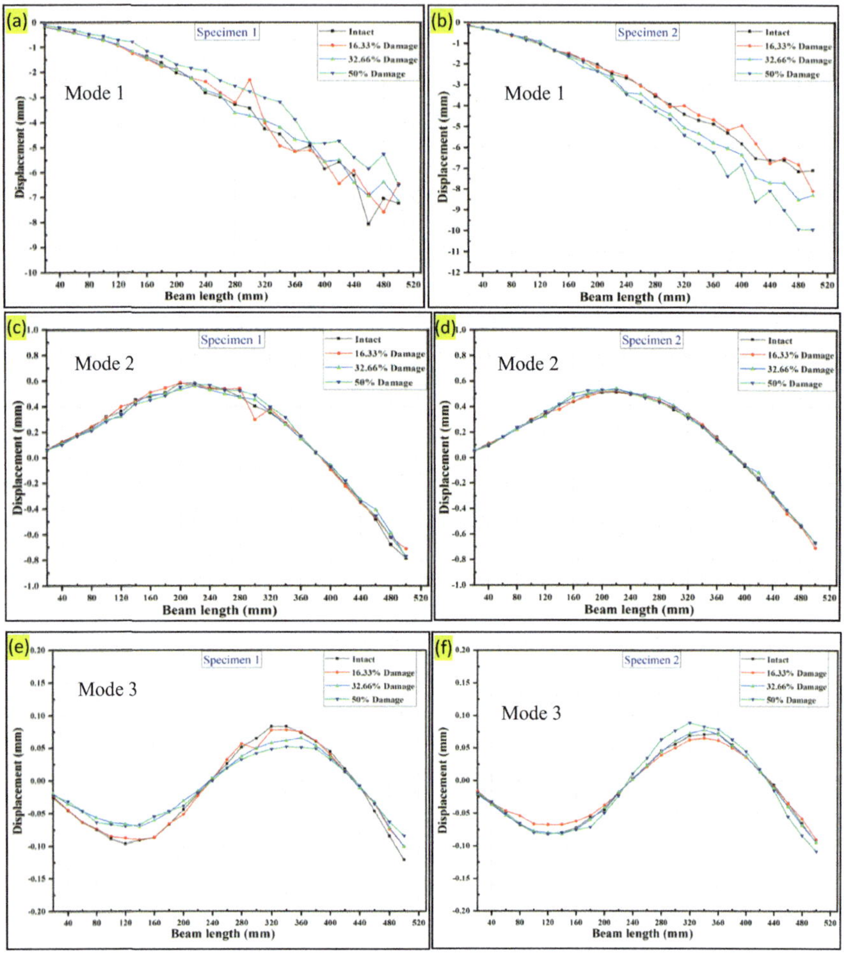

Fig. 6 Mode shape using raw modal data for specimen 1 and specimen 2

FRF yields a number of indications that can be used to forecast the occurrence of any structural changes. A number of FRF-based studies have shed light on the subject of damage identification and localization. FRF signatures such as shifting natural frequencies and changes in resonant peaks were used to study a two-story framed construction. The largest difference between undamaged and damaged FRF data was used to forecast the presence of damage and the location of damage in a beam [3]. The damage location is seen at all frequencies in a study based on normalized residual FRF shapes [5]. FRF data sets are created in diverse undamaged and damaged cases in the current investigation. The logarithmic FRF is shown in Fig. 5a, with damaged situations as intact, 16.33% damage, 32.66% damage, and 50% damage. It also demonstrates that, due to the structure's stiff behavior, the FRF variance between

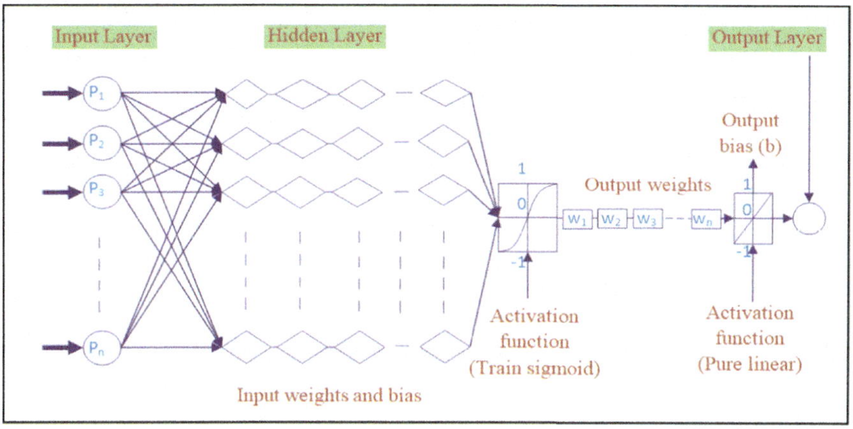

Fig. 7 Back-propagation neural network models

undamaged and 16% damage is quite modest. The imagined section of FRF has been shown in Fig. 5b, which indicates the variance even due to little damage. The larger plot (mode 2) reveals a distinct variance in damage statuses.

4.3 *Influence of Damage on Natural Frequencies*

The existence of flaws in the structures causes natural frequencies to change. Natural frequencies have been shown to be useful in the detection and localization of structural defects in a number of studies. The natural frequency shifts and FRF amplitude have been detected experimentally and numerically as a result of damage fluctuations [1]. Similar observations are made in the current investigation. With increasing damage depths, the response peak has changed from higher to lower frequencies, as seen in Fig. 5a, b. Tables 1 and 2 shows the variation in natural frequencies for the first four modes and their corresponding natural frequencies. The conditions of the structures and the degree of the damages are depicted by these frequency fluctuations. While the value does not change considerably in the first mode, the changes are larger in the following three modes. The natural frequencies are decreasing as the percentage of damage increases, as shown in Tables 1 and 2. It is also noted that the percentage fluctuations in the natural frequencies related to the second and third modes are substantially bigger, facilitating effective structural damage diagnosis.

The absolute mode shape curvature discrepancies reported in result Sect. 4.6 are used to determine the presence of structural problems and their severity, while their positions are determined using the results in Tables 1 and 2. In Tables 1 and 2, m1, m2, m3, and m4 indicated as mode 1, mode 2, mode 3, and mode 4 respectively.

Table 1 Specimen 1 contain damage between node 15 and 16

S. no.	Damage level	Natural frequency				Percentage age change in natural frequencies			
		m1	m2	m3	m4	m1	m2	m3	m4
1	Undamaged	18.5	113	314	612	–	–	–	–
2	16.33% damage	18.5	112	311	612	0	−0.892	−0.964	0%
3	32.66% damage	18.5	108	307	608	0	−4.629	−2.28	−0.657
4	50% damage	17.5	102	295	603	−5.71	−10.78	−6.44	−1.492

Table 2 Specimen 2 contain damage between node 9 and 10

S. no.	Damage level	Natural frequency				Percentage age change in natural frequencies			
		m1	m2	m3	m4	m1	m2	m3	m4
1	Undamaged	18.5	113	315	612	–	–	–	–
2	16.33% damage	18.5	112	312	609	0	−0.892	−0.961	−0.492
3	32.66% damage	17.5	109	307	605	−5.571	−3.669	−2.605	−1.157
4	50% damage	17	106	302	604	−8.82	−6.604	−4.304	−1.324

4.4 Analysis of Mode Shapes

The beam mode shapes are extracted using the experimentally obtained frequency response data set. A modified mode shape difference technique was utilized to find damage in a square-section mild steel tube in [13]. A beam structure's damage location was discovered utilizing operational deflection modes [16]. Figure 6a–f, which shows, how damage affects the appearance of both undamaged and damaged specimens.

In two-beam specimens, three-mode forms were first explored with varying damage depths and damage locations. To eliminate the possibility of measurement errors in the experimentally obtained FR, the data is further trained using an ANN technique to get error-free damage localization in the current study.

4.5 Modeling of Artificial Neural Network

A feed-forward neural network can be employed to express the non-linear functional mapping between a set of input and output modal variables. A non-linear function of multiple variables is described as a superposition of non-linear functions of single variables using an activation function [26]. As demonstrated in Fig. 7, a back-propagation neural network is a supervised feed-forward multi-layer neural network.

An input, hidden, and an output layers are the three layers of a back-propagation neural network. The input layer contains measured frequency response function data, which is then processed by hidden layers before being presented with accurate results by an output layer. The Levenberg-Marquardt (trainlm) algorithm and a multilayer perceptron with feed-forward back-propagation were used in this investigation, as described in [27]. A single hidden layer with a differentiable log-sig activation function is used to evaluate the results. Several studies have used hidden neurons with a number ranging from two to twenty-five to select the best network [28]. An ANN model is called to be optimal when statistical error lies within the ranges of MSE (Mean-square-error) < 0.001, MAPE (Mean absolute percentage error) < 5% and R (Regression) > 0.98. The statistical error can be determined using the expressions given in Eqs. (11–14). The training process stops when the validation error begins to rise, which doesn't allow over-learning of the model. Before training the network using an expression (14) as used in [29, 30], the FR data sets were normalized within the range of 0.1–0.9. Following normalization, 70% of the normalized data sets are randomly selected for training, 15% for validation, and the remaining 15% for testing.

$$MSE = \frac{1}{v} \times \left\{ \sum_{k=1}^{v} (e_k - p_k)^2 \right\} \tag{11}$$

$$MAPE = \left\{ \frac{1}{v} \times \sum_{k=1}^{v} \left| \frac{(e_k - p_k)}{e_k} \right| \right\} \times 100\% \tag{12}$$

$$R = \sqrt{\left[1 - \left\{ \frac{\sum_{k=1}^{v} (e_k - p_k)^2}{\sum_{k=1}^{v} p_k} \right\} \right]} \tag{13}$$

where, v = total output data; e_k and p_k = An experimental and predicted values.

$$N_k = 0.1 + \left\{ 0.8 \times \left(\frac{Z_k - Z_{k,min}}{Z_{k,max} - Z_{k,min}} \right) \right\} \tag{14}$$

N_k is the normalized data obtained from Z_k, while, $Z_{k,\,max,}$ and $Z_{k,\,min}$ are the maximum and minimum values in the data set.

4.5.1 Discussion on Predicted Results

As demonstrated in [31], the measured frequency response data is trained in the MATLAB software using the trainlm algorithm, which outperforms other algorithms such as trainbfg, trainlm, trainscg, trainrp, traingda, and traingdx. There are three sets of input data and one set of output data in this network. The position of force excitation, absolute value of damage depth, and damage depth as a proportion of total depth are the input data sets, while the measured displacement is the output data set. Two to twenty-five neurons are considered for the training task in the present work, as

Table 3 Optimal results using TRAINLM algorithm

	Topology	Regression coefficient				MSE	Percentage MAPE
		Training	Validation	Testing	Overall		
m1							
Specimen 1	3-10-1	0.99043	0.99883	0.9905	0.9919	0.00085	4.4791
Specimen 2	3-10-1	0.99764	0.99863	0.9948	0.9973	0.00027	3.072
m2							
Specimen 1	3-10-1	0.99836	0.99885	0.9979	0.9983	0.00016	1.9977
Specimen 2	3-8-1	0.99944	0.99959	0.9992	0.9994	0.00006	1.5349
m3							
Specimen 1	3-10-1	0.9981	0.9991	0.9983	0.9982	0.00016	3.0675
Specimen 2	3-10-1	0.9992	0.9995	0.9959	0.9990	0.00010	2.051

shown in [28]. Table 3, shows the most optimal findings. The network information for structural specimens 1 and 2 is shown in the table as the number of inputs, neurons, and outputs (3-10-1). Because the true objectives cannot be accomplished without removing the experimental mistakes from the frequency response data sets, the experimental errors are removed by training the data sets with ANN. The trained data sets that were determined to be capable of finding the damage positions were successfully used to construct the mode shapes illustrated in Fig. 8a–f.

The first three learned mode shapes, as shown in Fig. 8a–f, are used to assess the structure's current state. Using the mode shape curvature technique, the trained data sets are used to determine the damage location. Table 3 shows that the second mode fared better than the others since its MSE and MAPE values are the lowest.

4.6 Use of Absolute Mode Shape Curvature Difference in Damage Detection

Damage investigation of the numerically produced beam model utilizing the second-order derivative of mode shapes was given as a robust technique to show the fault site [10]. The first three mode shape curvatures in a composite beam were used to locate the fault [11]. The goal of this study's use of the 'absolute mode shape curvature difference' is to locate damage using mode shape curvature generated by a structure's displacement mode shapes. In this paper, a novel technique has been used to reduce error from observed FRF data sets using ANN training, which is useful for subsequent structural damage assessments. The curvature mode shape formed directly from measured frequency response data is incapable of finding damage, however the curvature mode shape generated using trained frequency response data is capable of tracking the existence and position of damage in the structure. The mode shape curvatures for specimens 1 and 2 are constructed using the first three mode

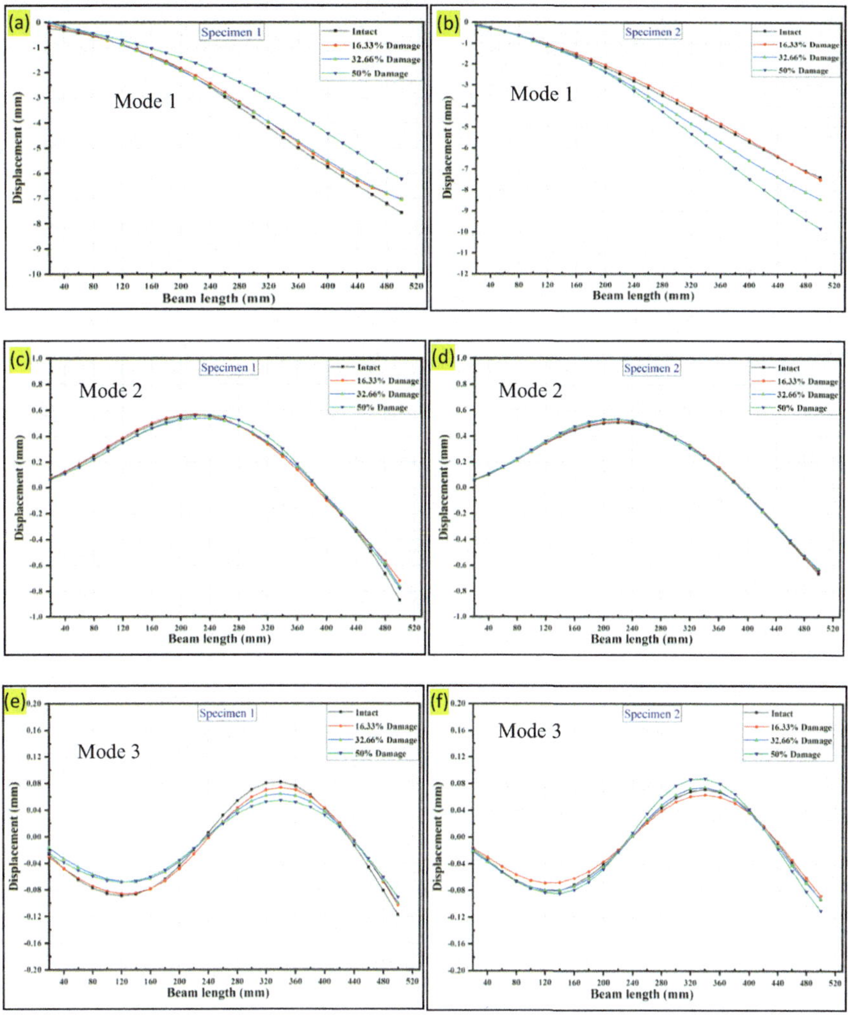

Fig. 8 Trained mode shapes of specimen 1 and 2

forms in this study. The damage location is indicated by mode shape curvatures, albeit not with the same accuracy. The curvature maxima for the first mode, as illustrated in Fig. 9a, b, are far away from the specific damage location. The damage site in Fig. 10a, b has been determined with great precision using second mode shape curvatures. In addition, in Fig. 11a, b, the third mode shape curvature has been seen to be less accurate than the second mode shape curvature. The experimental research, as well as ANN and AMSCD, were validated in this work using two separate damaged areas in two specimens.

In Figs. 10b and 11b, the damage has been recognized more precisely in specimen 2, where the damage is close to the support, than in specimen-1, where the damage

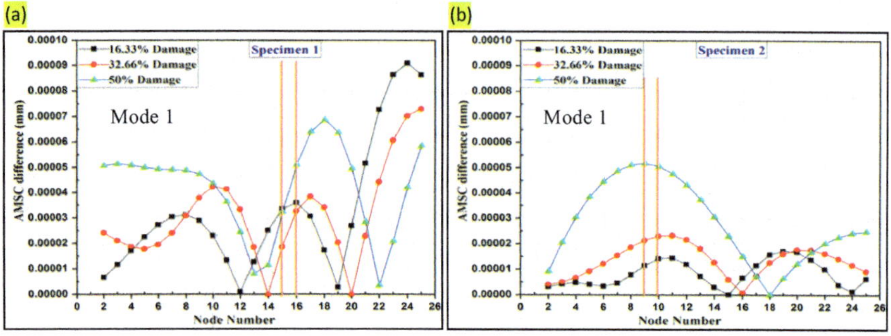

Fig. 9 **a** Absolute mode shape curvature difference for specimen 1 and, **b** absolute mode shape curvature difference for specimen 2

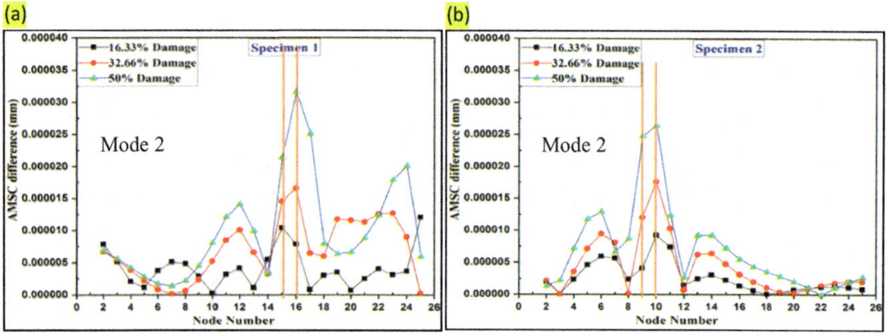

Fig. 10 **a** Absolute mode shape curvature difference for specimen 1 and, **b** absolute mode shape curvature difference for specimen 2

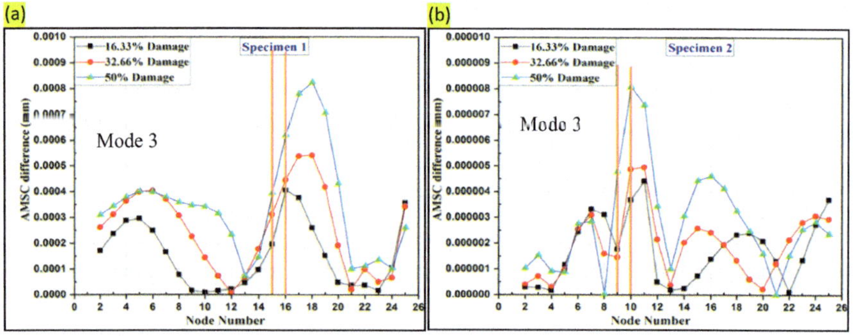

Fig. 11 **a** Absolute mode shape curvature difference for specimen 1 and, **b** absolute mode shape curvature difference for specimen 2

is far away from the support. The damage location was correctly detected in both situations.

5 Conclusions

When it comes to structural integrity, an early evaluation of the existence, location, and severity of damage is essential to estimate a structure's useful life and safety. Using frequency responses function, this research tries to forecast the existence, position, and severity of single damage position in a pair of identical cantilever beam model with varied damage locations. With varying degrees of damage, changes in natural frequency provide a clear indicator of the presence of damage. With increasing damage severity, the frequency response peaks shift toward lower natural frequencies. The frequency response data is used to produce the first three mode shapes. While earlier studies employed simulated modal data to detect structural deterioration, this study produces a modified set of frequency response data by training an artificial neural network on experimentally measured FR data to reduce errors. Following that, mod shape curvatures for both undamaged and damaged models are generated using the updated FR data. Finally, absolute mode shape curvature differences between undamaged and damaged structures that correspond to the identical modes are determined. The second of the mode shape curvature differences corresponding to the first three displacement mode shapes is found to provide the clearest information about the damage site, allowing the damage to be properly identified.

Conflict of Interest The authors declare that they have no conflict of interest.

References

1. Dackermann U, Li J, Samali B (2013) Identification of member connectivity and mass changes on a two-storey framed structure using frequency response functions and artificial neural networks. J Sound Vib 332(16):3636–3653
2. Maia NMM, Silva JMM, Almas EAM, Sampaio RPC (2003) Damage detection in structures: from mode shape to frequency response function methods. Mech Syst Signal Process 17(3):489–498
3. Limongelli MP (2010) Frequency response function interpolation for damage detection under changing environment. Mech Syst Signal Process 24(8):2898–2913
4. Bandara Rupika P, Tommy HT Chan, David P Thambiratnam (2014) Structural damage detection method using frequency response functions. Struct Health Monitor 13(4):418–429
5. Liu X, Lieven NAJ, Escamilla-Ambrosio PJ (2009) Frequency response function shape-based methods for structural damage localization. Mech Syst Signal Process 23(4):1243–1259
6. Cawley P, Adams RD (1979) The location of defects in structures from measurements of natural frequencies. J Strain Anal Eng Des 14(2):49–57
7. Hassiotis S, Jeong GD (1995) Identification of stiffness reductions using natural frequencies. J Eng Mech 121(10):1106–1113

8. Fan W, Qiao P (2011) Vibration-based damage identification methods: a review and comparative study. Struct Health Monit 10(1):83–111
9. Pan J, Zhang Z, Wu J, Ramakrishnan KR, Singh HK (2019) A novel method of vibration modes selection for improving accuracy of frequency-based damage detection. Compos Part B Eng 159:437–446
10. Pandey AK, Biswas M, Samman MM (1991) Damage detection from changes in curvature mode shapes. J Sound Vib 145(2):321–332
11. Lestari W, Qiao P, Hanagud S (2007) Curvature mode shape-based damage assessment of carbon/epoxy composite beams. J Intell Mater Syst Struct 18(3):189–208
12. Qiao P, Lu K, Lestari W, Wang J (2007) Curvature mode shape-based damage detection in composite laminated plates. Compos Struct 80(3):409–428
13. Elshafey AA, Marzouk H, Haddara MR (2011) Experimental damage identification using modified mode shape difference. J Mar Sci Appl 10(2):150–155
14. Pérez MA, Gil L, Oller S (2014) Impact damage identification in composite laminates using vibration testing. Compos Struct 108:267–276
15. Cao M, Radzieński M, Xu W, Ostachowicz W (2014) Identification of multiple damages in beams based on robust curvature mode shapes. Mech Syst Signal Process 46(2):468–480
16. Sampaio RPC, Maia NMM, Almeida RAB, Urgueira APV (2016) A simple damage detection indicator using operational deflection shapes. Mech Syst Signal Process 72:629–641
17. Wang Y, Liang M, Xiang J (2014) Damage detection method for wind turbine blades based on dynamics analysis and mode shape difference curvature information. Mech Syst Signal Process 48(1–2):351–367
18. Chaudhry Z, Ganino AJ (1994) Damage detection using neural networks: an initial experimental study on debonded beams. J Intell Mater Syst Struct 5(4):585–589
19. Masri SF, Smyth AW, Chassiakos AG, Caughey TK, Hunter NF (2000) Application of neural networks for detection of changes in nonlinear systems. J Eng Mech 126(7):666–676
20. Marwala T (2000) Damage identification using committee of neural networks. J Eng Mech 126(1):43–50
21. Dworakowski Z, Ambroziński Ł, Packo P, Dragan K, Stepinski T, Uhl T (2014) Application of artificial neural networks for damage indices classification with the use of Lamb waves for the aerospace structures. In: Key engineering materials, vol 588. Trans Tech Publications, pp 12–21
22. Brigham EO (1988) The fast Fourier transform and applications. Prentice Hall, Englewood Cliffs, NJ
23. Ewins DJ (1984) Modal testing: theory and practice, vol 15. Research studies press, Letchworth
24. Worden K, Farrar CR, Haywood J, Todd M (2008) A review of nonlinear dynamics applications to structural health monitoring. Struct Control Health Monit 15(4):540–567
25. Zhou Y-L, Figueiredo E, Maia N, Perera R (2015) Damage detection and quantification using transmissibility coherence analysis. Shock Vib 2015
26. Masri SF, Nakamura M, Chassiakos AG, Caughey TK (1996) Neural network approach to detection of changes in structural parameters. J Eng Mech 122(4):350–360
27. Chakraborty A, Roy S, Banerjee R (2016) An experimental based ANN approach in mapping performance-emission characteristics of a diesel engine operating in dual-fuel mode with LPG. J Nat Gas Sci Eng 28:15–30
28. Yusaf TF, Buttsworth DR, Saleh KH, Yousif BF (2010) CNG-diesel engine performance and exhaust emission analysis with the aid of artificial neural network. Appl Energy 87(5):1661–1669
29. Cay Y, Çiçek A, Kara F, Sağiroğlu S (2012) Prediction of engine performance for an alternative fuel using artificial neural network. Appl Therm Eng 37:217–225
30. Basheer IA, Hajmeer M (2000) Artificial neural networks: fundamentals, computing, design, and application. J Microbiol Meth 43(1):3–31
31. Bhowmik S, Panua R, Debroy D, Paul A (2017) Artificial neural network prediction of diesel engine performance and emission fueled with Diesel–Kerosene–Ethanol blends: a fuzzy-based optimization. J Energy Resour Technol 139(4)

Impact Assessment of Plastic Strips on Compressive Strength of Concrete

Snehal Kaushik, Sandip Shekhar Nath, and Titan Das

Abstract Solid waste is one of the many factors that negatively affect the environment. Plastic is an important type of solid waste with a strong environmental impact and is difficult to recycle and reuse. In order to decrease the plastic waste, many researchers have conducted various researches on concrete and have come up with innovative ideas of using plastic waste in concrete which not only helps in mitigating plastic waste problems but also helps in enhancing the compressive characteristics of a concrete. In this study, conventional concrete is studied under compression loading and compared with modified concrete samples having horizontally oriented plastic strips (MC-H). Three different arrangements of PET strips are considered. In the first arrangement, two layers of PET strips spaced at an equal distance of 50 mm between the concrete grades of M30. In the second arrangement, five layers of PET strips placed at an equal distance of 25 mm are placed. And lastly in the third arrangement, eleven layers of PET strips, each being separated by a distance of 12.5 mm. From load and deflection control tests, it is concluded that plastic (PET) strips positively contribute to concrete behavior in terms of increased stiffness and higher strength.

Keywords Solid waste · Plastic · Compressive strength of concrete · Environmental impact

1 Introduction

Concrete is the most commonly used construction material due to high durability and strength. Various researches have been carried out to reduce the negative influence of the cement industry and to develop sustainable concrete. In the past several attempts are made to develop sustainable concrete by replacing coarse aggregates by recycled solid waste aggregates, electronics plastic waste, plastic, glass, rubber, etc. [1]. Efforts

S. Kaushik (✉) · S. S. Nath · T. Das
Department of Civil Engineering, Girijananda Chowdhury Institute of Management and Technology, Guwahati, Assam, India
e-mail: snehal_ce@gimt-guwahati.ac.in

S. Mitra et al. (eds.), *Disaster Management and Risk Reduction: Multidisciplinary Perspectives and Approaches in the Indian Context*,
https://doi.org/10.1007/978-981-99-6395-9_14

have been made to tackle the dual problem of reducing the load on landfill sites with waste and cement consumption by using different class of materials derived from domestic wastes. Due to high strength and durability and easy processing plastic is widely used material around the world. Disposal of plastic waste is a huge problem due to absence of organic compound; it is non-decomposable material and a huge threat to the environment. Plastic has potential for enhancing concrete properties, which motivated researchers to explore the utilization of plastic waste in building sustainable concrete. The term plastics is a general one used to represent a category of materials, for example, low density polyethylene (LDPE), high density polyethylene (HDPE), polypropylene (PP), and polyethylene terephthalate (PET). In past researchers have found marginal reduction in strength of concrete and suggested the optimum result as 10–15% replacement of plastic for coarse aggregates. Most of the research found reduction in strength of concrete but support the use of plastic in non-structural concrete for the reason it shows higher workability and reduce environmental waste [2].

At present, production of cement, a key component of concrete, contributes approximately 5% of the global greenhouse gas (GHG) emissions [3]. Studies on utilizing crushed plastics as partial replacement of sand noted reduction in compressive strength with increasing plastic content [4]. A recent study observed that HDPE fibers used in structural concrete exhibited no impact on compressive strength and elastic modulus, whereas a slight improvement in flexural and tensile strength was reported [5]. The ability of PET fibers to reduce plastic shrinkage is generally acknowledged, but their influence on strength properties remains inconclusive. The reported data suggest that the addition of PET fibers can either increase or decrease compressive strength [6–9].

It is believed by the authors that plastics can be sustainably used in concrete only after a thorough understanding of their role in concrete prior to and during failure is achieved. To address the knowledge gap in this area, the present work represents an attempt to investigate the influence of adding PET strips on the strength characteristics of structural concrete. Three different arrangements of PET strips are considered. In the first arrangement, two layers of PET strips spaced at an equal distance of 50 mm are used. In the second arrangement, five layers of PET strips placed at an equal distance of 25 mm are used. And lastly in the third arrangement, eleven layers of PET strips, each being separated by a distance of 12.5 mm are used. The resultant concrete cubes are cured and tested under load and deflection control, and the observed behavior is analyzed to quantify the effect of the addition of PET strips.

Fig. 1 Plastic strip cut from used plastic water bottles added in concrete cubes

2 Preparation of Plastic Strips from Plastic Bottle

For preparation of strips, plastic water bottles are collected from different places and cleaned thoroughly with water. The top and bottom portion of the bottles is removed with the help of a cutter/knife. The remaining portion is cut to make a strip of length 150 mm and width 6 mm (Fig. 1).

3 Material and Experimental Details

3.1 Design of Concrete Mix

To achieve the desired strength of the concrete design of concrete mix is carried out. The right proportion of cement sand and aggregate is determined to get the desired strength. To study the effect of plastic strip on the compressive strength of concrete, cubes of 150 mm × 150 mm × 150 mm using a nominal mix of 1:1.36:2.42 are casted. A total 36 cubes were casted to determine the compressive strength at the end of 7 days, 14 days, and 28 days. The following parameters were used for the mixed design.

- Grade of concrete: M30
- Type of cement: OPC 43 grade confirming IS 8112
- Maximum nominal size of aggregate: 20 mm
- Minimum cement content: 340 kg/m3
- Maximum Water-Cement ratio: 0.40
- Workability: 75 mm (slump)
- Exposure Condition: Severe
- Method of Concrete Placing: Manually

- Type of Aggregate: Crushed angular aggregate
- Specific gravity of cement: 3.15
- Specific gravity of
  - *Coarse aggregate*: 2.39
  - *Fine Aggregate*: 1.58
- Water absorption of
  - *Coarse Aggregate*: 11%
  - *Fine Aggregate*: 34%
- Sieve Analysis
  - *Coarse Aggregate*: 20 mm crushed angular aggregate
  - *Fine Aggregate*: Confirming to grading zone III of table 4 IS 383
- Los Angeles Abrasion Test: 29.6%
- Slump Test: 60 mm

The details of the M30 mix are given in Table 1.

The casting of cube is done in a single stage. During casting of the cube mold are properly tightened to correct dimensions before casting. Care was taken that there are no gaps left from where there is any possibility of leakage of slurry. The coarse aggregate and fine aggregate were weighted first with an accuracy of 0.5 gm. The required quantities of materials required were weighed according to the mix proportion. The coarse and fine aggregate were mixed thoroughly and then the cement was added. The water was added carefully so that no water was lost during mixing. Then the concrete is poured in three layers in the cube mold compacting by hand and vibrator (Fig. 2). Three different forms of mixes in nine batches were prepared. Three different arrangements of PET strips are considered as shown in Fig. 3. In the first arrangement, two layers of PET strips spaced at an equal distance of 50 mm. In the second arrangement, five layers of PET strips are placed at an equal distance of 25 mm. And lastly in the third arrangement, eleven layers of PET strips are used each being separated by a distance of 12.5 mm. In all the above-mentioned layers, 5 numbers of PET strips were used at each location. The resultant concrete cubes are cured and tested under load and deflection control, and the observed behavior is analyzed to quantify the effect of the addition of PET strips.

Table 1 Mix proportion for M30 grade of concrete

Water cement ratio		Cement	Water	Coarse aggregate	Fine aggregate
0.45	Quantity of material (kg)	425.74	191.58	1030.95	582.34
	Ratio of material	1	0.449	2.421	1.36
	Mass for 1 bag of cement (kg)	50	22.49	121.05	68.4

Fig. 2 **a** Casting of concrete cubes; **b** curing of concrete cubes

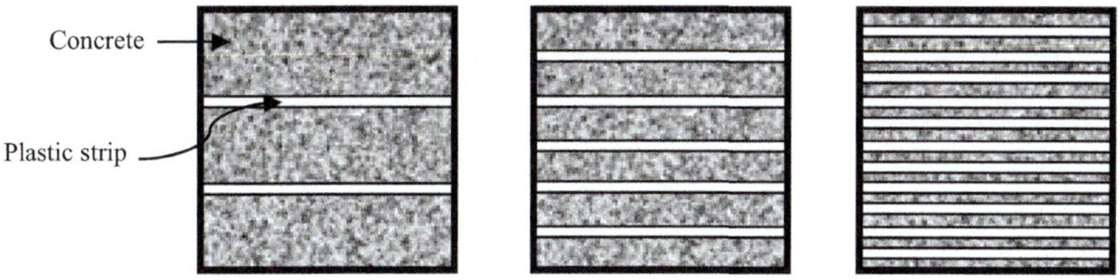

Fig. 3 Different arrangements of plastic strips: **a** spaced at 50 mm; **b** spaced at 25 mm and **c** spaced at 12.5 mm

3.2 Experiment Details

In this study experiments are carried out to understand the influence of PET strips on the properties of concrete. The basic properties of the cement and aggregates utilized in the study were checked to ensure that the choice of materials was in agreement with the recommendations of Indian standards (IS) [10–13]. Further, coarse aggregates corresponding to a maximum aggregate size of 20 mm were chosen to comply with Indian standard code [14]. Waste plastic strips derived from PET bottles were added to concrete. 36 numbers of cubes are tested to determine the compressive strength of concrete. Out of those, 9 numbers of cubes are of normal design mix and the remaining cubes are tested by adding plastic strips. The compressive strength of concrete is determined at the ages of 7, 14, and 28 days. Figure 4 shows the experimental set up of the concrete blocks to determine the compressive strength.

Fig. 4 Experimental set up for compression testing of concrete blocks

4 Results

It is observed from the experiment that the compressive strength of concrete initially increases with the increase of plastic content at 7 days and 14 days of curing but further curing for 28 days shows reduction in strength. Figure 5 shows the comparison of compressive strength of concrete tested after 7 days, 14 days, and 28 days of curing. The compressive strength values were significantly influenced by the amount of plastic strips and curing period. The optimum compressive strength of 42.66 MPa is obtained when the strips are laid at a distance of 50 mm c/c.

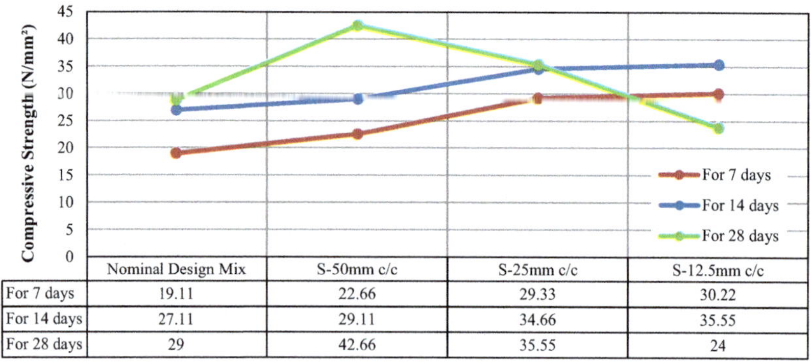

	Nominal Design Mix	S-50mm c/c	S-25mm c/c	S-12.5mm c/c
For 7 days	19.11	22.66	29.33	30.22
For 14 days	27.11	29.11	34.66	35.55
For 28 days	29	42.66	35.55	24

Fig. 5 Comparison of compressive strength of concrete considering different arrangements of plastic strip

5 Conclusions

In this study, conventional concrete is studied under compression loading and compared with modified concrete samples having horizontally oriented plastic strips (MC-H). Three different arrangements of PET strips are considered. In the first arrangement, two layers of PET strips spaced at an equal distance of 50 mm between the concrete grade of M30. In the second arrangement, five layers of PET strips placed at an equal distance of 25 mm are placed. And lastly in the third arrangement, eleven layers of PET strips, each being separated by a distance of 12.5 mm. From load and deflection control tests, it is concluded that plastic (PET) strips positively contribute to concrete behavior in terms of increased stiffness and higher strength.

- It has been noted that variation in compressive strength for modified concrete is much lower than that for conventional concrete. Further, the variation is observed to be higher if layers of plastic strips are increased and at a certain point the value rises to its extreme limit and then reduces gradually.
- The improved stiffness values accompanied by early crack initiation can be used to improve serviceability and measure damage accumulation in concrete.
- However, generalization of the proposed model will require extensive experimental investigation to confirm the role of the quantity and orientation of PET strips in concrete.
- Further, it was observed while experimenting that the compressive strength of concrete initially increases with the increase of plastic content at 7 days and 14 days of curing but further curing up to 28 days shows reduction in strength. The optimum compressive strength is obtained when the strips are laid at a distance of 50 mm c/c.

References

1. Pappu A, Saxena M, Asolekar S (2007) Solid waste generation in India and their recycling potential in building materials. Build Environ 42(6):2311–2320. https://doi.org/10.1016/j.buildenv.2006.04.015
2. Choi YW, Moon DJ, Chumg JS, Cho SK (2005) Effect of waste PET bottles aggregate on the properties of concrete. Cem Concr Res 35(4):776–781. https://doi.org/10.1016/j.cemconres.2004.05.014
3. Boden TA, Marland G, AndresRJ (2010) Global, regional and national fossil fuel CO_2 emissions. In: Carbon dioxide information analysis. Oak Ridge National Laboratory, Oak Ridge, TN, pp 425–441
4. Batayneh M, Marie I, Asi I (2007) Use of selected waste materials in concrete mixes. Waste Manage 27(12):1870–1876. https://doi.org/10.1016/j.wasman.2006.07.026
5. Pesic N, Zivanovic S, Garcia R, Papastergiou P (2016) Mechanical properties of concrete reinforced with recycled HDPE plastic fibres. Constr Build Mater 115:362–370. https://doi.org/10.1016/j.conbuildmat.2016.04.050

6. Irwan J, Asyraf R, Othman N, Koh H, Annas M, Faisal SK (2013) The mechanical properties of PET fiber reinforced concrete from recycled bottle wastes. Adv Mater Res 795:347–351. https://doi.org/10.4028/www.scientific.net/AMR.795.347
7. Kim SB, Yi NH, Kim HY, JayKim J-H, Song YC (2010) Material and structural performance evaluation of recycled PET fiber reinforced concrete. Cem Concr Compos 32(3):232–240. https://doi.org/10.1016/j.cemconcomp.2009.11.002
8. Fraternali F, Ciancia V, Chechile R, Rizzano G, Feo L, Incarnato L (2011) Experimental study of thermo-mechanical properties of recycled PET fiber-reinforced concrete. Compos Struct 93(9):2368–2374. https://doi.org/10.1016/j.compstruct.2011.03.025
9. Foti D (2011) Preliminary analysis of concrete reinforced with waste bottles PET fibers. Constr Build Mater 25(4):1906–1915. https://doi.org/10.1016/j.conbuildmat.2010.11.066
10. BIS (Bureau of Indian Standards) (1997) Specification for coarse and fine aggregate from natural sources for concrete. IS 383-1970. BIS New Delhi, India
11. BIS (Bureau of Indian Standards) (2013) Ordinary Portland cement, 43 grade-specification. IS 8112. BIS, New Delhi, India
12. BIS (Bureau of Indian Standards) (1959) Method of tests for strength of concrete. IS 516. BIS, New Delhi, India
13. BIS (Bureau of Indian Standards) (2000) Plain and reinforced concrete-code of practice. IS 456. BIS, New Delhi, India
14. BIS (Bureau of Indian Standards) (1963) Method of tests for aggregate for concrete. IS 2386-part 1. BIS, New Delhi, India

A Critical Socio-technical Insight of the 2014 Rainfall Induced Mudslide Disaster at Malin, India

Shubham Maurya and Arindam Dey

Abstract A massive catastrophic landslide wiped out the Malin village on July 30, 2014. The occurrence of such landslides cannot be prevented, but its effects can be minimized by taking timely preventive measures. This event has revealed the mismanagement of vulnerable hills of India. The 2014 Malin landslide has taught the lesson that the disaster management fraternity is yet to evolve in terms of pre-disaster mitigation strategies. There is an ardent necessity to start working on strategies to prevent a repetition of such costly disasters in nearby future. In this critical review, various aspects of the catastrophe are elucidated and the mitigating measures are described. A perspective towards the future direction of management is also provided. Such an attempt is important especially for the north-eastern region of India as the regional hillslopes are tremendously wetted by prolonged, incessant and recurrent rainfall. Hence, the lessons learned from the 2014 Malin landslide should help to develop an efficient preventive system against such similar occurrences in the future.

Keywords Malin village · Rainfall induced mudslide · Western ghat · Padkai scheme · Slope stability

1 Introduction

Malin is a small village in Ambegoan taluka located 110 km from the Pune city of Maharashtra, India. A massive catastrophic landslide occurred in the early morning (5 AM) of 30th July when the residents were still asleep. This tragedy affected 40 families and buried 45 houses. As the village was located in a remote area, the tragedy was known to the people and officials only when the driver of the first state

S. Maurya (✉) · A. Dey
Centre for Disaster Management and Research (CDMR), Indian Institute of Technology Guwahati, Guwahati, Assam, India
e-mail: smaurya@iitg.ac.in

A. Dey
e-mail: arindam.dey@iitg.ac.in

S. Mitra et al. (eds.), *Disaster Management and Risk Reduction: Multidisciplinary Perspectives and Approaches in the Indian Context*,
https://doi.org/10.1007/978-981-99-6395-9_15

transport bus reached the village a little after 7:30 AM. In a few reported articles, it is claimed that it was only known to the city officials when the bus driver drove back to the city. There was no early warning from the Indian Meteorological Department (IMD). According to the official records, the tragic event killed 151 people (59 men, 73 women and 21 children) and 300 cattle, along with more than 100s were missing. This event has revealed the mismanagement of vulnerable hills of India. There are many factors that are responsible for such massive landslide. This article would provide a walkthrough of the various aspects of the catastrophe and provide a discussion of the mitigating measures along with a perspective towards future direction.

2 Study Area

Malin village is located in the Western Ghats on the slopes of the 'Sahyadri'. Western Ghats consist of mainly fold mountains made of basaltic rocks and residuals of laterite and bauxite ores. These Ghats have a high level of biological diversity and endemism, and therefore it is internationally recognized as a UNESCO World Heritage Site. The geographical coordinates of the Malin village are latitude 19°9′46″ N and longitude of 73°41′2″ E, and is situated at an elevation of 796 m [1]. The schematic map showing the location of Malin village is shown in Fig. 1.

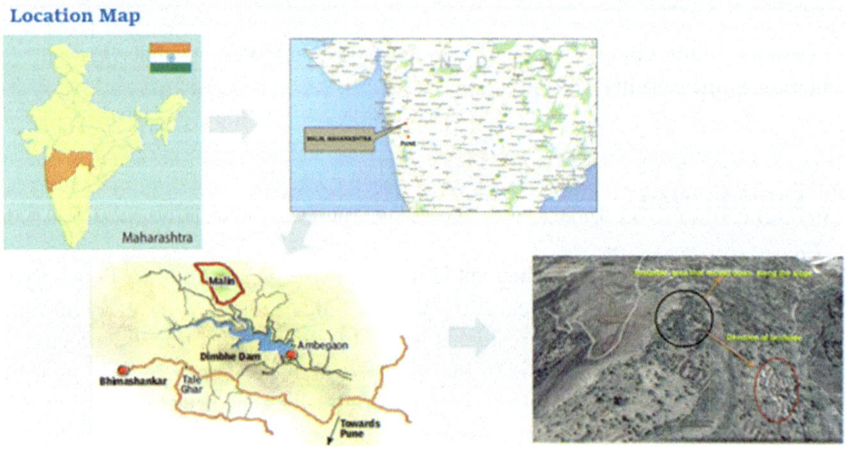

Fig. 1 Study area, map, location and topography of Malin, Maharashtra

3 Geological and Tectonic Setup

The village is situated in a geologically fragile region and is considered sensitive due to its location, geo-tectonic settings and high rainfall. The Malin area lies in the Deccan volcanic provinces, which is also called as Deccan trap, which was formed on solidified basaltic lava flows [2]. This Deccan lava flow is formed due to the fissure type of volcanic eruption and in this region, it is found mostly in the form of horizontal bed sheets. Two major types of soils are found in Ambegoan region: reddish-brown soils and dark-grey reddish-brown soils cover major areas of Ambegoan, predominantly along the hillside slopes. These soils are semi-fertile and have a loose texture, and thus was rapidly converted to agriculture land by local villagers. Hills in the Malin region have uniform and gentle slopes. The Google image-based interpretation shows that Malin region is bounded by faults from all four sides: ENE-WSW sinistral faults from the southeast and northwest and NW–SE dextral faults from the northeast and southwest. Hence, Malin block must be under geologic stresses due to bounding faults leading to warping and fracturing of the block [3]. The past occurrence of cracks in 2003 seen by the local people must be due to such deformations of tectonic faults. From the seismotectonic map, it was found that two seismic activities of smaller magnitude have occurred (within 100 km) in the past during 1964 and 1993.

4 Landslide Characteristics

The entire hill slope where the landslide occurred was divided into 4 Zones [4]. The dimension of landslide ranges from 45 to 62 m at the top and 134 m wide at the bottom. The runoff material extended over a distance of 512 m and the area affected is roughly 44,239 m^2. The depth or thickness of the material slid during landslide is approximately estimated as 7 m as most of the fills were removed in rescue operations before the forensic investigation. The zone of flow and accumulation unfortunately encompassed the resident zone of the Malin village, thus the effects compounded to a large catastrophe. It was found that the whole village, except a few houses and a primary school, was swallowed up by mounds of mud and debris. The debris material flow consisted of largely silty clay and big boulders. The size of boulders matched with those used for making bunds on the terraces for paddy cultivation and was, therefore, inferred to be the part of the destroyed bunds that existed along the slide zone. Three draining streams were observed in the slide zone: one of them running almost to the middle of the Malin landslide and other flowing closer [5]. In 2003, signs of cracks and distress were witnessed in the region by some NGOs. The villagers also noticed the cracks in the hills. After the detections, they were shifted to the adjacent village in specially erected shelters. However, they returned to Malin village again after some time without understanding the risk associated and

(a) (b)

Fig. 2 Malin village **a** before, **b** after landslide. *Source* BBC News

probably this act led to large-scale casualties during the event. Figure 2 highlights the scenarios of the Malin village before and after the 2014 mudslide event.

5 Chronology of Disaster

The region experienced very heavy rainfall from the 22nd July 2014 itself. As Malin village does not have rain-gauge station in its area, the rainfall data were collected and analyzed from the nearby tehsil region's station. Hence, the assessed rainfall might not represent the actual rainfall experienced by Malin village; however, the same can be considered for a forensic understanding under all practical circumstances. It was observed that from 22nd to 28th July, the expected normal rainfall was in occurrence, but the nearby rain-gauge station recorded a whopping 108 mm of rainfall on 29th July, or even more (as shown in Fig. 3).

The rainfall recorded by the National Aeronautics and Space Administration of US's (NASA) Tropical Rainfall Measuring Mission (TRMM) on July 29th shows the Malin region as a very susceptible-zone [6] which suggested the rainfall exceeded 175 mm. The continuous rainfall since, 5–6 days before the event had saturated the hill slope of the area. This developed mound of mud and debris that swallowed the entire tribal village of around 50 families. On investigating the rainfall data, it was found that in the year 2013, the place received more rainfall than 2014 (as shown in Fig. 4). However, no landslide was recorded in Malin or surrounding region in the year 2013. Hence, it was comprehended that other than natural factors, anthropogenic contributions including faulty housing practices, engineering projects, unscientific development and land-use practices as well as deforestation have contributed significantly to the occurrence of such huge devastating disaster.

Poor socio-economic conditions, non-access to goods and services, literacy, and ignorance has contributed to the vulnerability of the people. Lack of these resources and amenities forced the people to construct settlements in such hazardous and

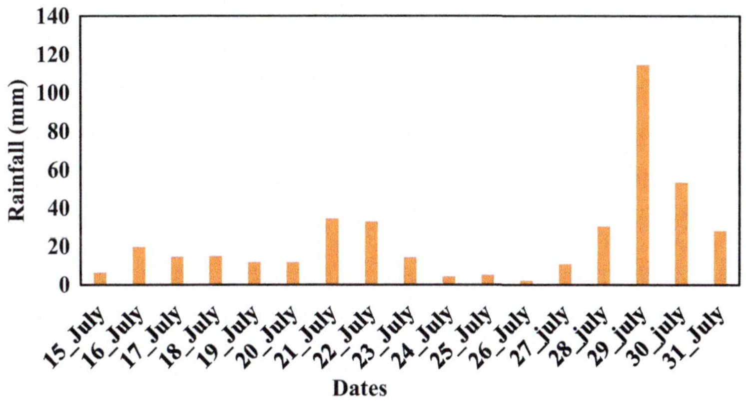

Fig. 3 Rainfall data collected at nearby rain-gauge station

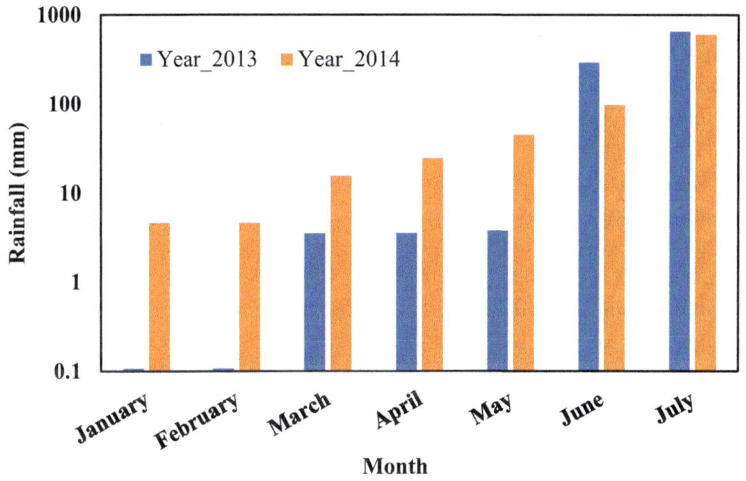

Fig. 4 Comparison of cumulative rainfall record of the year 2013 and 2014 from January to July

vulnerable areas. People living in Malin area are poor farmers and are mainly dependent on agricultural activity. To improve the socio-economic status of the villagers, income paddy cultivation was promoted by the Government of Maharashtra with **Padkai scheme** for tribals employment under Mahatma Gandhi National Rural Employment Guarantee Act (MNREGA). This project-initiated flattening of hill slopes. Trees were uprooted and land was levelled which, in turn, loosened the soil. Different machineries [7] were used to make plots favorable for rice cultivation under this project. Several investigations also claimed that the project was undertaken without discussing and ignoring the soil scientist's opinions.

As for paddy cultivation, large amount of water is kept in hold. When such storage is effectuated for several days, the water is able to percolate down and make passage through the pores and spaces in basaltic rocks along the shear seams and weak plane. Hence, in case of occurrence of incessant rain, there is a good chance of initiation and triggering of slope failure due to the infiltrating water. The roads of the village had been constructed under Pradhan Mantri Gram Sadak Yojana (PMGSY). This construction led to the movement of different earth moving equipment [8] which possibly caused destabilization of the slopes. Excessive deforestation for windmill project and irrigation projects have also contributed in destabilization. Some researchers have also claimed that the back water of Dimbhe dam might have played a significant role in instability, as it is well documented by some researchers that standing water with fluctuating levels can induce landslide [9].

6 Aftermath of the Disaster

The landslide hit early in the morning (5 AM) when residents were still asleep. Due to the remoteness and in the absence of tele-connectivity in the village, the tragedy came to light a little after 7.30 AM. Rescue and relief operations were launched immediately after getting information. 9 relief and rescue teams of National Disaster Response Force (NDRF) were immediately put into action and they carried out operation from 30 July 2014 to 6 Aug 2014. The continuous rainfall, adverse weather and narrow road condition made the rescue operation extremely challenging. On the first day of the rescue operation, the NDRF team successfully rescued 8 live victims, and subsequently the rescue operation was continued for continuous 7 days and nights. According to the official records given later, the tragic event affected 40 families and killed 151 people (59 men, 73 women and 21 children), 300 cattle while more than 100s went missing [10]. The team (including doctor, nurses and volunteers) and close to 400 officers were involved in the biggest rescue and relief operation of landslide in the country [11]. Bank of Maharashtra came forward and arranged various essential relief materials such as food, drinking water and medical kits at the spot. It also arranged few workers for facilitating the rescue operation and coordinating with district officials and NDRF team. Few telecom companies came forward and established a temporary network at the site to facilitate the relief and recovery operation. A group of villagers (around Malin village) started recognizing, marking, and updating the list of the people present in the village and also those who have been away from the village at the time of the landslide. Maratha Bhawan (a common house) had been provided as the immediate shelter to the villagers. Later on, the administration took the decision to construct pucca houses for the people who survived. Further, the administration prepared a list of affected villagers in order to provide employment opportunities to those who survived.

Various financial assistance was provided to the victims. The then Maharashtra Chief Minister announced a financial aid of Rs five lakh to the family of victims from the CM relief fund [12]. The then Home Minister of India announced Rs. 2

lakhs for family of the victims. The cost of the medical treatment of the victims was covered by the state government. Construction of school was done later in the nearby area through Sakal Relief Fund [13] which had provided educational facilities to the children of the Landslide-hit Malin village. Different programs were run by various organizations and local bodies to encourage people to donate for the victims. For example, Maratha Relief Fund has collected about Rs 1.11 crore for supporting the rehabilitation of the disaster-struck villagers of Malin.

7 Possible Mitigation and Associated Monitoring

Mitigation measures are taken up to reduce the impact of similar future disasters on human and financial consequences by taking up technical and technologically viable actions. The occurrence of the landslide cannot be prevented, but its effects can be minimized by taking timely preventive measures.

7.1 Mitigation Ways in Malin Landslide

A myriad of techniques was adopted at different times from the catastrophic event in the Malin village and associated area to prevent any lurking or similarly impending slides and flows. These include:

- Land assessment
- Generation of slope stability maps
- Preparation of susceptibility database
- Slope stabilization and development
- Drainage
- Redirecting debris channel
- Retaining walls
- Soil reinforcement
- Vegetation plantation for stabilizing slopes
- Telecommunication system setup
- Installation of rain gauge station
- Early warning systems
- Afforestation
- Participatory rural appraisal

Land Assessment: As the area is geologically very fragile and environmentally degraded and thus experiences number of landslides [14]; aerial surveys should be preferred for such areas. Through aerial surveys, a proper representation of existing vegetation can be obtained which can be indicative of potential landslide region exhibited by slow growth of vegetation over a disturbed soil mass. Growth of blackberry bushes indicates disturbed soil or occurrence of even a small magnitude of landslide that might have occurred at the place in the past. Any development activity planned for the area should be taken only after a detailed study of the region and slope protection measures should be provided according to the code developed for the region.

Generation of Slope Stability Maps: Regional level slope stability maps should be prepared and monitored continuously. These maps should be regularly updated based upon the development activities in the surrounding. Monitoring of sensitive landslide zones should be done very carefully.

Preparation of susceptibility database: Landslide susceptibility map identifies areas which are prone to landslide [15]. In this, different areas are ranked based on the degree of actual or potential hazard from landslide. Landslide susceptibility map should be prepared by combining the prevailing factors in the region. The various parameters considered are mainly rainfall intensity, slope angle, slope aspect, elevation, soil type, distance to roads, distance to rivers, lineament density, road density, and river density. The more the number of parameters considered, a more complex but realistic analysis can be made to replicate real scenario of susceptibility of the area.

Slope Stabilization and Development: Slope stabilization is one of the crucial factors that need to be considered while performing any kind of activity in a critical or potent landslide region. Slope stabilization comes in the picture when the slope is very steep or the fill material is not well compacted. One major reason of destabilization of a slope is the accumulation of water that leads to an increase in pore pressure which, in turn, plays a major role in initiating landslide events. Hence, surface and sub-surface water management on the slopes is the most effective remediation measure for controlling many landslides (as shown in Fig. 5). Management of surface runoff and sub-surface water is done through the construction of efficient drainage networks and, hence, the mass movement can be minimized through this activity. At Malin village, appropriate site-specific solutions need to be incorporated in order to avoid future slides.

(a) (b)

Fig. 5 Slope stabilization with the use of drainage. *Source* https://youtu.be/At0G2HEgVDo

Drainage: The pre-existing conduit capacity of the natural drainage system in the hillsides should be increased to cater to heavy rainfall. Prior to this event in Malin village, water accumulated in the upstream due to insufficient passage. This led to percolation of rainwater at the greater depth through rock soil interface and through major joints [16], and this contributed to one of the major reasons of the mudslide.

Retaining Walls: Support retention walls should be constructed along the periphery of the area where hill slopes were cut and a very high slope is exposed. Weep holes also should be provided at the proper locations.

Telecommunication system: For better preparedness, telecommunication system should be developed in such vulnerable remote areas. As mentioned earlier, in Malin region, the incident came to light only when the transport bus reached the village and thereafter it drove back to the city and informed officials. The catastrophic nature of a disaster depends on the time elapsed in its rescue response. The chances of survival of the victims is high if relief operations are started immediately and survival quotient decreases rapidly with the passage of time.

Installation of rain gauge station: The Malin and associated area comes under heavy rainfall zone and environmentally, it is heavily degraded and vulnerable to landslides for many decades. The rainfall data collected for the investigation was from the Tehsildar office, Ambegoan as Malin village does not have rain gauge station. This puts limitation on the actual analysis. Hence, installation of rain-gauges and monitoring stations need to be installed in such areas, which can be achieved without spending a significant amount.

Early Warning System (EWS): The scale of the disaster in case of Malin landslide could have been reduced if an efficient early warning system had been in place. EWS are helpful to reduce the intensity of disaster [17] or to avert disaster by providing various timely warnings, which has the capability to reduce the associated risk to the people's life in the region. EWS can be used to monitor inclination, deformations, soil moisture content, rainfall etc., which would help in reducing the potential hydrodynamic time lag, a limitation of the conventional instruments.

Afforestation: Plantations can largely help in reducing the chances of sliding of hillslopes [18]. However, plants used for this purpose need to be selected only after consultation with the experts. Selection of suitable plants can help in reducing the stress of the terrain. In Malin region, some trees can be seen to play a major role in limiting the spread of debris and landslide extent. Few houses which were located adjacent to the trees were saved from being buried, although the trees were uprooted.

Participatory Rural Appraisal (PRA): Participatory rural appraisal practice should be done in such communities (as shown in Fig. 6) since only through community participation, it is possible to apprise the dangers and significance of disasters to the community [19]. This will help in the accurate exchange of information (i.e., risk communication). This will also ensure the involvement of the people in risk management right from the beginning. Various ways of doing it include regular conduct of

Fig. 6 Rapid rural appraisal practice. https://tamannakalim.wordpress.com/2013/12/02/95/

workshops, community engagement, and also disseminating some basic information about safety tips for landslide-prone areas that may be posted and distributed among the public in such places.

8 Measures to Reduce the Impact of Such Disasters

Several measures can be adopted to attempt reducing the intensity and impact of such disasters, and these may include the following:

- Alert system should be devised and implemented across sensitive zones. Such areas should be closely monitored.
- The best option to avert the impact of such landslides would be to create awareness among the people by organizing programs about changing agriculture practices and climatic factors that can initiate the landslide.
- Explain the basics of good hill slope practices in the simplest and generalized way to the people.
- Ensure that the natural drainage is not disturbed or blocked due to slope modification and improper terracing. Furthermore, ensure that water is not accumulating or percolating on slope, as it will reduce the safety factor of the hill slope.
- Cutting of slope to expose a very high vertical face should be strictly avoided.
- A check on illegal constructions and deforestation should be strictly undertaken.
- Protecting the environment is one of the best ways to reduce the risk of such disasters.
- Using appropriate engineering technology such as construction of retaining walls around the periphery of the area where Padkai Scheme is developed.
- Generating alerts and transferring people to safe zones should be done within proper time.
- Increasing vegetative growth in the sloping areas to avoid soil erosion.

- Training the people and making them capable to how to take basic safety measures through Participatory Rural Appraisal (PRA).
- Project/Schemes related to land use should be done in consultation with experts such as geologists, environmentalists, and Geological Survey of India before implementation.
- Those crops should be cultivated which requires less water in the slope.
- Poor construction of houses has led to many number of casualties. It was observed that everything, except a few houses and a primary school, were buried, under debris flow. Minor cracks were observed in the school building which was made of bricks and cement. Hence, pucca house construction with scientific measures should be adopted in the region

9 Conclusion

The landslide that occurred in Malin has taught the lesson that more work needs to be done to yet evolve in terms of pre-disaster mitigation strategies. There is an ardent necessity to start working on strategies to prevent a repetition of such costly disasters in nearby future. Following are some of the important observations and suggestions derived from the Malin landslide that can be used to minimize the impact of landslides in an area.

- Landslide is a complex phenomenon and to decipher the underlying physics of each such landslides to answer the question 'why did it happen?', a lot more is yet to be investigated. A major limitation is not having a framework that incorporates analysis of unsaturated soil with traditional slope-stability analysis, which can overcome the inherent limitation of conventional stability analysis that assumes saturated behaviour of soil.
- Very heavy rainfall is considered as one of the main triggering factors for the landslide disaster. Yet, there are several other factors as well which are equally important such as deforestation, stone quarrying, and large-scale land-use modification. A threshold limit for regional scales should be developed by formulating a relationship between rainfall intensity, duration and occurrence of landslide.
- It is mainly seen that occurrence of landslide is more in the monsoon season. Hence, surface and sub-surface water management can play a key role in preventing the occurrence of the associated tragedy. This management of water can be done through establishing proper drainage connections.
- Timely soil health monitoring and zoning will help in critical analysis of the slope condition.
- Hazard zonation and risk assessment should be mandatorily performed for such areas to understand the risk associated. In case of Malin incident, signs of cracks were observed by the villagers earlier in 2003, but they moved back to the village, which is an indirect reason of concurring such large-scale casualties.

- Develop people-centric early warning system. Efficient communication system should be developed between the local bodies and the people living in the hazardous regions. Together with the credibility of the alarming system and collaborative team effort, such disasters can be inadvertently averted in the future.
- Afforestation program should be planned with a scientific perspective (keeping in view the appropriate vegetation for the terrain) to increase the green belt.
- Promoting awareness about insurance policy will help to reduce the financial impact on vulnerable people.

References

1. Ering P, Kulkarni R, Kolekar Y, Dasaka SM, Babu GS (2015) Forensic analysis of Malin landslide in India. IOP Conf Ser Earth Environ Sci 26:012–040. IOP Publishing
2. Singh TN, Singh R, Singh B, Sharma LK, Singh R, Ansari MK (2016) Investigations and stability analyses of Malin village landslide of Pune district, Maharashtra, India. Nat Hazards 81(3):2019–2030
3. Ramasamy SM, Muthukumar M, Subagunasekar M (2015) Malin-Maharashtra landslides: a disaster triggered by tectonics and anthropogenic phenomenon. Curr Sci 108(8):1428–1430
4. Dey N, Sengupta A (2018) Effect of rainfall on the triggering of the devastating slope failure at Malin, India. Nat Hazards 94(3):1391–1413
5. Mishra M, Dugesar V, Raju KP, Sarkar S, Vishwakarma M (2017) Post-Disaster Investigation of the Malin Slope Failure Deccan Plateau, India. Natl Geograph J India 63(03):1–7
6. Advancing Earth and Space Science Blogosphere. https://blogs.agu.org/landslideblog/2014/07/31/malin-landslide-1/. Accessed 10 June2022
7. The Indian Express. http://indianexpress.com/article/india/india-others/heavymachinery-created-farms-in-vulnerable-zone/. Accessed 10 June 2022
8. BBC News, India. https://www.bbc.com/news/world-asia-india-28579668. Accessed 10 June 2022
9. Iqbal J, Tu X, Gao W (2019) The impact of reservoir fluctuations on reactivated large landslides: a case study. Geofluids. Meshram S (2016) Investigations of the causes of landslide at Malin and some preventive measures. J Geotech Stud 1(2):1–14
10. The Hindu https://www.thehindu.com/news/national/Malin-landslide-Rescue-operations-draw-to-a-close/article60089904.ece. Accessed 10 June 2022
11. Chitre SH (2022) The landslide threats in Maharashtra: lessons to be learnt from the Malin Landslide. J Emerg Technol Innov Res 9(6):163–166
12. The Economic Times. https://economictimes.indiatimes.com/news/politics-and-nation/malin-landslide-toll-climbs-to-63-rs-5-lakh-aid-to-victims/articleshow/39403681.cms. Accessed 10 June 2022
13. Hindustan Times. https://www.hindustantimes.com/india-news/a-new-village-a-new-life-malin-landslide-survivors-start-afresh-after-tragedy/story-Yuko0EREm9dCr93MVN41IJ.html. Accessed 10 June 2022
14. National Disaster Management Guidelines—Management of landslides and snow avalanches, National Disaster Management Authority (NMDA, 2009), New Delhi. https://ndma.gov.in/sites/default/files/PDF/Landslide/landslidessnowavalanches.pdf. Accessed 10 June 2022
15. Mersha T, Meten M (2020) GIS-based landslide susceptibility mapping and assessment using bivariate statistical methods in Simada area, northwestern Ethiopia. Geo Environ Disast 7(1):1–22
16. Shah CR, Sathe SS, Bhagawati PB, Mohite SS (2021) A hill slope failure analysis: a case study of Malingoan village, Maharashtra, India. Geol Ecol Landscapes 5(1):1–6

17. Guzzetti F, Gariano SL, Peruccacci S, Brunetti MT, Marchesini I, Rossi M, Melillo M (2020) Geographical landslide early warning systems. Earth Sci Rev 200:102973
18. Niraula RR, Gilani H, Pokharel BK, Qamer FM (2013) Measuring impacts of community forestry program through repeat photography and satellite remote sensing in the Dolakha district of Nepal. J Environ Manage 126:20–29
19. Zubir SS, Amirrol H (2011) Disaster risk reduction through community participation. WIT Trans Ecol Environ 148:195–206

Seismic Risk Mitigation of RC Frame Building in North-East India Using Buckling Restrained Braces

Aakash Kumar and Needhi Kotoky

Abstract North-East (NE) India has experienced several devastating earthquakes in the past decades. The seismic susceptibility of reinforced concrete frame buildings has been exposed by these earthquakes. In this region, most of the existing buildings are non-ductile as they are built before the introduction of modern seismic design codes. Seismic safety and resilience of these non-ductile structures can be increased by the use of a reliable retrofitting scheme. Therefore, buckling-restrained braces (BRBs) are designed to withstand earthquake-induced cyclic lateral loadings as a passive control system. In this study, the influence of BRBs in reducing the seismic risk of non-seismically designed frame is discussed. A finite element model of a 3-story 3-bay low ductility moment-resisting RC frame is developed. Nonlinear time-history analyses are carried out using a suite of ground motions to incorporate record-to-record variability. For both as-built and BRB retrofitted frame seismic fragility curves are obtained. For seismic risk assessment, the site-specific seismic hazard curve of Guwahati city giving relationship between annual probability of exceedance and peak spectral acceleration at 1.0 s is obtained. The estimation of seismic risk reduction as a result of the application of the BRBs retrofit within the bare frame is done by convolution of the seismic fragility curves with the regional seismic hazards for the Guwahati region. The findings of this study present the significance of BRBs on the seismic performance of the building, as well as the efficacy of a BRB retrofit for the NE region.

Keywords Buckling-restrained braces · Seismic risk mitigation · Seismic retrofitting · Reinforced concrete frame · North-East India

---

A. Kumar · N. Kotoky (✉)
National Institute of Technology Meghalaya, Shillong, Meghalaya, India
e-mail: needhikotoky@nitm.ac.in

A. Kumar
e-mail: p22ce002@nitm.ac.in

S. Mitra et al. (eds.), *Disaster Management and Risk Reduction: Multidisciplinary Perspectives and Approaches in the Indian Context*,
https://doi.org/10.1007/978-981-99-6395-9_16

1 Introduction

North-East (NE) India has experienced several devastating earthquakes such as 1897 Shillong earthquake (M_w 8.1), 1934 Bihar Nepal earthquake (M_w 8.1), 1950 Assam earthquake (M_w 8.7), and 2011 Sikkim earthquake (M_w 6.9). The seismicity of NE India is the combined effect of the Burmese arc in the east and the Indian–Eurasian Plate boundary in the north [1]. This region is among the highly earthquake-prone regions of the world. In this region, most of the existing buildings are non-ductile as they are built before to the introduction of modern seismic design codes. Due to the limited ductility capacity, these buildings are significantly vulnerable to severe ground shaking. Demolishing these seismically deficient structures and replacing them with modern structures is not feasible. As a result, seismic retrofitting is seen as a reliable technique for reducing seismic risk. For seismic design and retrofit of structures, steel braces are widely used. However, these bracings are prone to collapse when experiencing high cyclic or dynamic loads such as earthquakes. Restraining devices are added outside or within the braces, modifying them to buckling-restrained braces (BRBs) to mitigate the buckling effect as passive control system [2]. Passive control systems have proven to be very efficient devices for newly constructed as well as seismically retrofitted existing structures [3, 4]. The addition of a bracing system to a low-ductility frame enhances the structure's collapse modalities as well as the probabilistic characteristics of its seismic response [5]. Along with additional stiffness and strength to seismic deficient buildings, BRBs provide good energy absorption efficiency as a seismic-resistant structural element. These braces contribute to the additional path for lateral loads caused by earthquakes, improving the seismic performance of the frame. Therefore, this retrofitting technique may lead to a reduction in seismic vulnerability and losses of RC frame buildings in case of future earthquakes.

Addressing the gap, this study evaluates the influence of BRBs in seismic risk reduction of older non-seismically designed RC frame buildings. The next section introduces a benchmark 3-story low ductility RC frame representative of typical design details before the introduction of modern seismic design codes is chosen and BRB retrofit design is carried out. Next, the finite element modeling strategy of the case-study frame is discussed followed by a discussion on the design of BRB retrofit. To obtain seismic fragility curves, non-linear time-history analyses (NLTHA) are performed on the detailed finite element model of the as-built and retrofitted frames. Finally, for seismic risk assessment, the site-specific seismic hazard curve of Guwahati city giving the relationship between the annual probability of exceedance and peak spectral acceleration at 1.0 s [6] is convolved with seismic fragility curves to obtain lifetime seismic risk. The findings of this study present the significance of BRBs on the seismic performance of the building, as well as the efficacy of a BRB retrofit for the NE region.

2 Case Study Building–Finite Element Modelling of Bare and BRB Retrofitted Frame

2.1 Description of Case Study Frame

A 3-story 3-bay low ductile moment-resisting RC frame has been considered for the present study. In this frame, the bay width is taken as 5.49 m and each story height within the frame is taken as 3.66 m. Concrete compressive cube strength of 24 MPa and Grade 40 steel reinforcing bars with a yield strength of 276 MPa have been used for the frame modeling. The beam size is 260× 460 mm, and the column size is 300× 300 mm. Before the introduction of modern seismic codes, buildings were mostly designed for gravity loads alone, without taking seismic design into account.

2.2 Finite Element Modeling of Bare Non-Ductile Frame

The case study building frame is modeled in the finite element package OpenSees [7]. The *beamWithHinges* element, which consists of 2 plastic hinge regions at the element end and a central elastic element defined by fiber sections, is used to model the nonlinear flexural hysteretic response of beams and columns [9]. The plastic hinge lengths for columns and beams are obtained as per Panagiotakos and Fardis [10]. A moment–curvature analysis of the section is used to estimate the elastic component of effective flexural stiffness while taking into account the axial force level introduced by dead loads. *FiberSection* is used to define the ends of column and beam cross-sections with rectangular concrete patches and reinforcement layers. *uniaxial-Material Hysteretic* model is used to model the longitudinal reinforcements whose controlling parameters like degraded unloading stiffness, pinching, and damage are calibrated using experimental results. The nonlinear degrading *Concrete02* material model is being used to model confined and unconfined concrete within fiber sections. The contribution of the slab is modeled in the beams using T-sections with effective width 4 times the beam width. A *zerolength* shear spring placed at the column top is used to model the shear response of columns in older non-ductile buildings. The *uniaxialmaterial limit-state* created by Elwood [11] is used to designate the characteristic attributes of this shear spring [11]. Building joints (interior and exterior) are modeled using a multilinear response envelope and a trilinear unloaded path is implemented using *Pinching4* material [12]. High axial stiffness is assigned to the beam for the modeling of diaphragm of rigid floor. Gravity loads are distributed across the beams, whereas the masses are concentrated in the lumped form at the beam-column joints.

2.3 BRB Design for Case-Study Frame

In this study, a case study frame has been taken and retrofitted with BRBs at each story, BRBs are introduced in the central bay. The elastic brace exhibiting sufficient over-strength is attached in series with an elastoplastic dissipative device to form the dissipative braces. This configuration gives, autonomous calibration of the strength ($F_c{}^i$) and stiffness ($K_c{}^i$) of the dissipative diagonal braces. At each story, $K_c{}^i$ is designed to maintain the first mode shape of the as-built frame after retrofitting in order to prevent significant changes to the internal force distribution in the frame within the elastic behaviour. Moreover, the $F_c{}^i$ distribution is intended to achieve yielding of the devices across all stories at the same time. As a result, the overall ductility of the bracing system is equivalent to the ductility of the individual braces. For different strengths proportion coefficient (α) values, the bracing system can be designed. It is defined as the ratio of seismic base shears associated with BRB frame and bare frame. For the current study, α is taken as 1. One more significant parameter that affects the design is the dissipative brace ductility (μ_{BRB}) which has been considered as 15. Table 1 shows the properties of dissipative braces $K_c{}^i$ and $F_c{}^i$ at each story together with the material's yield strength ($f_{y,BRB}$), the area (A_{BRB}), and length (L_{BRB}) of the BRB device. The device behavior is modeled using *SteelBRB* material model in OpenSees.

3 Influence of BRB Retrofitting on Seismic Risk

This section presents the influence of BRBs in seismic risk mitigation of RC frame. First, seismic fragility curves of as-built and retrofitted frames are developed for different damage states (DS) using *NLTHA* considering a suite of ground motions. Next, seismic risk reduction has been evaluated as a result of the implementation of the BRBs retrofit within the bare frame due to the convolution of the seismic fragility curves with the regional seismic hazards for the Guwahati region.

Table 1 Design properties of the BRBs distributed across the floors of the retrofitted frame

Floor no	$F_c{}^i$ (kN)	$K_c{}^i$ (kN/m)	$f_{y,BRB}$ (MPa)	A_{BRB} (mm^2)	L_{BRB} (mm)
1	207.9	45,967.4	250.0	831.6	2799.3
2	178.9	30,940.0	250.0	715.7	3579.2
3	103.0	28,242.4	250.0	412.0	2257.4

3.1 Seismic Fragility Comparison of Bare and Retrofitted Frame

Seismic fragility curves representing vulnerability under seismic shaking constitute critical precursors to seismic risk assessment of building structures. The seismic fragility curves for as-built and BRB retrofitted frames are developed using a two-step approach. In the first step, probabilistic seismic demand models (PSDMs) for critical are developed. Seismic fragility curves are developed in the second step using the demand model (obtained in first step) and limit state capacity model. PSDMs that relates the median peak engineering demand parameter (*EDP*) of the RC frame with the intensity measure (*IM*) of the ground motion is developed by conducting NLTHA of the RC frame using a suite of ground motion representative. The NLTHAs employ a synthetic suite of 150 unscaled strong ground motions so that the structure experiences behavior ranging from the linear to the non-linear domain, hence encompassing the various DS defined. Following *NLTHA*, *PSDMs* are developed for as-built and retrofitted BRB frames to estimate median seismic demand using the relation given:

$$\ln(EDP) = \ln a + b\ln(IM) \tag{1}$$

where, *EDP* is the engineering demand parameter, ln *a* and *b* are the linear regression coefficients, and *IM* is ground motion intensity measure. This study considers maximum inter-story drift ratio ($IDR_{\max}$) as *EDP* to estimate structural damage due to earthquakes. The bare and retrofitted frame has distinct fundamental time periods (1.2 s for the bare frame and 0.6 s for the BRB retrofitted frame), therefore, to allow comparison of PSDMs and fragility curves spectral acceleration at 1.0 s [$S_a(1.0\text{ s})$] is considered as *IM*. Figure 1a shows the comparison of PSDMs of as-built bare frame and BRB retrofitted frame. The figure shows the data points ($S_a(1.0\text{ s})$– $IDR_{\max}$ pairs) obtained from *NLTHA* simulations, and regression lines represent the median estimate of seismic demand for bare and retrofitted frame. Comparison of *PSDMs* results indicates a reduction in median seismic demand for BRB retrofitted frame as compared to as-built frame.

Following the development of *PSDMs* for bare and retrofitted frame, limit state capacity of different *DS* –slight (S), moderate (M), extensive (E), and complete (C) in terms of $IDR_{\max}$ are obtained using nonlinear static pushover analysis. Figure 1b shows the pushover curve of bare and retrofitted frames and markers corresponding to the attainment of different DS. These markers are obtained by measuring local *EDPs* such as material strains of steel and concrete in beams and columns and accounting for shear failure of columns. A detailed description of various *DS* is given in Table 2.

Finally, *PSDMs* and limit state capacity of different *DS* are utilized to develop seismic fragility curves. The seismic fragility for a particular *DS* follows lognormal distribution and is estimated as:

$$P[DS|S_a(1.0s)] = \Phi\left(\frac{\ln(S_a(1.0s)) - ln(med)}{\zeta}\right) \tag{2}$$

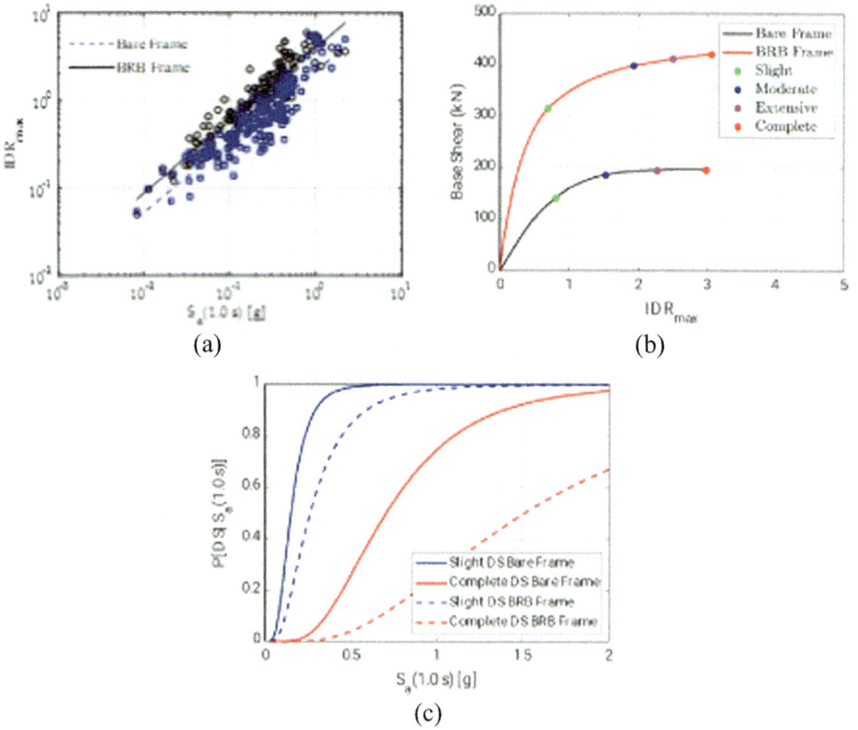

Fig. 1 **a** Nonlinear time history response and fitted PSDMs for as-built bare frame and BRB retrofitted frame, **b** Pushover curves of bare and BRB retrofitted frames showing markers corresponding thresholds of different DS, **c** Seismic fragility curves of as-built and retrofitted frames for slight and complete DS.

Table 2 Description of damage states and IDR limits

Damage states	Description of failure	As-built bare frame–IDR_{max}	BRB retrofitted frame–IDR_{max}
Slight (S)	Yielding of 50% of columns at one story	0.81	0.72
Moderate (M)	Crushing/Spalling of concrete in 50% of columns at one story	1.53	1.81
Extensive (E)	Average of M and C *DSs*	2.27	2.40
Complete (C)	Initiation of shear failure in 50% of columns at one story	2.98	2.99

where P[DS| S_a(1.0 s)] is the probability of damage state exceedance given a S_a(1.0 s) intensity, *med* is the median of lognormal distribution for fragility and ζ is the dispersion of fragility. Figure 1c shows the comparison of seismic fragility curves of as-built and retrofitted frame for S and C DSs. Results show that the failure probability of the BRB retrofitted frame is significantly lower than that of the as-built frame. The median S_a (spectral acceleration corresponding to 50% probability of failure) for S, M, E, and C damage states of bare frame is 0.14, 0.32, 0.51, and 0.70 g. On the other hand, the median S_a for BRB retrofitted frame is 0.26, 0.82, 1.15, and 1.51 g, respectively. Clearly, across all DS implementation of BRB results in a significant reduction in vulnerability. The next section utilizes the developed seismic fragility curves for seismic risk assessment.

4 Seismic Risk Reduction of Retrofitted Frame

For seismic risk assessment, the site-specific seismic hazard curve of Guwahati city giving relationship between annual probability of exceedance and peak spectral acceleration at 1.0 s is obtained from past literature [6] and shown in Fig. 2a. Using the seismic fragility curves for the as-built or BRB retrofitted frame, as well as site-specific seismic hazard information for Guwahati region, the lifetime (T = 50 years) probability of DS exceedance, P_{Tf}, can be calculated as follows.

$$P_{Tf} = 1 - (1 - P_{Af})^T \tag{3}$$

Hence, using the convolution of seismic fragility and site-specific hazard [$H(IM = im)$], the annual probability of DS exceedance, P_{Af} can be obtained as follows:

$$P_{Af} = \int_{im} [Fragility|IM = im]\left|\frac{dH(im)}{d(im)}\right|d(im) \tag{4}$$

Figure 2b shows the $T = 50$ years probability of exceedance comparison for as-built and retrofitted frames. The results show that the failure probability of the BRB retrofitted frame is significantly lower than that of the as-built frame. The BRB retrofit leads to 39, 71, 72, and 73% reduction in seismic risk estimates of S, M, E, and C damage states.

5 Conclusions

In the present study, the influence of BRBs in reducing seismic risk of non-seismically designed frame in NE India is discussed. The proposed methodology's capability and effectiveness are evaluated by considering a realistic benchmark RC frame with

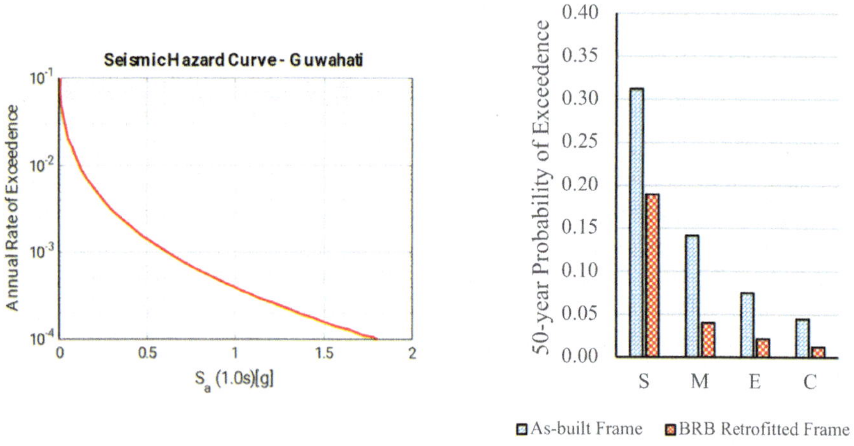

(a) (b)

Fig. 2 **a** Seismic hazard curve of Guwahati region obtained from Nath and Thingbaijam [6] **b** Comparisons of seismic risk for the as-built and BRB retrofitted frames for S, M, E, and C damage states

limited ductility capacity retrofitted with BRBs. The retrofitted frame using BRBs significantly reduces seismic fragility, as indicated by the seismic fragility curves for the as-built and retrofitted frames. The estimation of seismic risk reduction as a result of the application of the BRBs retrofit within the as-built frame is due to the convolution of the seismic fragility curves with the regional seismic hazards for the Guwahati region. Results reveal 73% reduction in seismic risk estimates of complete damage state highlighting the influence of BRBs retrofit for non-seismically designed RC building in the NE region. As a future scope of the study, the work can be extended to different retrofit levels, and also variations in the case study frame can be introduced.

Acknowledgements The first author's fellowship is granted by the Ministry of Education (MoE), Government of India (New Delhi).

References

1. Kumar A, Harinarayan NH, Verma V, Anand S, Borah U, Bania M (2018) Seismic site classification and empirical correlation between standard penetration test N value and shear wave velocity for guwahati based on thorough subsoil investigation data. Pure Appl Geophys 175(8):2721–2738
2. Zhou Y, Shao H, Cao Y, Lui EM (2021) Application of buckling-restrained braces to earthquake-resistant design of buildings: A review. Eng Struct 246:112991

3. Freddi F, Ghosh J, Kotoky N, Raghunandan M (2021) Device uncertainty propagation in low-ductility RC frames retrofitted with BRBs for seismic risk mitigation. Earthquake Eng Struct Dynam 50(9):2488–2509
4. Dall'Asta A, Ragni L, Tubaldi E, Freddi F (2009) Design methods for existing RC frames equipped with elasto-plastic or viscoelastic dissipative braces. In: Proceedings of XIII Convegno Nazionale ANIDIS. Bologna, Italy
5. Freddi F, Tubaldi E, Ragni L, Dall'Asta A (2013) Probabilistic performance assessment of low-ductility reinforced concrete frames retrofitted with dissipative braces. Earthquake Eng Struct Dynam 42(7):993–1011
6. Nath SK, Thingbaijam KKS (2012) Probabilistic seismic hazard assessment of India. Seismol Res Lett 83(1):135–149
7. McKenna F, Fenves GL, Scott, MH, Jeremic B (2000) Open system for earthquake engineering simulation (OpenSees). Pacific Earthquake Engineering Research Center, University of California, Berkeley, CA
8. FEMA (2003) HAZUS-MH MR4 Technical manual—Earthquake model. Federal Emergency Management Agency, Washington, DC
9. Scott MH, Fenves GL (2006) Plastic hinge integration methods for force-based beam–column elements. J Struct Eng 132(2):244–252
10. Panagiotakos TB, Fardis MN (2001) Deformations of reinforced concrete members at yielding and ultimate. Structural J 98(2):135–148
11. Elwood KJ (2004) Modeling failures in existing reinforced concrete columns. Can J Civ Eng 31(5):846–859
12. Lowes LN, Mitra N, Altoontash A (2003) A beam-column joint model for simulating the earthquake response of reinforced concrete frames

Extreme Events, Resilience and Disaster Management: Lessons from Case Studies

Rajarshi Majumder

Abstract Disaster Management is thought to start only after a disaster has struck. But that is only a part of the overall management strategy. We are increasingly waking up to the fact that Disaster Management should start much ahead and avoiding disasters is the best way to manage them. This calls for building a resilient society. For the last few decades environmental irregularities have become more frequent across the globe, which scientists claim to be the signs of an irreversible climate change. Scientists are clamouring for the need to build mitigating and adaptive measures in local, national and global policies to face this challenge. In this paper, we start with a broad outline of disasters, resilience, damage costs and avoidance costs. We also explain why we find resilience planning to be generally underfunded in developing countries. We then discuss the consequences of such under-preparedness using three case studies of cyclone management in India to understand the complexities of resilience planning and extreme event management policies. We also examine how the pandemic had affected resilience activities during one of the extreme events chosen for study. In the final section we outline the lessons learned from this experience and the policy response that may be put in place to deal with such situation—which looks to become increasingly regular in near future. The paper helps in bringing to limelight the problems faced during multiple extreme events.

Keywords Climate change · Extreme events · Resilience · Tropical cyclone · Disaster management · Pandemic · Multiple shocks · COVID19

1 Introduction

Disaster Management is thought to start only after a disaster has struck. But that is only a part of the overall management strategy. We are increasingly waking up to the fact that Disaster Management should start much ahead and avoiding disasters is the

R. Majumder (✉)
Department of Economics, University of Burdwan, Burdwan, West Bengal, India
e-mail: rmajumder@eco.buruniv.ac.in

S. Mitra et al. (eds.), *Disaster Management and Risk Reduction: Multidisciplinary Perspectives and Approaches in the Indian Context*,
https://doi.org/10.1007/978-981-99-6395-9_17

best way to manage them. This calls for building a resilient society. However, it is also a sad truth that developing countries generally spend less on resilience building and hence are under-prepared while dealing with disasters. In addition, in recent times we have seen multiple disasters striking simultaneously, making the task of disaster management more complex. In this paper, we try to portray the fundamentals of disaster management from a social science perspective and put forward few case studies to strengthen our argument that a more concerted and comprehensive disaster management policy should be put in place in developing countries in general, and in India in particular to reduce the economic losses. We also argue that the focus should be more on disaster avoidance than its management post-facto.

The paper has four sections. The next section deals with the concept of resilience and the costs of shocks and their prevention. In the third section, we describe the case studies to understand how resilience towards was compromised due to a simultaneous occurrence of another shock—the biological hazard of pandemic. In the final section, we outline the lessons learned from this experience and the likely policy suggestions.

2 Disaster Management And Resilience

2.1 Disaster & Its Management

A disaster is defined as a massive disruption to the normal working of human society for a short or moderately long time period with associated loss of life or property or both. Disasters can be natural (flood, earthquake, cyclones, etc.) or man-made (fire, building collapse, oceanic oil spill, etc.). Many of these disasters are sudden and unforeseen, but modern scientific developments have enabled us to foresee a host of disasters. The use of gadgets and instruments, past experiences, simulation of data using modern modelling techniques, and a thorough analysis of geographic and social data helps us identify vulnerable regions and population. Thus, we have ample warning of the likely turn of events and therefore have greater scope and duration to 'manage' such disasters. Consequently, Disaster Management now starts much early in the form of avoidance & mitigation.

There are four pillars of Disaster Management as practised internationally. They are Mitigation, Preparedness, Response, and Recovery. The first phase of mitigation aims at avoiding disasters by taking appropriate steps (e.g. not building homes in fragile/seismic zones, keeping flood channels free of construction/obstruction, etc.). The second phase looks to plan and impart training & awareness for actions to be taken if a disaster that cannot be mitigated strikes (e.g. building cyclone shelters, holding drills & mock exercises simulating fire/earthquake, etc.). The third phase is just immediately after a disaster strikes when the administration and civil society respond to the shock based on the planning and training of the second phase. In this phase, the normal activities of the society are compromised and the focus is on human safety (e.g. evacuation, search & rescue, ensuring water, food and medical supply, etc.).

The fourth phase aims at restoring the normal activities of the society as much and as fast as possible (e.g. rebuilding damaged structures, helping evacuees get back to normal livelihood, restarting economic processes, etc.). Thus Disaster Management is a comprehensive process of foreseeing & preventing hazards, planning to minimise disruptions & loss from them, taking appropriate steps in case of emergency, and bringing back society to its normal state.

In India too, the Disaster management Act of 2005 defines Disaster Management as.

> .... a continuous and integrated process of planning, organising, coordinating and implementing measures which are necessary or expedient for–(i) prevention of danger or threat of any disaster; (ii) mitigation or reduction of risk of any disaster or its severity or consequences; (iii) capacity-building; (iv) preparedness to deal with any disaster; (v) prompt response to any threatening disaster situation or disaster; (vi) assessing the severity or magnitude of effects of any disaster; (vii) evacuation, rescue and relief; (viii) rehabilitation and reconstruction;
>
> GoI [1]

Thus what we call Disaster Management is actually building up capacity in the society to foresee, avoid, and prepare against extreme events so that the society faces al little loss as possible in its normal functioning. This is often referred to as Resilience which we discuss next.

2.2 *Extreme Event, Damage Cost and Resilience*

Disasters or extreme events are shocks to society that threaten and jeopardise normal functioning and lead to loss in terms of human and animal life, structure, production and money. The measures taken by human society to tackle disasters can be categorised into two components—*Avoidance* by taking preventive measures, and *Remedy* by minimising negative impact of disasters. The capacity of a society to combat shocks is called Resilience. As the IPCC puts it, Resilience is:

> The capacity of social, economic, and environmental systems to cope with a hazardous event or trend or disturbance, responding or reorganizing in ways that maintain their essential function, identity, and structure, while also maintaining the capacity for adaptation, learning, and transformation.
>
> IPCC [15]

Thus there are four key elements of a resilient society. The first one is how it prevents hazards to happen in the first place. The second element is planning so that in case of an unavoidable extreme event, the impact on our built environment is minimum. The third element is to maintain functionalities as much as possible. The fourth element is to restore the normal functionality of society and economy and get back to the pre-shock level as early as possible. A resilient society can reduce the cost of damages arising out of an extreme event or a shock. Figure 1 helps us in understanding this further.

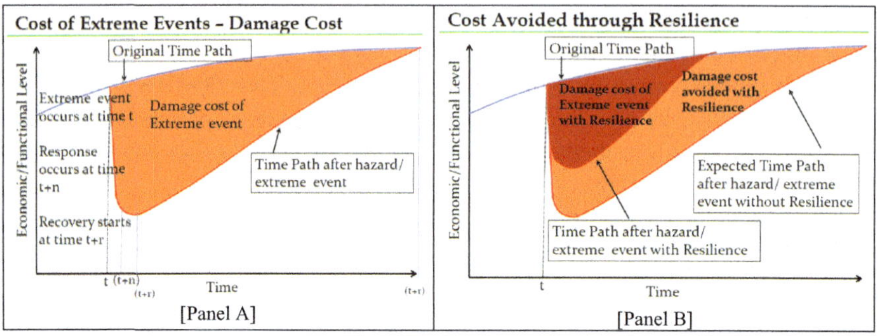

Fig. 1 Understanding resilience. *Source* Author's depiction

In Panel-A we draw the likely time path of a society/economy that faces a shock at time t. The blue line is the normal time pasth without shock. As the extreme event (or disaster or shock) strikes at period t, immediately the functional level of the economy nosedives from its normal time path (shown by the red line). Response actions that arrest the declining functionality start at time t + n and the sharp decline is checked. Recovery starts at period t + r when society starts bouncing back towards its normal level once more. Finally, at t + s period the society once more comes back to its pre-shock time path. Thus the shaded area is the loss in functionality (or economic loss, if measured in terms of GDP, etc.) or Damage Cost due to the extreme event.

In Panel-B we superimpose the pathway followed by a resilient society on the earlier time path. A resilient society follows the dark red line. It is obvious that with resilience initial damage/loss is smaller and response is quicker. As a result, the U-turn comes earlier and the rate of recovery is also faster. So society reaches the normal time path, or the blue line, much earlier than what would have happened without any resilience. The damage cost now is the area shaded dark red. It is thus clear that Damage Cost is lower with Resilience. The area shaded light red in this panel is a measure of the Damage Cost Avoided due to Resilience.

The response time, recovery time and the recovery rate are dependent on the degree of resilience of the society. If resilience is low, response will be late, recovery time will be long and recovery rate will be slow. As a result, Cost Avoided will be low and actual Damage Cost will be high. On the other hand, a highly resilient society will have a fast response, short recovery period, and fast recovery rate. So it will get back to its normal path quite quickly (and may even surpass that). Cost avoided in this case will be high and actual Damage Cost will below.

It is thus clear that there is a trade-off between investment in resilience (also called Avoidance Cost) and Damage Cost. In other words, to reduce Damage Cost later, one has to spend more on Avoidance Cost now. Increasing degree of resilience is associated with decreasing Damage Cost and increasing Avoidance Cost (Fig. 2). In most cases, total cost would come down as the degree of resilience increases, indicating that a society should try to build as much resilience as physically possible.

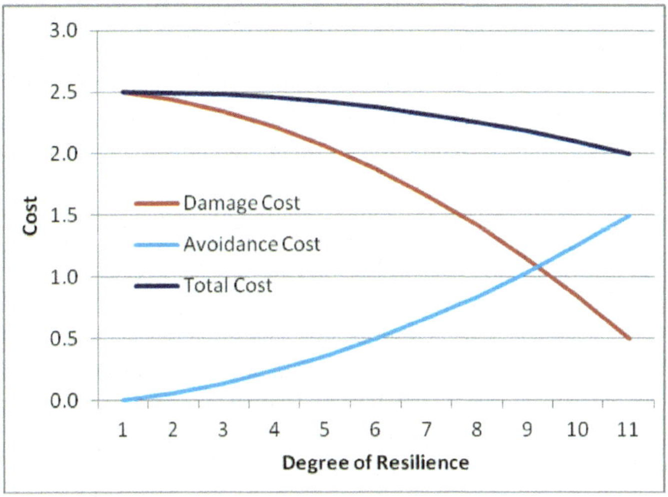

Fig. 2 Degree of resilience and costs. *Source* Author's depiction

However, the situation is not that simplistic for four reasons. First, there is always financial constraint and alternative demands on expenditure by the State. Hence one cannot simply go for maximum possible resilience pathway. Second, the extent and frequency of extreme event and hence that of damages are uncertain. Third, life time of resilience mechanism (structures like dams, embankments, shelters, etc., or fiscal instruments like taxes) is finite. Fourth, value of money today is not the same as value of money in the future, i.e. money depreciates at a rate called the discount rate.

An example will explain matters further. Suppose there is a river that frequently tops its banks during rainy season (a good example is River Brahmaputra for this audience) and causes damage to life and property in the floodplain. The extent of damage depends on the margin by which it tops its banks which in turn depends on the extent of excess rain. This last factor is uncertain and so the height of embankment to be constructed to prevent flooding cannot be ascertained exactly. For that, we look at past experiences i.e. historical data on flooding to find out frequencies or probabilities of different levels of topping margin. An embankment of level 'h' will prevent flooding up to level 'h' but not floods beyond that level. If the embankment has a lifetime of 't' years, then the damage cost saved (S) would be damage cost for all 'h' (or below) topping floods multiplied by their expected frequency of occurring over a 't' year period and discounted by the discount rate. This has to be compared with the construction and maintenance cost (C) of an embankment of height 'h'. The difference (S-C) gives us the Net Savings from building the embankment. Now, the higher the embankment the higher would be C. The damage cost saved would also be higher, but not monotonically increasing as frequencies/probabilities of progressively higher levels of flood are lower. This is illustrated in Table 1 where we have assumed a 30 year life span of the embankment and a discount rate of 7% per year. For example, a 10 cm high embankment will save about Rs. 0.25 million per year in avoided damage costs which adds up to a discounted present value of Rs. 3.10 million over the 30 year life of the embankment. Cost of building such an embankment is Rs. 1.75 million resulting in a Net Saving of Rs. 1.35 million over its life. Discounted expected damage cost saved increases as height of embankment height increases but the rate of increase dips after a time while cost of building the embankment increases monotonically. As a result Net savings initially increase as we raise the height of the embankment to 30 cm, but thereafter the gain from avoided damage cost is less than the increase in building cost and net savings start decreasing. A 50 cm or higher embankment results in a negative net savings since cost of building is very high but such a high river topping & devastating flood is extremely rare. A typical net saving from resilience building curve therefore looks like that in Fig. 3. It is obvious that the optimum policy should be to build a 30 cm high embankment as Net Saving is maximum at that point. A lower degree of resilience would increase damage cost while a higher one will lead to avoidance cost increasing more than proportionately to damage cost.

Two other factors that determine the optimum degree of resilience are the lifespan of the structure and the time discount rate, which were assumed to be 30 years and 7% in our illustration. A rise in life span leads to higher savings of damage cost and hence increases Net Savings while a rise in discount rate (valuing present more compared to future) will reduce the present value of savings. As illustrated in Fig. 4, a structure with a life span of 30 years in a society that discounts time at 5% will have a substantially higher Net Savings compared to a structure of lifespan 10 years in a society with 15% discount rate (Panel A). But more importantly, in the former case the optimum, i.e. maximum Net Savings occurs at an embankment height of 30 cm while in the latter, the optimum is at 15 cm (Panel B).

Developing countries are typically short of resources and try to build structures with shorter lifespan. Being low-income countries, society has a very high time discount factor as they value the present substantially higher compared to the future. Thus with a low life-span and high discount rate, the optimum degree of resilience is also low for these countries. This explains why developing countries are in general underprepared in terms of resilience and typically under-invest in Disaster Management.

2.3 Resilience Planning

Resilience planning, specifically for environmental extreme events like Tropical Cyclones, is an extensive long run phenomenon. The actionables can be divided into two broad categories Pre-event or Pre-emptive Actions, and, Post-event or Restorative Actions. Pre-emptive actions include Hazard Mapping & Monitoring, Warning systems & Forecast, Building Codes, Awareness & Insurance, Response Training, Evacuation Planning, and Stocking Essential Commodities including Food and Medicines. Restorative actions include Ensuring Drinking & Sanitation Services, Health Management, Restoring Critical Infrastructural Services, Recover/repair

Table 1 Net saving from resilience building—An illustration

Topping height/ Embank ment height (cm)	Frequency (once every)	Damage (million ₹)	Probability (%)	Expected annual damage (million ₹)	Cumulative expected annual damage cost saved (million ₹)	PV of annualised damage cost saved (million ₹)	Building cost (million ₹)	Net saving (million ₹)
(A)	(B)	(C)	(D)	(E)	(F)	(G)	(H)	(I = G–H)
10 cm	4 years	1	25	0.25	0.25	3.10	2.75	1.35
15 cm	7 years	2	15	0.30	0.55	6.82	4.25	3.57
20 cm	20 years	6	5	0.30	0.85	10.55	6.25	5.30
25 cm	50 years	20	2	0.40	1.25	15.51	8.75	7.76
30 cm	100 years	40	1	0.40	1.65	20.47	11.75	9.72
40 cm	200 years	50	0.50	0.25	1.90	23.58	19.25	5.33
50 cm	500 years	100	0.20	0.20	2.10	26.06	28.75	(−) 2.69
60 cm	750 years	125	0.13	0.16	2.26	28.08	40.25	(−) 12.17
75 cm	1,000 years	150	0.10	0.15	2.41	29.94	61.25	(−) 31.31

Source Author's depiction

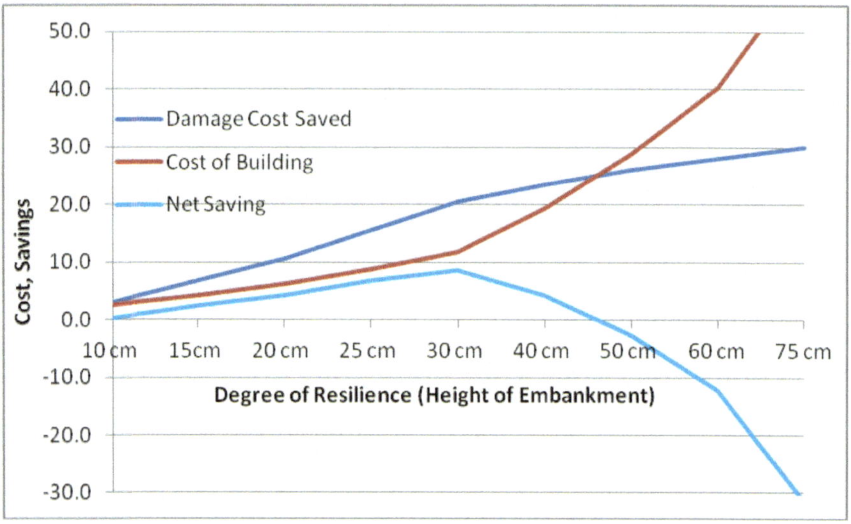

Fig. 3 Degree of resilience and net savings. *Source* Author's depiction

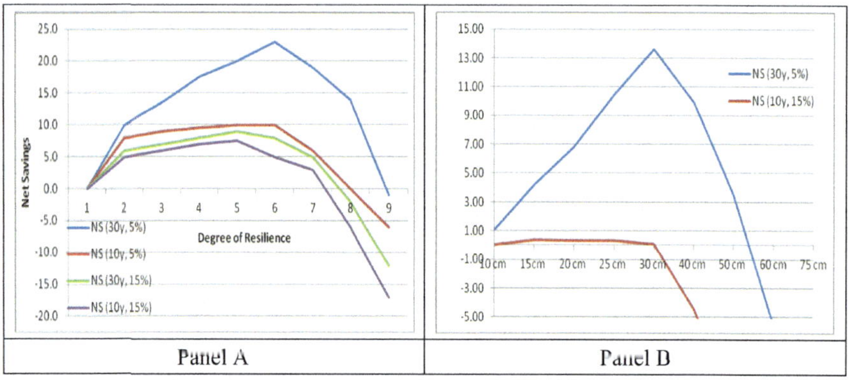

Fig. 4 Net savings for different lifespan and discount rate combinations. *Source* Author's depiction

damaged assets, Rehabilitation of people, Restart economic activities and Fast insurance pay-outs. For large countries like India, this planning has to happen not only at the macroeconomic level but also at the State and Local level. Another important aspect is coordination of such actions across geographical boundaries and jurisdictions of several agencies[5]. It is generally observed that the pre-emptive actions are mostly capital intensive in nature and involve investment in environmentally resilient infrastructures and buildings, capacity building for relocation and shelters, investment in scientific and technological breakthrough for advanced monitoring and warning, and maintaining stocks of essential commodities including medicines. Most of the restorative actions are on the other hand labour intensive in nature and

Table 2 Comparing the impact of some recent cyclones in India

Year	Event	Loss of life	Estimated economic damage
1999	Odisha Super Cyclone (250–275 kmph)	9,658	US$ 2.5 billion (0.54% of GDP)
2019	Very Severe Cyclone 'Fani' (185–210 kmph)	64	US$ 2.4 billion (0.09% of GDP)
2020	Very Severe Cyclone 'Amphan' (185–210 kmph)	108	US$ 13.8 billion (0.54% of GDP)

Source GoI

involve transport & distribution of essentials, repair & restoration of infrastructures, and rehabilitation of people. As already mentioned, resilience planning in developing countries is generally poor because of capital shortages and lack of foresightedness.

Against this backdrop, we focus on three disasters over the last quarter century to have struck eastern India to understand the components of disaster management plans that worked and those that need overhauling (see Table 3).

3 The Case Studies

3.1 Case Study I: Orissa Super Cyclone 1999

(a) Background

Tropical Cyclones are pretty common in the Indian subcontinent. They start as low pressure areas in the North Indian Ocean and move towards the Arabian Sea or the Bay of Bengal. The first set strikes the western coast of India while the latter strikes the eastern coast of India/Bangladesh coast. Oldest cyclone in recorded history was in 1737 which coincided with an earthquake of moderately high intensity and devastated the Bengal delta, killing more than 300 thousand people. Since then 20 of the world's 25 major cyclones in terms of lives lost have taken place in the Indian subcontinent. Tropical cyclones are generally associated with high velocity winds, storm surge, and unusually heavy rainfall. As a result, coastal and low-lying areas are most vulnerable and witness the most damage in terms of destruction of buildings, bridges, dams, and power & communication installations; submerging of standing crop; and loss of future income due to increased salinity of cultivable land following ingress of sea water and loss of cattle & farm machinery. The human and economic damage caused by tropical cyclones is considered to be one of the highest among all natural disasters, affecting more than 20 million people annually and causing economic damage averaging US$ 30 billion every year [2].

However, cyclone management in India focussed mainly on post-disaster relief and rehabilitation till about 1970s. The Cyclone Distress Mitigation Committee (CDMC) set up in 1970 recommended a paradigm shift in cyclone management from relief to prevention. It talked of the installation and modernisation of warning systems, identifying vulnerable areas, construction of cyclone shelters, and mass awareness campaigns. Many of these were followed up and taking a cue from Bangladesh Cyclone Shelters were started to be constructed in vulnerable coastal areas. However, when a Super Cyclone struck Orissa on 28/29 October 1999 the ground realities and lacunae were brought to light.[1]

(b) Impact

The Super cyclone had a landing wind speed of about 260 kmph. It affected close to 18 thousand villages with a population of close to 20 million. 1.9 million households were affected, 350 thousand houses collapsed completely and another 250 thousand houses collapsed partially. More than 10 thousand human lives were lost of which Erasma block of Jagatsinghapur district accounted for more than 8000 as 6 villages were completely washed away by storm surge and tidal wave as high as 8 m that continued for almost two days. The estimated loss was US$2.5 billion which was more than 0.5% of India's GDP at that time.

(c) Lessons

The first lesson learnt was that while meteorological forecasts and warning system had improved, on ground preparedness was low. The warnings reached the villages in time but were not unambiguous or specific regarding evacuation The disaster management infrastructure was also inadequate as Orissa had just 23 specialised Cyclone Shelters at that time compared to more than 1,000 in Andhra Pradesh [3], 4. Second, public perception of the danger involved was low and there was no training of local communities or mock drill regarding the Standard Operating Protocol that was to be followed in case of a severe cyclone. People simply did not know how big the threat was or what specifically to do during that time. It was compounded by the fact that another cyclone warning was sounded 12 days before but that cyclone turned direction at the last moment and made a weak landfall, affecting only the Ganjam and Gajapati districts mildly. As a result, people were complacent and did not take the warnings seriously. Third, there was no back-up for the power or communication systems which depended on high-tension power transmission towers and radio masts. The cyclone had literally mowed down the towers and masts and Orissa was incommunicado for about 4 days.

The relief operations however, were exemplary, and rehabilitation efforts were also comprehensive. This cyclone, perhaps for the first time, brought into sharp focus the economic losses associated with disasters—most of which could be avoided with proper planning and resilience building. Administrators and planners started to look at disaster management as investment for prevention and mitigation rather than post-disaster relief & rehabilitation.

3.2 *Case Study II: Extremely Severe Cyclone 'Fani' 2019*

(a) Background

As mentioned, the Orissa Super Cyclone of 1999 brought to focus the huge economic loss associated with disasters and planners started to look at investing for resilience. Orissa State Disaster Mitigation Authority (OSDMA) & Orissa Disaster Rapid Action Force were immediately formed after the cyclone and started to look at disaster mitigation in a comprehensive manner. In 2005 (after the 2004 tsunami) National Disaster Management Authority (NDMA) & National Disaster Response Force were formed at the union government level and OSDMA became Odisha State Disaster Management Authority [5]. In 2006 a National Cyclone Risk Mitigation Project (NCRMP) was initiated with support from Word Bank [6]. In 2010 District Disaster Management Authorities were formed in 84 coastal districts of the country. Alternative power and communication channels through underground cables, satellite communication network and ham radios were increasingly deployed and a dedicated National Disaster Communication Infrastructure was put in place to connect control rooms, vulnerable region operation centres and NDMA/SDMA. At the technological front, the warning system had improved substantially and in 2009 a State-of-the-art Cyclone Early Warning System was operationalised [7, 8]. Cyclone management became a pre-emptive action based on early warning and tracking and evacuation (see Table 2 for a flow chart).

(b) Impact

Extremely severe cyclone '*Fani*' made landfall near Puri in Ganjam district of Odisha on 3rd May, 2019, about twenty years after the super cyclone of 1999. The wind speed at landfall was about 180 kmph with gusting wind velocity of 205 kmph. It affected close to 18 thousand villages with a population of about 17 million and 64 lives were lost. Power & communication system were badly damaged but restored

Table 3 Flow chart of cyclone management in India

Year	Initiatives
1979	National Committee of Science and Technology (cyclones)
1990	World Bank-assisted Cyclone Emergency Reconstruction Project (CERP), GoAP
1999	Orissa State Disaster Mitigation Authority & Orissa Disaster Rapid Action Force
2005	(after 2004 tsunami) National Disaster Management Authority & National Disaster Response Force
2006	National Cyclone Risk Mitigation Project (NCRMP) with World Bank assistance
2009	State-of-the-art cyclone early warning system
2010	District Disaster Management Authorities in 84 coastal districts; National disaster communication infrastructure connecting with operation centres

Source GoI [5, 6, 7, 8]

within a reasonable time. Considerable damage was caused to transport infrastructure, dwelling houses, crops, and livestock, including fisheries. The economic loss was estimated to be about US$ 2.4 billion (about 0.1% of India's GDP at that time).

(c) Lessons

What had helped bring down the loss of life in a span of twenty years? The key game-changer was the Odisha government's zero-tolerance policy towards mortality. First, the infrastructure was augmented and during the interim period, with assistance from the NCRMP, Odisha had constructed more than 800 multipurpose cyclone shelters and more than 8000 other designated shelters. Second, a Community Based Disaster Preparedness (CBDP) programme was initiated to provide awareness and training related to potential natural disasters in the region, actions to be taken for their mitigation, and operating protocol immediately before, during and after the disaster. Involving school children in awareness building and mock-drill enabled the messages to reach every home even in the hinterland. Third, the warning system from the IMD was translated to locally understandable simple language and alerts were sent out through radio/television inserts, SMS and public address system from Day-Zero-minus-Four. These messages were changed to Warnings from Day-Zero-minus-Two. HAM-radios played an important role in information dissemination. Fourth, the state government undertook a massive evacuation programme in which close to 1.6 million people including 25,000 tourists were evacuated from vulnerable areas. Fishing activities were suspended and the district & block operation centres were well-stocked with medicine, food, polythene sheets and other essentials. Fifth, the search & rescue operations after the landfall were well coordinated and there was a clear chain of command.

However, the economic loss was substantial. A part of this was inevitable in terms of physical damage to structures and inundation of agricultural land. But the magnitude was multiplied simply because there were more vulnerable structures and economic activities going on in the areas that came in the path of the cyclone. For example, the strict building standards and Coastal Zone Regulations were not adhered to because of economic reasons and the hotelier tourism construction lobby. The coast became a hotbed for economic activities and these suffered a huge blow. The same is true for the urban slums, semi-pucca houses in rural and urban areas, and old dilapidated rural schools. Being poor, informal, unorganised, small-scale implied that most of these assets and economic activities were outside the insurance coverage and hence the losses were personal, private and catastrophic for the poor and vulnerable sections of the society. While the State did pick up a part of the reconstruction cost, the impact is long-lasting.

3.3 *Case Study-III: Very Severe Cyclone 'Amphan' in West Bengal 2020*

(a) Background

A particular previously unknown strain of influenza/pneumonia combination (later named COVID19) spread worldwide during December 2019–February 2020 affecting more than 200 countries, making over 115 thousand people sick and causing 4263 deaths by the end February. WHO declared COVID19 to be a pandemic in March, 2020 WHO [9]. Since the strain was unknown previously, there was no vaccine or proven therapy to combat the disease and the only recourse left was prevention. It was observed that the virus spread through oral/nasal discharges of droplets during speaking/breathing/coughing/sneezing, preventing people from coming to close physical proximity of one another was thought to be the only strategy. As a result, countries resorted to suspension of economic and social activities, mandating people to stay indoors and wearing masks. Globally termed as *Lockdown* in the media, it was imposed in India on 24th March. 2020. The century-old Epidemic Diseases Act (1897) was invoked and orders were given to arrest persons who violated the lockdown orders. The only exceptions were fire, police and emergency services, transportation of essential goods, banking services, and petrol pumps. Under such situation, very severe cyclone 'Amphan' made landfall at Bakkhali, part of Indian Sunderbans on 20th May.

(b) Impact

Very severe cyclone *Amphan* had an average speed of 185 kmph, gusting to 210 kmph. Close to 30 million people were affected in India. 3 million houses were damaged, 1.7 million hectares of crop land were inundated and there was a substantial damage to power and communication infrastructure as also to mangroves, forests and urban tree cover. The economic damage was estimated to be about US$ 14 billion, about 0.5% of GDP, making it the costliest cyclone in recorded history in terms of economic loss [10]. Close to hundred human lives were lost. However, there were secondary impacts as well. Incidence of COVID19 increased by more than 50% in the cyclone devastated areas compared to pre-cyclone period, presumably because of close proximity of people in the cyclone shelters [11].

(c) Lessons

The damage caused by '*Amphan*' was much higher than the cyclone '*Fani*' which struck the previous year though the severity of them in terms of wind speed etc. were less for '*Amphan*' compared to '*Fani*'. Both loss of life and economic loss were much higher. This was caused by several factors. First, the pandemic and associated lockdown posed severe constraints in terms of preparedness, evacuation, search and rescue, and restoration & rehabilitation activities. Though the scientific & technical aspect of detection and EWS were on target, dissemination of information and preparing for the disaster was lax. With public transport completely shut down (Indian

Railways came to a standstill for the first time after independence), movement of officials, disaster response teams, volunteers, along with transport of materials were slow and could not achieve the target quantum. Second, most of the multipurpose cyclone shelters were already being used as Quarantine Centres for the people affected/ suspected to be affected with COVID19. In addition, the 'social distancing protocol' put in place meant that structures that could accommodate a thousand people were supposed to provide shelter to only 200. Though more than 2 million people were evacuated, this was not even half of the vulnerable population as the focal path of the cyclone was densely populated. Third, even the field staff necessary for evacuation and restoration activities could not be fully deployed due to lack of public transport facilities and scarcity of accommodation. Fourth, the cyclone damaged power & communication infrastructure substantially and in absence of electricity even wireless communication and ham-radio operations broke down for more than 3 days. Fifth, relief and rehabilitation were inadequate as the economic activities were already closed for close to 2 months when the cyclone struck and materials were hard to come by. The government distributed food, tarpaulin and medicines but the reconstruction work was left to individuals. Sixth, the cyclone showed the vulnerability of densely populated semi-urban areas which are generally unplanned and without any standards/regulations as regards construction of buildings or putting in place a drainage system.

4 Conclusion

The three case studies bring out the importance of building up resilience for disaster management. While the first one showed us the missing elements in terms of Warning system and Awareness building, the second showed how preparedness can go a long way in minimising damages from disasters. The third showed how convergence of natural and biological disasters can undo disaster management planning. Since scientists predict simultaneous occurrence of multiple disasters to become more frequent, a relook into our disaster management programme and protocol become imperative (see Table 4 for a summary). So far our Disaster Management plans to operate in silos and we have separate plans for each of the major natural disasters that the country may face Flood, Cyclone, Drought, Earthquake, and Landslide. But so far there are no plans for tackling biological or epidemiological disasters. Neither is there any convergence in tackling multiple disasters at the same time. Our Disaster Management programme should therefore be reformulated along the following lines:

(i) Include biological disasters in our disaster management plans;
(ii) Redesign DM plans to allow for multiple disasters occurring simultaneously;
(iii) Take into account the changed socioeconomic realities like long distance movement of men and materials, commuting to work as a common phenomenon and rapid urbanisation;

Table 4 Lessons for disaster management from the case studies

Event	Enabling factors	Failures
Odisha Super Cyclone, 1999	Warning system	Lack of awareness; Ineffective Dissemination of warning; No training; Lack of evacuation infrastructure; No coordination among different agencies; Poor communication and power back up system
Extremely Severe Cyclone 'Fani', 2019	Early warning system; Awareness building & training of local communities; Effective Dissemination of warning; Evacuation infrastructure; Coordination among agencies; Communication and power back up system	Economic activities in Coastal Zone disregarding regulations; Poverty leading to unsafe buildings; Construction lobby disregarding standards
Very Severe Cyclone 'Amphan', 2020	Early warning system; Awareness building & training of local communities; Effective Dissemination of warning;	Densely packed economic activities & Unplanned urbanisation in Coastal Zone disregarding regulations; Construction lobby disregarding standards; Pandemic induced shortage of infrastructure, materials, staff and coordination activities; No plan to deal with multiple disasters

Source Author (based on references in text)

(iv) Make vulnerability mapping and resilience building an integral part of all sectors, especially town planning, construction, transport, power, communication, education & health;
(v) Make basic disaster management courses a compulsory part of our education system to increase awareness.

It is also to be noted that global warming and climate change-induced sea surface temperature rise was one of the factors that led to a fast transformation of the cyclone Amphan from a depression to a severe cyclone within a short period of time [12]. Since global warming is now an accepted phenomenon, frequency and severity of cyclonic storms are also expected to increase in the region. Similarly, few scientists have also related emergence of new strains of virus with global warming [13]. Thus it is highly likely that future disasters will not strike alone and unless we take appropriate mitigating measures and revamp our disaster management plans to be operational all throughout the year, we would be stuck between the devil and the deep sea.

Note

[1] Any tropical cyclone that develops in the North Indian Ocean is monitored by the IMD on a 7-scale category based on 3-min sustained wind speeds. The lowest category with windspeeds between 31–49 kmph are labelled as Depressions. Further classifications are: Deep Depression (50–61 kmph); Cyclonic storm (62–88 kmph); Severe Cyclonic Storms (89–117 kmph); Very Severe Cyclonic Storms (118–166 kmph); Extremely Severe Cyclonic Storms (166–220 kmph); and the most severe being Super Cyclonic Storm (over 222 kmph).

References

1. GoI (2005) The disaster management Act, 2005 (No. 53 of 2005). Ministry of Law & Justice (Legislative Department). [www.cdn.s3waas.gov.in/s365658fde58ab3c2b6e5132a39fae7cb9/uploads/2018/04/2018041720.pdf]
2. Guha-Sapir D (2017) EM-DAT: The emergency events database. Université catholique de Louvain (UCL) - CRED, Brussels, Belgium [available from www.public.emdat.be, accessed on 12 July 2020]
3. Dhara S (2000) Deaths foretold. Down to Earth:39–42
4. Samal KC (2006) Facing sudden impact: experience of Orissa super cyclone of 1999. Man & Dev
5. GoI (2008a) National disaster management guidelines: Management of cyclones. National Disaster Management Authority, Government of India
6. GoI (2008b) Environment and social management framework: The National Cyclone Risk Management Project (NCRMP). National Disaster Management Authority, Government of India
7. GoI (2015) Strategies and lessons for preparing better & strengthening risk resilience in Coastal Regions of India. National Disaster Management Authority, Government of India
8. GoI (2019) National disaster management plan 2019. National Disaster Management Authority, Government of India
9. WHO (2020) Timeline of WHO's response to COVID-19 [available from https://www.who.int/news/item/29-06-2020-covidtimeline, Accessed on 15 Jun 2020]
10. Sud V, Rajaram P (2020) Cyclone Amphan caused an estimated $13.2 billion in damage in India's West Bengal: government source. CNN. [Available from https://edition.cnn.com/2020/05/22/weather/cyclone-amphan-damage-intl-hnk/index.html, accessed on 20 Aug 2020]
11. Kumar S, Lal P, Kumar A (2021) Influence of super cyclone "Amphan" in the Indian Subcontinent amid COVID-19 pandemic. Remote Sens Earth Syst Sci 4:96–103. https://doi.org/10.1007/s41976-021-00048-zu
12. Koll RM (2020) Chennai braces for heatwave due to super cyclone Amphan, mercury breaches 40-degree mark. The New Indian Express. [Available from https://www.newindianexpress.com/cities/chennai/2020/may/19/chennai-braces-for-heatwave-due-to-super-cyclone-amphan-mercury-breaches-40-degree-mark-2145375.html, accessed on 20 Aug 2020]
13. Curseu D, Popa M, Sirbu D, Stoian I (2010) Potential impact of climate change on pandemic influenza risk. In: Dincer et al. (eds.), Global Warming: Engineering solutions. Springer
14. GoI (undated) Cyclone related disasters. Indian Meteorological Department Ahmedabad. [Available from http://www.imdahm.gov.in/cycdisasters.htm, accessed on 15 Jun 2020]
15. IPCC (2014) Summary for policymakers. In: Field CB, Barros VR, Dokken DJ, Mach KJ, Mastrandrea MD, Bilir TE, Chatterjee M, Ebi KL, Estrada YO, Genova RC, Girma B, Kissel ES, Levy AN, MacCracken S, Mastrandrea PR, White LL (eds) Climate change 2014: impacts, adaptation, and vulnerability. Part A: global and sectoral aspects. Contribution of working group II to the fifth assessment report of the intergovernmental panel on climate change. Cambridge University Press, Cambridge, United Kingdom and New York, NY, USA, pp 1–32

Blue-Green Infrastructure: A Possible Connect to Guwahati Smart City

Debdut Sengupta, Sudip Mitra, and Rajib Shaw

Abstract In the context of flood management, Blue-Green Infrastructure (BGI) can be defined as the network of natural & designed landscape components that include blue and green spaces, and green spaces are designed to turn blue during extreme rainfall & flood events. BGI provides multifunctional benefits like less urban water discharge, better climate change adaptation, less water pollution, increased biodiversity, and water storage. Even at an individual level, green space promotes good health, well-being, and better physical & mental development. Developed countries like Europe & North America have promoted the use of Green Infrastructure (GI), emphasizing more on environmental sustainability. Copenhagen mandated the use of green roofs as a BGI measure in 2010 which absorbs 50–80% of their annual rainfall, delays discharge & reduces the urban heat island effect. In Portland City, the Downspout Disconnection project (1993–2011) alone can attribute to 1.2 billion gallons of stormwater discharge reduction. In Toronto City, building with greater than or equal to 2000 m^2 area must apply green roofs in 20–60% of their area. Japan mandated green roofs in all the newer constructions. The developed countries are taking benefits both in terms of cost & ecology by shifting from grey to green. But in India, the main focus is on housing & transportation development. In the Indian context, it is challenging to implement BGI because of the rapid urbanization and the hierarchical model. Thinking & implementing BGI had already begun in metropolitan cities like Ahmedabad and Mumbai. But, until now, no proper general framework has been available. Our study area is focused mainly on Guwahati City. The city is a major commercial and educational hub of Assam and the gateway for the north-eastern region of India. The city witnesses flooding as a recurrent event. The focus of our work is to identify the possible improvements for the Guwahati

D. Sengupta · S. Mitra (✉)
Center for Disaster Management and Research (CDMR), Indian Institute of Technology Guwahati, Assam, India
e-mail: sudipmitra@iitg.ac.in

D. Sengupta
e-mail: debdut.sengupta@iitg.ac.in

R. Shaw
Graduate School of Media and Governance, Keio University, Fujisawa, Japan

S. Mitra et al. (eds.), *Disaster Management and Risk Reduction: Multidisciplinary Perspectives and Approaches in the Indian Context*,
https://doi.org/10.1007/978-981-99-6395-9_18

Smart City by analyzing the existing infrastructural gaps in the city, the objectives of the Smart City Programme & how Guwahati plans to develop its own Smart City. The study aims to see what could have been the critical GI implementations in the Smart City Project to obtain its objectives & whether the BGI implementation was done fractionally or not.

Keywords Guwahati Smart City · Blue-green infrastructure · Landscape · Multifunctional benefits · Green infrastructure

1 Introduction

Blue-Green Infrastructure (BGI) has become a widespread term in the context of land conservation & land development [1]. Whereas traditional land conservation approaches oppose development, BGI, on the other hand, focuses on sustainable development without compromising the natural ecosystem. The adjective Blue-Green reminds practitioners to consider sustainable natural ecosystems while making infrastructure. The meaning of this term with respect to the context may change—for some, it refers to trees that provide ecological benefits in urban areas; for others, it refers to environmentally friendly engineered structures (such as stormwater management or water treatment facilities) [1]. A strategically planned and managed network of wilderness, parks, greenways, conservation easements, and working lands with conservation value that supports native species, maintains natural ecological processes, sustains air and water resources, and contributes to the health and quality of life for communities and people are called as green infrastructure [1].

BGI provides multifunctional benefits like less urban water discharge, better climate change adaptation, less water pollution, increased biodiversity, and water storage provision [2]. Even at an individual level, green space promotes good health, well-being, and better physical & mental development.

Developed countries like Europe & North America have promoted GI use, emphasizing environmental sustainability more [2]. Implementation of BGI should consider the conservation of water dynamics at its source, conservation & maintenance of the urban open space (to facilitate infiltration, accumulation, evapotranspiration, flood zones for the coming rainwater) & finally, space for urban development [2]. The successful consideration of the first two objectives is critical to facilitate sustainable infrastructure growth.

One of the main focuses of BGI is to mitigate flood risks. The implementation of Grey Infrastructure (dam, concrete pipes) for flood risk management & to provide urbanization opportunities are no longer helpful. The construction work for such infrastructure also causes more surface water runoff, increasing the risk of urban flooding. Although we cannot completely erase our dependency on grey infrastructure, we need to limit it, focusing more on the natural solutions available. Water is always seen as an external element required to be accommodated in the city's spatial planning management. It should be considered an intrinsic element [2].

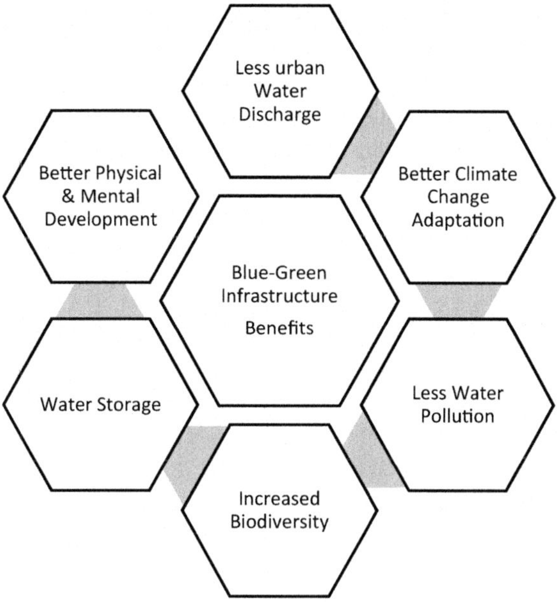

Fig. 1 Blue-green infrastructure benefits

Developing cities with higher economic growth, infrastructural & service facilities, and transportation facilities tend to have more people migrating, which further increases the risk of environmental degradation. Metropolitan cities like Ahmedabad, Mumbai, Pune have already addressed the issue & working on the Blue-Green Infrastructure aspect. From 2006–2021 only 212 Research Articles were found on Elsevier's Scopus database using the terms: "blue-green infrastructure," "blue and green infrastructure," "green–blue infrastructure," and "green and blue infrastructure" [2]. Of this, only 17% of the publications are from developing countries, which shows the apparent need to promote more & more research in this regard as developing countries are at higher risk. In India, the focus is on housing, and transportation development, i.e., satisfying the basic needs of people [2]. Unlike the developed countries, since India is a developing country, the government and planners are struggling to balance urbanization & protecting the environment (Fig. 1).

2 Guwahati City Profile

Guwahati is the largest city in the North-Eastern region & also the only metropolis having a population greater than 1 million in this region. Guwahati is spread over an area of 264 sqkm [3]. The city is also a major commercial and educational hub of Assam and the north-eastern region of India. It is home to premier institutions such as the Indian Institute of Technology (Guwahati), Gauhati University, and Cotton College (Fig. 2).

Guwahati lies between the banks of the Brahmaputra River and the foothills of the Shillong plateau, with LGB International Airport to the west and the town of Narengi to the east [4, 5]. The geographical extension of the present-day city may be considered to be between 91°33′18.141” E and 91°10′41.005 “E Longitudes and 25° 59′ 33.183” N and 26° 15′ 50.945” N Latitudes [4, 5]. The central part of the city has small hillocks such as Sarania hill, Nabagraha hill, Nilachal hill, and Chunsali Hill. The major water bodies are Deepor Beel, Silpukhuri, Dighali Pukhuri, Borsola Beel, and Silsako Beel [4, 5]. The Guwahati Metropolitan Area is divided into six drainage basins which finally drain into the Brahmaputra River. These six basins are Bharalu Basin, Dipar Basin, Silsako Basin, Foreshore Basin, North Guwahati Basin, and Kalmoni Basin. Guwahati comes under the seismic zone five, making it highly vulnerable to earthquakes & landslides. Also, the existence of the mighty Brahmaputra makes it susceptible to flooding. One of the main objectives of the Guwahati Smart City program is to deal with disasters more efficiently so as to reduce the damage to the decadal progress.

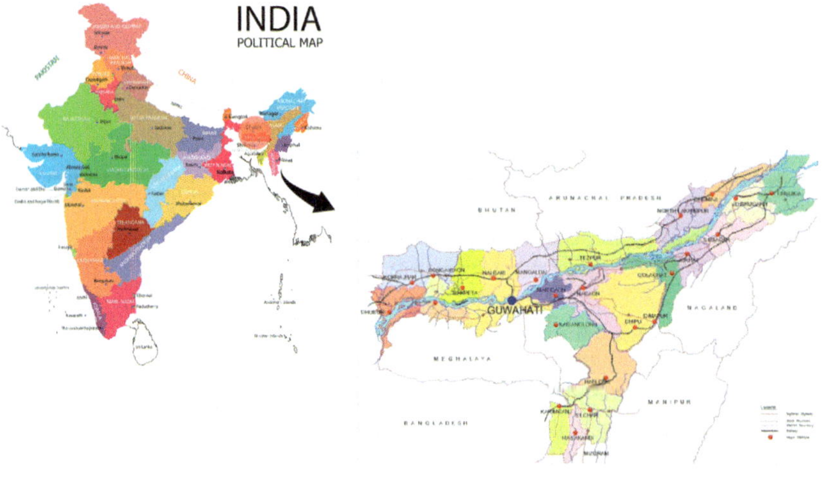

Fig. 2 Location of Guwahati City in Assam (*Source* Master Plan for Guwahati Metropolitan Area–2025)

3 Literature Review

3.1 Guwahati City Problems

Guwahati is the biggest city in the state of Assam, India, holding a population of over one million. It is situated at the bank of the mighty Brahmaputra River. Due to its geographical location, along with the other north-eastern parts, each year, Guwahati witnesses flooding events. In recent decades the devastation due to flooding has increased few folds. In the last 20 years, the average annual rainfall for Guwahati City has been 1607 mm. Very few cases of urban flooding were reported before the 1980s in Guwahati City [6]. Unplanned urbanization & changes in land use patterns are the main reasons for flood recurrence. The main reasons for the rampant changes in land use are increased employment opportunities due to growth in the industrial sector & immigration and encroachments by economically weaker sections of society over hills for better urban service, education & medical facilities [7]. Encroachment of hills & cutting down of trees causes higher soil loss due to surface flow over exposed slopes: 188.09 gL^{-1} (exposed slope) compared to 3.062 gL^{-1} (vegetated slope) of surface flow [6].

Guwahati witnessed a 65% increase in population growth from 1981–1991 due to the greater population migration for better infrastructural facilities in the education, health & employment sectors [6]. Although the population growth currently stands at 8.7% only, earlier from 1971–2001, high population growth was observed [3]. To compensate for this population growth, real estate activities have also increased. The climatic conditions of the area in and around Guwahati are changing rapidly, and in 2009 dusty sky was observed. Earthwork of NHAI, cutting of hills in nearby regions of Meghalaya, and Construction work undertaken by the Real Estate Industry were responsible for this [6].

3.2 Smart City Programme

"Smart City is defined as a city where structure and function of various urban systems are clearly defined, simple, highly responsive and malleable via contemporary technology and design" (Smart Cities, 2010; MoUD, 2014). Smart cities should perform smartly in all the urban sectors like energy, sanitation, and infrastructure. In light of India's growing urban footprint, more than 40% of its population is expected to live in cities by 2030 [8]. Urbanization causes significant challenges regarding climate change adaptation, more frequent & intense hazardous events, more energy requirements, emission of higher amounts of greenhouse gasses, etc. In recent decades climate change & low carbon concepts, along with a better quality of life for citizens, have made the Smart City Programme implementation more challenging. But on the other hand, urbanization provides opportunities for economic growth. Urban areas produce close to 60% of India's GDP and are estimated to rise beyond 70%

Urban Rejuvineation Programmes						
Smart Cities Mission (SCM)	Swachh Bharat Mission Urban (SBM-U)	Pradhan Mantri Awas Yojana-Urban (PMAY-U)	Deendayal Antyodaya Yojana National Urban Livelihoods Mission (DAY-NULM)	Atal Mission for Rejuvenation and Urban Transformation (AMRUT)	Heritage City Development and Augmentation Yojana (HRIDAY)	Urban Transport with a focus Metro Rail Systems

Fig. 3 Urban rejuvenation programmes (*Source* Ministry of Housing and Urban Affairs: Making a city smart learning from the smart cities mission)

by 2030 [8]. To avail the economic growth opportunities, provide an overall better quality of life to its citizens & tackle the emerging challenges due to urbanization, the Government of India initiated the Urban Rejuvenation Programme in 2015 [8] (Fig. 3).

Under the Smart Cities Mission, 100 cities were selected in a two-phase manner. Of these 100 cities, 8 North-Eastern Cities were selected. The cities were to propose projects based on three significant aspects: **1. Sustainability, 2. Liveability & 3. Economic Ability**. Core infrastructure elements of the Smart City Programme include adequate water supply, assured electricity supply, sanitation, including solid waste management, efficient urban mobility and public transport, affordable housing, especially for the poor, robust IT connectivity and digitalization, good governance, especially e-Governance and citizen participation, sustainable environment, safety and security of citizens, particularly women, children and the elderly and health and education (MoHUA, Smart Cities).

There are mainly three models of Area-Based Smart City Development: **1. Retrofitting** (makes the existing built-up area more efficient & liveable, **2. Redevelopment** (replacing the current built-up environment and installing enhanced infrastructure using mixed land use), **3. Greenfield** (introduces most of the Smart Solutions in a previously vacant area) (MoHUA, Smart Cities). In addition to the above three, pan-city focuses on Smart Solutions for the existing city-wide infrastructure for better infrastructure & services. Smart solutions in the transportation sector to reduce commute time or wastewater recycling are good examples. Each selected city is to frame its own concept, vision, mission, and plan (proposal) for Smart City. Guwahati Smart City Limited has listed Borsola Beel (Lake), Morabharalu River, Bharalu River, Brahmaputra Riverfront, and DeeporBeel (Lake/Wetland) Development as Area Based Development (ABD) Projects. The ABD Projects do include activities like removing sludge/cleaning of Borsola Beel, Morabharalu River, Bharalu River; construction of Sewage Treatment Plant (STP) using Sequential Batch Reactor (SBR) technology for Borsola Beel, Morabharalu River, Bharalu river; making sewage interception & collection drain (Drain, sewage bypass arrangement) & development of landscape (five senses garden, water sports, seven sister's walkway) for Borsola

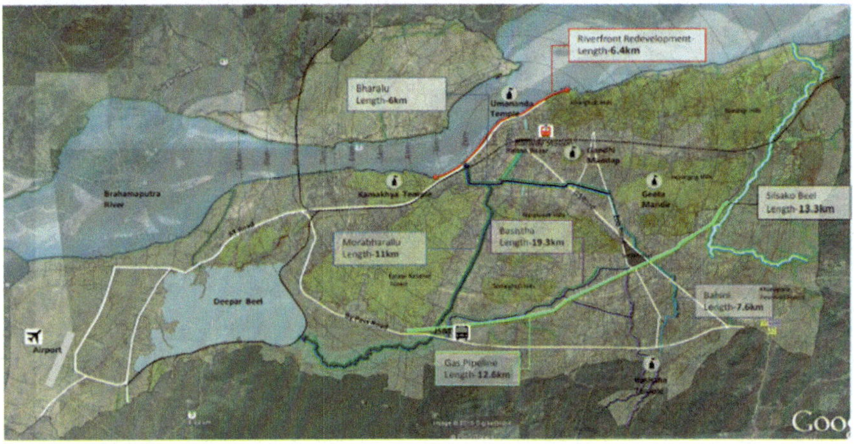

Fig. 4 Strategic plan for Guwahati Smart City limited (*Source* Guwahati Smart City Limited: Guwahati Smart City Draft Proposal)

Beel. The Brahmaputra Riverfront Development does include some critical future improvements for flood management based on the feasibility report for the riverfront design—1. Upgrading of outflow structures of stormwater and drainage system within the topographical scope riverfront, 2. Hydrological data survey & design of integrated storm & drainage system of the city (GOA, Guwahati Smart City Limited). Deployment of various sensors for traffic violations, environment and weather conditions, and flood warning systems comes under the Pan-City activities. A command & communication center (COC) will be established at the center of Guwahati City. All smart solutions and key components viewing, centrally monitoring, storage of data, and operation of the entire components & solutions installed will be done through COC (GOA, Guwahati Smart City Limited) (Figs. 4 and 5).

3.3 Guwahati SWOT Analysis

From the Approved Proposal of Guwahati Smart City, we can find the SWOT analysis report of the city [9] (Table 1).

The SWOT analysis indicates that the primary focus areas should be—1. Environmental Sustainability & 2. Better Transportation System Development. BGI can mitigate the flood challenges, soil erosion, and degradation of air & water quality. Although only restoring the natural ecosystem will not be enough to reduce disastrous events, hence careful planning considering the objectives of the Smart City Programme is needed. For the rest part, the existing grey infrastructure needs to be modified, or new infrastructure needs to be developed. In transportation, other than providing footpaths & equitable distribution of road spaces, we should focus more on the public transportation medium as it reduces the space required to accommodate

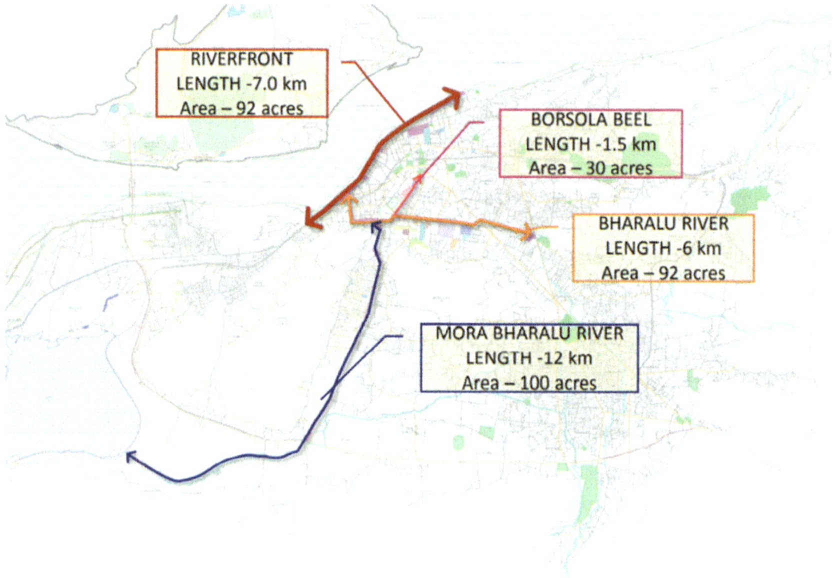

Fig. 5 Scope of work for area based proposal (*Source* Guwahati Smart City Limited: Guwahati Smart City Draft Proposal)

passengers compared to private cars. This reduced space further reduces the load on road construction activities. This causes a reduction in siltation problems, better infiltration of stormwater, lower greenhouse gas emissions & fewer flooding problems, indirectly mitigating sustainability-related issues. Sustainability maintains the balance between natural & built ecosystems in a city to prevent issues like **1. Rising temperatures, 2. Deteriorating air quality, 3. More frequent floods and droughts, and 4. Declining urban biodiversity** [8]. All these issues can be addressed using Blue-Green Infrastructure. We need to integrate BGI into the smart city mission carefully.

3.4 *Infrastructural Gaps*

Before implementing Smart City thinking, we should first identify the existing infrastructural gaps & should invent smart solutions to minimize them. A study on the existing infrastructural gaps w.r.t water, sanitation & drainage systems (critical systems for urban life) has been made in Guwahati City [7]. The city's primary water supply source is from the Brahmaputra River & groundwater sources. People cannot drink the water from the Brahmaputra directly as it is highly contaminated & needs a high level of purification before consumption [10]. The underground water quality is also poor because of leakages & presence of metals like iron, fluoride, and

Table 1 Guwahati SWOT Analysis (*Source* Guwahati Smart City Limited: Approved proposal of Guwahati Smart City)

Strength	Gateway to the North-Eastern region. Fast-growing metropolis. Presence of hills, basins, rivers, and beels adds to its scenic beauty, boosting the tourism economy. Called the "City of Temples" for its historical, cultural & religious heritage.
Weakness	Yearly flash floods caused due to heavy rains & degradation of the natural ecosystem. Frequent landslides due to hill encroachment & cutting down of trees. Higher soil loss due to exposed surface & higher surface runoff. Traffic congestion issues. Lack of pedestrian infrastructure causes higher road accidents. Road accidents cause 50.4% of deaths in Assam, more significant than the national average of 36.4% in 2013.
Opportunity	Guwahati has approximately 388 primary schools and 101 secondary & high schools. The 18 higher educational institutes include an IIT. The higher education institutes attract students from the North-Eastern States and other parts of the country. The strategic location & connectivity with the North-East provides the opportunity to serve as a commercial and IT capital of the East. To become an eco-tourism destination by preserving and conserving its ecological and cultural treasures. 43.02% of the trips in Guwahati are shared by public transport (30.2%) + intermediate public transport (13%). Walk trips consist of 23%. But the city lacks properly designed footpaths, equitable distribution of road spaces, etc.
Threat	Unauthorized urbanization causes landslides. Soil erosion reduces the holding capacity of storm drains causing flood risks. Shrinking of natural wetlands to provide urbanization opportunity. Growing number of vehicles cause air & water pollution. Sewage & waste discharge into the stormwater drains affect the environment.

arsenic. Out of the 110 MLD installed potable water capacity, only 73 MLD (33.6% less) is available [7]. The per capita water supply is 75 Lpcd which is 44% less than the MoUD standards of 135 Lpcd [7]. The city has no integrated sewage systems other than certain pockets of railway colonies, IOCL refineries, and defense establishments. The small drains installed along the roadsides are not efficient enough and mostly remain waterlogged during the monsoon month. Siltation & encroachment activities make the Bharalu Basin the most flood-prone of this region.

The main gaps w.r.t mentioned water, sanitation & drainage system can be significantly improved by BGI installation. Increased water infiltration will help in groundwater recharge & will reduce surface runoff, reducing water body pollution. This will protect water bodies like the Brahmaputra River (a primary source of water for the city) from pollutants, heavy metals, toxic chemicals, etc. Less surface discharge causes less water to be treated & less risk of waterlogged roads. So BGI alone can elevate much of the loads on the infrastructural gaps.

3.5 Urban Sustainability Index & Ease of Living Index

Urban Sustainability Index (USI) checks the state of sustainability of the urban growth pattern of a city & it is measured based on three aspects: 1. Economic, 2. Environmental and 3. Social [11]. Guwahati, revealed as one of the fastest-growing cities in the world by a UK media outlet, has a USI of 0.451, which is far below the ideal 1. The reduction of natural & semi-natural vegetated areas over the years (at an annual rate of 2.2 km^2) causes a decrease in environmental sustainability, where Guwahati is only 35% sustainable [11].

In the Ease of Living Index 2020, Guwahati was ranked 46th out of the 49 cities under the million-plus category, while Bengaluru topped, followed by Pune & Ahmedabad [12]. The Index aims to assess the ease of living of citizens living in the cities across three pillars: quality of life, economic ability, and sustainability where the pillars include 14 categories (Education, Health, Housing & Shelter, WASH & SWM, Mobility, Safety & Security, Recreation, Economic Development, Economic Opportunities, Gini coefficient, Environment, Green spaces & building, Energy Consumption and City Resilience) [13]. Although the performance breakup based on the quality of life, economic ability, and sustainability couldn't be found but based on the environmental sustainability is 35% [11], in the USI, Guwahati is expected to perform somewhat similar here also. The weightage distribution of these three pillars is **1. Quality of Life (35%)**, **2. Economic Ability (15%)** & **3. Sustainability (20%)**. The rest is Citizen Perception (30%).

3.6 Portland & Copenhagen: Blue-Green Cities

Portland has one of the oldest & successful GI programs in the U.S. The city is an internationally renowned leader for its GI implementation for sustainable development & climate change adaptation. Johnson Creek is a 40 km long tributary of the Willamette River in Portland. Johnson Creek watershed is divided into numerous smaller catchment areas and almost 75% of the stormwater discharge without treatment was used to enter Johnson Creek through piped networks. During the 1980s this discharge caused a high concentration of heavy metals, *E. Coli* & a higher stream temperature. Investments were done for extensive river restoration, riparian planting, and installation of BGI elements to reconnect the creek to its remaining floodplains and reduce flood risks. The city invested 1.4 billion dollars in combined sewer overflows (CSO) control through "Big Pipe Projects" which eliminated 94% of CSOs for the Willamette River. The parallel 55 million dollars of GI projects not only reduced the size of the required pipes but also saved a huge amount of money. Till October 2020 the city witnessed 56,000 downspout disconnections from the combined sewer system which redirects the stormwater to citizens' gardens for infiltration [14].

Copenhagen has a combined sewer system in most of the city area & has a separate sewer system close to the port. For reduction in CSOs, the city had built reservoirs that lowered overflow from 20–70 per year to 2–6 per year & discharge volume was reduced from 1.6 million m^3year^{-1} to 300,000 m^3year^{-1}. The city had kept mainly three possible options for heavy rainfall events: 1. Installing newer sewer pipes to increase sewer capacity which costs DKK 10–15 billion if laid across the city. DKK 3–5 billion is additionally needed to separate rainwater from domestic water; 2. BGI installation to improve water quality & decrease water quantity. This costs DKK 3–5 billion which is much lower compared to traditional grey infrastructure development; 3. Ensuring flooding takes place only where it causes less damage. Making dedicated flooding places is needed like—playing fields & parks. In Copenhagen, green roofs absorb 50–80% of their annual rainfall; furthermore, green roofs delay stormwater discharge & reduce the urban heat island effect. The city is also planning to mandate downspout disconnection with 5 years prior notice to the locals [15].

4 Discussion & Research Gap

4.1 Water

The ABD projects for Guwahati City put a lot of focus on the water bodies' (lake, wetland, river) wellness restoration. Activities like cleaning of sludge, sewage treatment plant installation, and sewage bypass arrangement are expected to restore the water ecosystem. Water is an ancient city constituent & we need to consider the water dynamics for the sustainable development path of our cities. The drainage system of Guwahati City is heavily dependent on the existing natural drains, which get clogged due to waste material, garbage, and sewage. In addition, siltation problem due to hill cutting adds to the problem. The city doesn't have a planned drainage system [10]; hence the development of sewage interception & collection drain, and stormwater drains were much needed. The tertiary treated wastewater from the sewage treatment plant will be used for flushing & gardening purposes in Governmental Buildings as per the ABD project promoting better water usage.

4.2 Flood Resilient City

One of the primary visions of Guwahati Smart City is to become a "Flood Resilient City" (GOA, Guwahati Smart City). As a long-term sustainable measure, the city needs to focus on watershed management & basin-specific stormwater management plans. As a part of ABD projects, installing grey infrastructure such as collection drains, stormwater drains, and cleaning natural drains are expected to lessen the flooding problems in the future. Pan-City projects like Flooding Monitoring System will also be developed, which will be monitored centrally at City Operation Centre (COC) in an integrated manner. Approximately 50 sqkm of land is required by Guwahati Metropolitan Area for various urban activities by 2025. This indicates a drastic increase in residential, commercial, public, and semi-public spaces [7]. Hence a clearly defined allocated urban green space was necessary. On the Guwahati Smart City Limited Website, as well as the Guwahati Smart City Draft proposal, we couldn't find any planning specifically for Green Infrastructure. The Smart City proposal of each city was expected to encapsulate either a retrofitting or redevelopment or greenfield development model, or a mix thereof and a Pan-city feature with Smart Solution(s) (MoHUA, Smart City Guidelines). The cities were free to formulate their concept, vision, mission, and planning for a Smart City that is appropriate to its local context, resources, and levels of ambition. But the benefits of Green Infrastructures (GI) such as downspout filters, permeable pavements, green roofs & walls should have been considered based on their relevance to the context of flooding. Urban GI schemes are expected to face four major challenges for any metropolitan city—**1. Design Uncertainty** (how to plan, implement, and maintain. Site-specific planning based on threats & resources is essential), **2. Equity** (Certain communities, based on age, gender, income, and race, might not enjoy BGI benefits), **3. Regulatory Body** (Distribution of responsibilities to different bodies & commitment to sustained maintenance work), **4. Micro-Scale** (Fines for non-compliance with water, air, and soil quality for private & public lands. Triggers broad individual level BGI usage) [16].

4.3 Permeable Pavement

The installation of permeable pavement aims at reducing the stormwater runoff problem at the source. It reduces the cost of water treatment & also facilitates the water to get soaked into the soil. It also comes under GI which is missing from the smart city planning. The permeable paving temporarily stores the water into the underlying base course & then slowly releases it into the subgrade or underlain. This causes a reduction in peak stormwater discharge & also an increased lag time in the hydrograph. In addition, an underground tank system treats, stores & supplies the stormwater for non-potable applications like irrigation, toilet, vehicle washing, etc. A recent study was done on Guwahati City for the feasibility check of permeable pavements to control flash flooding [17]. The study reveals the need for more research work in this regard as porous asphalt; porous concrete pavement systems are vulnerable to frequent clogging, which hinders their usability. Under the Pan City Projects for Guwahati Smart Limited, we could find smart traffic and smart parking mentioned (GOA, Guwahati Smart City Limited). But the further elaboration of the terms couldn't be found. On the Google Scholars website, very few papers are related to the terms "permeable pavement guwahati" and "permeable road Guwahati." The installation of such pavements needs more research to provide hardcore conclusions for practitioners to consider in Pan-City Projects.

4.4 Transportation

The SWOT analysis of Guwahati City indicates the clear need for better transportation system development. Guwahati witnesses high traffic congestion issues. Lack of pedestrian infrastructure causes higher road accidents. Road accidents cause 50.4% of deaths in Assam, more significant than the national average of 36.4% in 2013. The city lacks properly designed footpaths and equitable distribution of road spaces. 200 busses are available under the Jawaharlal Nehru Urban Renewable Mission (JNURM) scheme & at present, 2500 total city busses are available [10]. Most city buses are private, causing frequent halts of busses to acquire more passengers. The location of the wholesale market, trade center, inter-district, and interstate bus terminus inside the city region caused traffic congestion issues & hence shifted to the outskirts of the city [10]. The ABD project for the Brahmaputra Riverfront Development includes the development of walkway, cycle track, and jogging path along the full Riverfront (6 km stretch Riverfront) from Rajbhawan to Kamakhaya Temple. Various other infrastructure works like ramps, stairs, jetties, ferry terminals, and bridges over outflow structures are to be designed.

5 Learnings from a Preliminary Study

We can plan the BGI flood management of Guwahati City into mainly two parts—1. Man-made features like a green roof, blue roof, green street, green buildings, etc.; 2. Natural Features like wetlands, ponds, rivers, lakes, etc. The man-made features are to arrest the stormwater, delay stormwater discharge, prevent stormwater from mixing with contaminants, let stormwater get infiltrated into the soil, enhance bio-diversity and so much more. The man-made features mainly work as an on-source treatment. The natural features work as detaining & retaining the rest of the discharge where the bio-retention systems remove the contaminants & sediments from the discharge. Together these two parts can reduce yearly CSOs significantly as can be seen in the case of Copenhagen & Portland. We cannot fully remove grey infrastructure but we should focus on substituting grey with green infrastructure. Green infrastructure not only comes with added ecological & social benefits but also saves a lot of money. In Copenhagen, cost of BGI construction is 3–5 times less than installing newer sewer pipes. Most of the cities in the world have a combined sewer system that collapses during heavy rainfall events. Downspout disconnection is a great way to stop stormwater from entering the sewer system. Cities like Portland have already made 56,000 downspout disconnections till 2020 whereas Copenhagen is on the verge of mandating downspout disconnection. Few of the cities have also considered provisions for flooding zones where the flood water will be directed & will cause very less damage.

Considering the importance of on-source treatment for flood management, we are working on green roof development for Guwahati City. Although its successful implementation comes with many benefits including social & economic, its design is a challenging part. As mentioned in the literature [18], what will be the thickness of the soil stratum, what should be the vegetation, how much water is to store, what are the geographic & climatic conditions of the region, and how much dead-load the buildings can withstand—all are very important questions in the design process. Sedum Mosses is a very popular option worldwide as vegetation for green roofs since they can withstand different climate conditions, can continue photosynthesis metabolism even after having a long duration (4 months) of water scarcity, and can be used in shallow stratum (about 7 cm is more than sufficient). But its availability is limited & hence it is better to start with the locally available vegetation & soil stratum. In the UK most of the buildings can withstand an additional dead-load of about 8 to 10 KN which is enough for the building to be safe from Green Roof retrofitting. But the same study is not available in Guwahati & based on the dead-load idea the soil stratum needs to be prepared. Another important aspect is the presence of pollinators: it is important to see that pollination is happening. On the rooftop availability of nutrients can be an issue for which the emphasis should be on vegetation that can sustain with limited nutrition.

6 Conclusion

The ABD Programmes have intensely focused on the water bodies, which is supposed to improve the overall sustainable development of the city by restoring the lost water ecosystem. It will also help in the eco-tourism part of the city & will mitigate flood risks. But critical improvements could have been added to the proposed planning. The ABD projects are more focused on runoff management not on the on-source treatment. Green roofs & downspout filter disconnections can save a lot of water from entering the sewer system. But the implementation of such BGI elements other than extensive research on their design & functionality, requires detailed cost–benefit analysis for practitioners to adopt green over grey infrastructure.

List of Abbreviations

Abbreviations	Definition
BGI	Blue-Green Infrastructure
GI	Green Infrastructure

GOI	Government of India
IOCL	Indian Oil Corporation Ltd.
LGB	Lokpriya Gopinath Bordoloi
CDP	City Development Plan
MoUD	Ministry of Urban Development
MoHUA	Ministry of Housing and Urban Affairs
ABD	Area Based Development
COC	Command & Communication Centre
SWOT	Strength, Weakness, Opportunity, Threat
USI	Urban Sustainability Index
GOA	Government of Assam

References

1. Benedict MA, McMahon ET (2012) Green infrastructure: linking landscapes and communities. The Conservation Fund, Island Press
2. Tardin-Coelho R, Cynthia T, Miguez MG (2021) Water dynamics and blue-green infrastructure (BGI): Towards risk management and strategic spatial planning guidelines. J Cleaner Prod. https://doi.org/10.1016/j.jclepro.2021.129993
3. Census Tables (2011) Ministry of home affairs, Government of India: https://censusindia.gov.in/census.website/data/census-tables. Accessed on 11.07.2022
4. City Development Plan Guwahati (2006) https://www.yumpu.com/en/document/view/7548447/city-developmentplan-guwahati-jnnurm. Accessed on 11.07.2022
5. Master Plan Guwahati 2025 (2009) Guwahati Development Department, Government of Assam: https://gdd.assam.gov.in/portlets/master-plan-guwahati-2025. Accessed on 11.07.2022
6. Bora MC Growth of infrastructure and climate change in Guwahati city. In: International conference on infrastructure finance-2010, vol 2. Indian Institute of Technology, India. https://doi.org/10.13140/2.1.2771.0406
7. Kaushik A, Dhingra M, Singh MK, Parikh JK (2015) Smart city expo world congress 2015: Indian cities towards smartness: a case study of Guwahati City. In: Conference: Smart city expo world congress 2015 at Barcelona. Climate resilient smart cities
8. Government of India, Ministry of home and urban affairs, Smartnet: making a city smart: learning from the smart city mission
9. Government of Assam, Guwahati Development Department, Guwahati Smart City Limited: The Smart City Challenge Stage 2 (Smart City Proposal)
10. Basumatary MG, Anand S Sustainable urban infrastructural development for Smart City in Guwahati, India. UDC 911.375.4:316.422] (540.35–21Guw). https://doi.org/10.26565/2076-1333-2018-25-05
11. Borah B, Borah C (2021) Urban sustainability index of Guwahati City. J Contemp Issues Bus GovMent 27(1). P-ISSN: 2204–1990. E-ISSN: 1323–6903
12. Ease Of Living Index 2020: Guwahati Ranks 46 Out Of 49 Cities, The Guwahati Times, Published March 4, 2021
13. Government of India, Ministry of Home and Urban Affairs, Smartnet: Ease of Living Index 2019 Assessment Framework
14. Sustainable flood risk and stormwater management in blue—Green Cities. An Interdisciplinary case study in Portland, Oregon. JAWRA J Am Water Resour Assoc 56(26). https://doi.org/10.1111/1752-1688.12854

15. Brears RC Blue and green cities: The role of blue-green infrastructure in managing urban water resources.
16. Satiram BG, Chakraborty I, Banerjee S (2020) Challenges in implementation of green infrastructure in metropolitan cities—case of Ahmedabad. https://doi.org/10.37628/jaip.v6i2.641
17. Das CK, Saikia MD (2018) Feasibility study for permeable pavements to control flash flood in Guwahati city. Int Res J Eng Technol (IRJET) 05(05)
18. Alibaba HZ (2019) GREEN roof benefits, opportunities and challenges. December 2019; Project: He assessment of indoor thermal comfort in a building: a case study of Lemar, salamis road, Famagusta, Cyprus. Eastern Mediterranean University
19. Wescoat JL, Rawoot S (2020) Blue-green urban infrastructure in Boston and Bombay (Mumbai): A macro-historical geographic comparison. ZARCH. https://doi.org/10.26754/ojszarch/zarch.2020154857
20. Government of Assam, Guwahati Development Department, Guwahati Smart City Limited: Guwahati Smart City Draft Proposal

Precipitation Analysis and Rainfall Forecasting for Kamrup Rural District

Arnab Paul Choudhury, Debaditya Gupta, and Sudip Mitra

Abstract A considerable portion of the farmers and people in rural areas, especially in Assam, involved in agricultural practices are solely dependent upon climatic factors for producing crops. Crops, too, are sensitive to climatic conditions and other environmental factors and failing to provide the right environment at the right time eventually leads to loss of productivity. Hence, it is prudent that a system is devised in the form of an early warning system such that farmers can be made aware of any irregularities in rainfall in advance, thereby avoiding massive losses to crops. A long-term precipitation analysis over 120 years is performed for the Kamrup Rural region using gridded data provided by the Indian Meteorological Department. The MK-test, Sen's Slope and EMD method which is proposed as an effective alternative for trend identification of series, has been used for trend analysis. Additionally, an attempt is made to develop a rainfall forecasting model using recurrent neural networks such as Long Short-Term Memory.

Keywords Empirical mode decomposition · MK-test · LSTM · Rainfall forecasting · Precipitation forecasting · Trend analysis

---

A. P. Choudhury · D. Gupta · S. Mitra (✉)
School of Agro and Rural Technology, Indian Institute of Technology Guwahati, Assam, India
e-mail: sudipmitra@iitg.ac.in

A. P. Choudhury
e-mail: arnab.paul@iitg.ac.in

D. Gupta
e-mail: dgupta@iitg.ac.in

S. Mitra
Centre for Disaster Management and Research, Indian Institute of Technology Guwahati, Assam, India

S. Mitra et al. (eds.), *Disaster Management and Risk Reduction: Multidisciplinary Perspectives and Approaches in the Indian Context*,
https://doi.org/10.1007/978-981-99-6395-9_19

1 Introduction

A majority of the Indian population is involved in agricultural practices. Approximately 70% of Indian rural households still depend primarily on agriculture for their livelihood, with 82% of farmers being small and marginal. Water is one of the chief components required for the cultivation of any major crop. Additionally, water is required in the correct quantity and at the right time to ensure a good harvest. However, as per the World Bank, in 2015, only about 38% of the total agricultural land was irrigated. The numbers are even lesser for the state of Assam, where according to the Assam State Rural Livelihood Mission, only 5.4% of the gross cultivated area of the state of Assam is irrigated. The numbers clearly show that a considerable portion of the farmers and people in rural areas, especially in Assam, involved in agricultural practices are solely dependent upon climatic factors for producing crops. Crops, too, are sensitive to climatic conditions and other environmental factors and failing to provide the right environment at the right time eventually leads to loss of productivity.

Recently, data-driven models have gained much more popularity over knowledge-based models [13]. Several data driven-models based upon Deep BLSTM-GRU [1], artificial neural networks, Adaptive Network based Fuzzy Inference System optimized with Particle Swarm Optimization and Support Vector Machines [16], deep neural networks [15, 18], seasonally-integrated autoencoders using Long Short-Term Memory (LSTM) autoencoders [17], generative adversarial networks [24] have been used to forecast rainfall. Additionally, hybrid models such as Multilayer Perceptron with Whale Optimization Algorithm [2], complementary ensemble empirical modal decomposition(CEEMD)-LSTM [26], Convolutional Neural Network(CNN)-LSTM model [11], Least Square Support Vector Machine with Multi-Factor Integration [10] based models have also shown a good capability to model a nonlinear and complex process like rainfall. In particular, models based on LSTM have found good use in sequence modelling and have proved useful in rainfall forecasting [9, 23].

A comprehensive literature review revealed that minimal studies had been carried out to understand the precipitation trend as well as rainfall forecasting for the Kamrup Rural district. Since a majority of the farmers in the Kamrup Rural district are dependent upon rainfall as the primary source of water for their agricultural fields, it is, therefore prudent to develop an early warning system that can predict/forecast rainfall and warn farmers if rainfall is about to get delayed or is arriving early for any season.

In the preceding sections, trend analysis of the rainfall time series over the Kamrup Rural district is conducted using established statistical techniques, and further, an attempt is made to develop a model that can forecast and predict rainfall that can be used as an early warning system by local farmers.

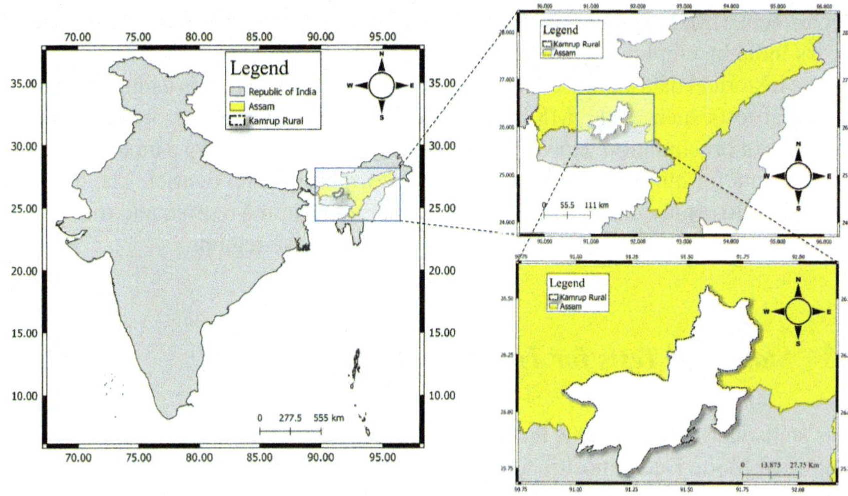

Fig. 1 Study area

2 Materials and Methods

2.1 Study Site

The study area is the entire district of Kamrup Rural, which is located in the eastern part of the state of Assam in India. The district is roughly contained within the bounding box having coordinates (90.875, 26.625), (91.875, 26.625), (91.875, 25.625), (90.875, 25.625) in decimal degrees. Figure 1 shows the study area and its location with respect to India and the state of Assam.

As per data released by Govt. of India for census 2011, the total population of Kamrup Rural is 1517542, of which 51.3% are males and 48.7% are females. Also, 90.6% of the total population resides in rural areas. The census categorizes the working population, which comprises 41.4% of the total population, into three categories, namely, cultivators, agricultural labourers, household industry workers and other workers. Of these, cultivators and agricultural labourers together constitute 45.3% of the working population clearly indicating agriculture and allied activities to be a major source of livelihood among the masses.

2.2 Data Used

The study is carried out using a 120-year-long gridded dataset of daily precipitation with a resolution of $0.25^\circ \times 0.25^\circ$. The data is developed by the Indian Meteorological

Department (IMD) using daily rainfall from 6,955 hydro-meteorological stations across India by employing Shepard's interpolation scheme [14].

The historical dataset is used for analysing the significant trends using established statistical tests such as the MK-test and EMD method, thereby providing a broad overview of precipitation. However, the primary goal is to develop a model using the gridded data for all the 13 grids that cover the Kamrup Rural district. The methods utilized in this study can be replicated for other areas in the region, and the models can be used as an early warning system by farmers in the region.

2.3 *Statistical Tests for Trend Analysis*

The statistical methods like the Mann–Kendall test, Sen's slope estimator, and Empirical Mode Decomposition, as described in the forthcoming paragraphs, were used for analysing the estimating the significant trend and magnitude of change in precipitation

Mann–Kendall Test

The rank-based non-parametric Mann–Kendall is highly appropriate for trend detection in hydrological variables for several reasons:

i. it does not require data to be normally distributed
ii. it supports multiple observations per time period,
iii. it allows missing values and censored observations in the time series and
iv. it is low sensitive to abrupt breaks due to inhomogeneous time series.

The MK-test is essentially limited to testing the null hypothesis that the data are independent and identically distributed [7, 12]. The test for trend in a series without specifying whether the trend is nonlinear or linear.

For a Given series x(t) with the length of n, the null hypothesis of no trend assumes that the series x(t) is independently distributed. It is based on the test statistic S given by Eq. 1:

$$S = \sum_{i=1}^{n-1} \sum_{j=i+1}^{n} \text{sgn}(x(j) - x(i)) \tag{1}$$

where,

$$\begin{aligned} sgn(x) &= 1\, if\ x > 0 \\ sgn(0) &= 0\, if\ S = 0 \\ sgn(x) &= -1\, if\ x < 0 \end{aligned} \tag{2}$$

A positive (negative) value of S indicates an upward (downward) trend. S is approximately normally distributed when $n > 8$, with the mean E(S) and variance Var (S) as follows:

$$E(S) = 0$$
$$\mathrm{Var}(S) = \frac{1}{18}\left[n(n-1)(2n+s) - \sum_{i=1}^{m} t_i i(i-1)(2i+5)\right] \quad (3)$$

where ti is the number of data in the tied group and m is the number of tied groups. The standardized test statistic Z follows the standard normal distribution:

$$Z = \begin{cases} (S-1)/\sqrt{\mathrm{Var}(S)} & \text{if } S > 0 \\ 0 & \text{if } S = 0 \\ (S+1)/\sqrt{\mathrm{Var}(S)} & \text{if } S < 0 \end{cases} \quad (4)$$

The null hypothesis of no trend gets rejected if the absolute value of Z is bigger than the theoretical value $Z_{1-\alpha/2}$; here, α is the concerned statistical significance level.

Sen's Slope Estimator

The MK test identifies the existence of a trend in a negative or positive direction. Trend strength is calculated using Sen's slope [20] as follows

$$\beta = \mathrm{Median}\left(\frac{Y_i - Y_j}{i - j}\right) \textit{for all } j < i \quad (5)$$

where Y_i and Y_j are data values at time steps i and j, respectively, the test statistic β denotes the median of all slope estimates. The negative value of β indicates a decreasing trend and a positive value indicates an increasing trend

Empirical Mode Decomposition and Trend Significance Testing

The significant advantage that EMD has over other time series decomposition is methods like Fourier Transform is that it has no underlying assumptions of the given time series data being stationary or linear.

However, for applying EMD decomposition on a time series data, the data is assumed to have the following characteristics [6]:

i. the series has at least two extrema (one maximum and one minimum)
ii. the characteristic time scale is defined by the time lapse between the extrema
iii. if the data were totally devoid of extrema but contained only inflection points, then it can be differentiated once or more times to reveal the extrema.

The EMD method decomposes an entire time series as the summation of so-called Intrinsic Mode Functions (IMF) along with a residue. Mathematically, this is represented as,

$$X(t) = \sum_{j=1}^{n} C_j + R_n \tag{6}$$

where X(t) is Time series, C_j are IMF components of X(t) and R_n is the residue of X(t).

The IMF components, C_j, are generated in an adaptive and iterative sifting process that continues until an end criterion is achieved. The end criterion in EMD is the generation of a monotonic signal, referred to as the residue, R_n.

In order to separate signal from noise, Wu and Huang [25] further studied the characteristics of white noise and identified a relationship between energy density and mean period of IMFs of white noise empirically using a Monte Carlo experiment. The relation identified is as follows,

$$E_n T_n = constant \tag{7}$$

where E_n is the energy density of nth IMF and T_n is the mean period of nth IMF.

They derived the energy distribution of the IMFs and the corresponding spread function, which can be used to obtain lines for various confidence levels. Based on these confidence levels, they proposed two methods, the "a priori" and "a posteriori" test, to identify IMFs with a higher probability of having some information. Yan-Fang [19], further investigated the latter of the two to identify the trend in a hydrological time series where the residue of the decomposed time series was assumed to be the trend component.

They compared the performance of the EMD method with the widely used MK test for finding trends in hydrological series. The MK tests were performed before and after data whitening to see if the performances varied. The results indicated that EMD is more powerful for trend identification and can be an effective alternative for trend identification of a series.

In this study, to find the trend of the rainfall time series, we first mean normalize the time series and decompose it into its IMF components and residue using EMD. On the residue, we applied the "a priori" statistical significance test with a 5% level of significance on the residue R_n, and if R_n fails the test, then the "a posteriori" test was employed. If the time series passed any of the tests, it was then decomposed using EMD, and the magnitude of the residue, which is taken to be the trend, is found using the Sen's Slope estimator. Figure 2 shows the methodology applied for statistical trend analysis in the form of a flowchart.

From the literature review, it was deduced that ANN models perform better compared to traditional statistical methods and the neural networks which are capable of sequence modelling can be utilized for time-series modelling. The section below describes the methods applied for developing a rainfall forecasting model

Long Short-Term Memory

A Long Short-Term Memory (LSTM) is a kind of a Recurrent Neural Network (RNN). An RNN has a recurrent hidden unit, where the output of a succeeding cell

Fig. 2 Methodology flowchart for statistical trend analysis

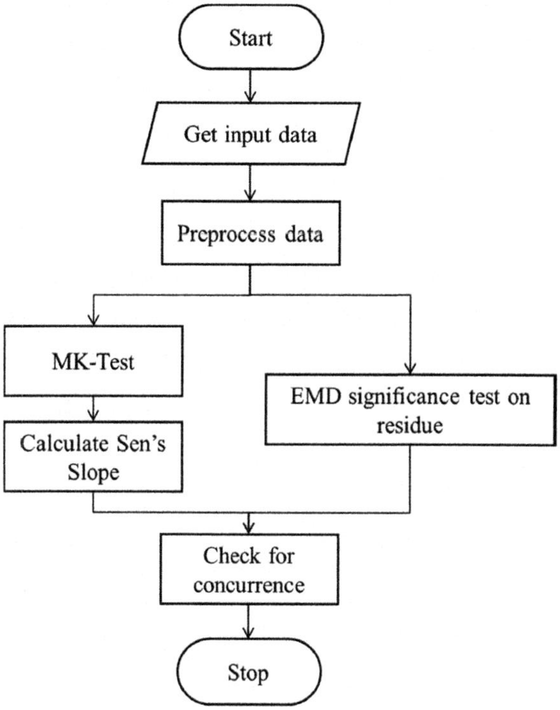

is given as input to the preceding cell in order to maintain information about all the past elements in the sequence, [3]. In an RNN model, a cell receives an input, x_t at the current cell and also another input from the hidden state, h_{t-1} of the previous cell. The RNN maps the inputs x_t and h_{t-1} to the current hidden state, h_t by applying a non-linear activation function which is then forwarded to the next RNN cell and can also be used as an output, y_t of the current cell. An illustration can be seen in Fig. 3. as to how information is forwarded in an RNN model.

$$h_t = f(U_h h_{t-1} + w_i x_t + b_h) \tag{8}$$

$$y_t = f(W_o h_t + b_o) \tag{9}$$

where, U_h, w_i, W_o are the respective weights, and b_o and b_h are the biases

RNNs are a powerful tool for sequence modelling, but they struggle with the problem of exploding or vanishing gradients. The problem was investigated, and a solution was proposed in the form of LSTM by Hochreiter and Schmidhuber [5]. They proposed a gated structure instead of the recurrent unit in the RNN cell. A number of modifications have since been proposed to further improve the LSTM model.

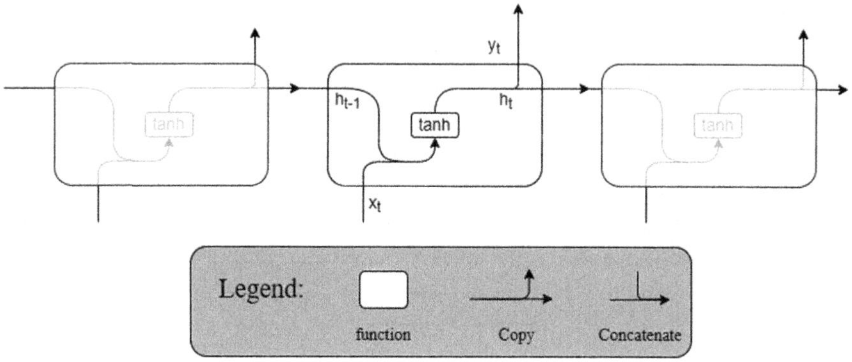

Fig. 3 Architecture of RNN cell

The core components of an LSTM are gates that control the flow of information or the cell state within the LSTM cell. The gates in an LSTM cell are:

i. Forget gate: The forget gate is a sigmoid activation that ensures all values are within "0" and "1". A value of "1" ensures that all the information from the previous state is kept, while "0" represents that the information is got rid off.
ii. Input gate: The input gate is similar to the forget gate as it is also a sigmoid activation. However, instead of using it to retain or remove information, it is used for adding any new information to the current cell state.
iii. Output gate: The output gate controls the flow of information to the next hidden state.

An illustration of LSTM can be seen in Fig. 4. From the Fig. 4. we have,

$$\text{Forget gate}, f_t = \sigma\left(W_f x_t + U_f h_{t-1} + b_f\right) \tag{10}$$

$$\text{Input gate}, i_t = \sigma(W_i x_t + U_i h_{t-1} + b_i) \tag{11}$$

$$\text{Potential Cell state}, c' = \tanh(W_{c'} x_t + U_{c'} h_{t-1} + b_{c'}) \tag{12}$$

$$\text{Cell state}, c_t = f_t \otimes c_{t-1} + i_t \otimes c' \tag{13}$$

$$\text{Output gate}, O_t = \sigma(W_o x_t + U_o h_{t-1} + b_o) \tag{14}$$

$$\text{Hidden state}, h_t = \tanh(c_t) \otimes O_t \tag{15}$$

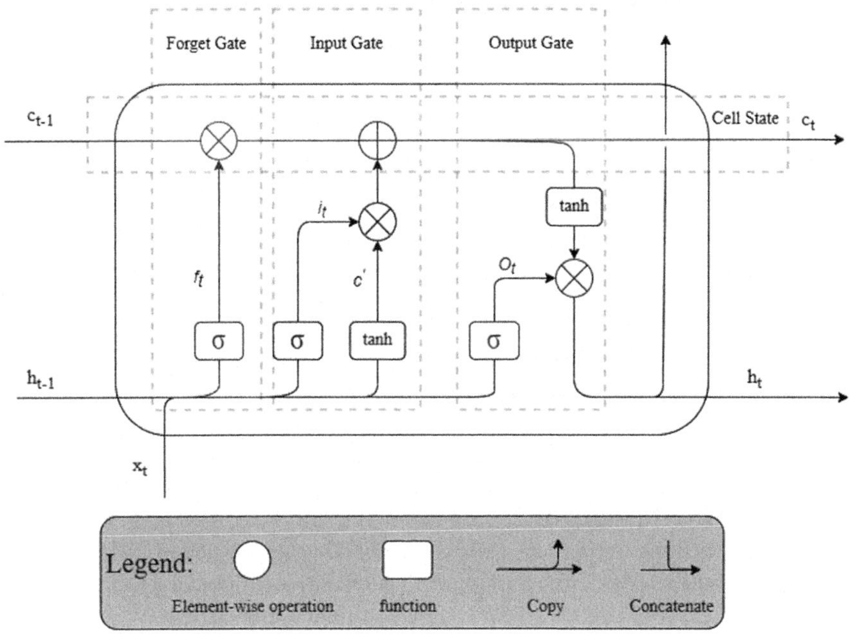

Fig. 4 Architecture of LSTM cell

where, σ(.) is the sigmoid function, tanh(.) is the hyperbolic tangent function and ⊗ denotes element-wise multiplication; the variables W_f, U_f, W_i, U_i, $W_{c'}$, $U_{c'}$, W_o, U_o, b_f, b_i, $b_{c'}$, b_o are the respective weights and biases that can be trained.

In this work, two methods were used to train models for each of the 13 grids that cover the entire Kamrup Rural district. In the first method, stacked LSTM was used, and the second set of models were trained using a stacked LSTM with an exception that the data is preprocessed using EMD with the IMF components as input feature vectors. The results of LSTM and EMD+LSTM methods are compared and presented in the report. Unlike various other sequence models utilizing EMD and similar decomposition methods [21, 26], where each IMF is trained using a separate LSTM and the final predictions are made by summing the results, a many-to-one LSTM model has been implemented as training multiple models for IMF components of all the grids will be more resource-heavy.

For both the methods, the time-series data were preprocessed using a square-root transformation which was shown to improve the predictability of forecasts in the Indian monsoon by Stephenson et al. [22]. Additionally, this also ensures that the predicted values are always greater than zero after the inverse transformation is applied to the predicted values. The square root transformation is followed by a min-max normalization technique as given in Eq. (16) to ensure that measurement units do not impact model training and to improve model efficiency and performance [4].

$$x' = \frac{x - x_{min}}{x_{max} - x_{min}} \tag{16}$$

where, x' is the normalized data, x is the original data, x_{max} is the maximum value in the data and x_{min} is the minimum value in the data.

Among multiple hyperparameters which can be tweaked to ensure that the model is properly trained, one key hyperparameter in sequence modelling is the time step to lookback. A well-known method, the Partial Auto-Correlation Function (PACF), was used to find the time steps to look back. The confidence level for selecting the significant lags was set to 99.95%. Though this method only considers linear correlation and rainfall may have a non-linear relationship with its lags, this is a well-known parsimonious method for selecting the time steps rather than arbitrary trial and error [9]. Figure 5 shows the PACF plot for location (91.0, 25.75), which is a representative PACF plot for all the other 13 locations.

For developing the model, the time-series data for each grid needs to be divided into training, validation and testing datasets. However, there is no thumb rule for data division, and it depends on the problem of interest [9]. In this work, for training, testing and validating the models, approximately 30 years of gridded data were used, and the first half was used for training, of which 30% was used for validation and the latter half i.e. 50 % was used for testing the trained model.

Once the data is preprocessed, it was then re-shaped as per the dimensions of the model. Additionally, the EMD+LSTM method has multiple input vectors which is equal to the number of IMF components generated by EMD. However, the number of IMF components generated by EMD is not fixed and depends upon the input data, so in order to ensure that the input data dimension is consistent, the last i IMF components are summed together such that the number of input feature vectors, say

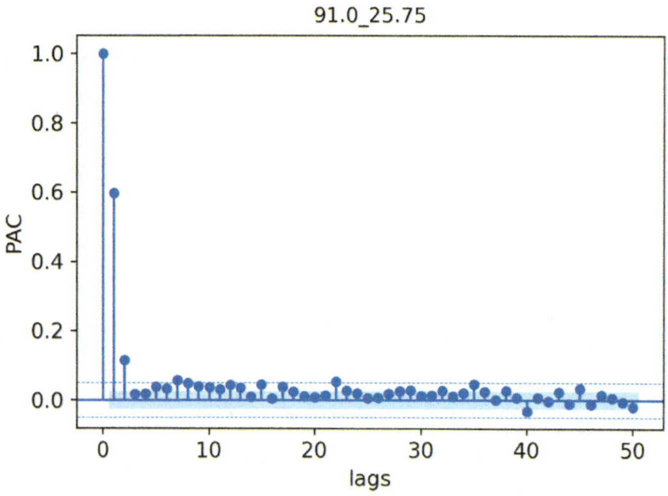

Fig. 5 PACF plot for location location (91.0, 25.75), time-steps = 23, Confidence level: 99.95%

n in testing and training data as well as during prediction is constant. The value of n is found by first decomposing both the training and testing dataset separately using EMD. If the decompositions generate unequal number of IMFs, then the decomposition where the number of IMF components is lesser, say m, is used to set the input dimension such that,

$$n = m - k \tag{17}$$

else if the decompositions generate equal number of IMFs, say p then,

$$n = p - k \tag{18}$$

The value of, k, used in this study is 2.

Additionally, the various other hyper-parameters and configurations used for training the model are given in Tables 1 and 2, respectively.

Figure 6 shows the flowchart used for training and testing the models for both the LSTM and the EMD+LSTM model.

Table 1 Hyper-parameters for LSTM

Sl. No.	Hyper-parameter	Value
1	Time-steps	As found by PACF
2	Number of LSTM layers	2
3	LSTM units in layer 1	175
4	LSTM units in layer 2	Number of input IMFs
5	Number of units in dense layer	1
6	Activation function	Leaky Relu, alpha: 0.01
7	Learning rate	0.001
8	Number of epochs	10,000
9	Batch size	Training size/50

Table 2 Configurations for training LSTM

Sl. No.	Configurations	Value
1	Early stopping	Monitor: Validation Loss
		Patience: 40 epochs
		Minimum change: 0.001
		Restore best weights: True
2	Reduce learning rate	Factor: 0.4
		Patience: 15
		Cooldown: 5
3	Optimizer	Adam
4	Loss function	Mean absolute error

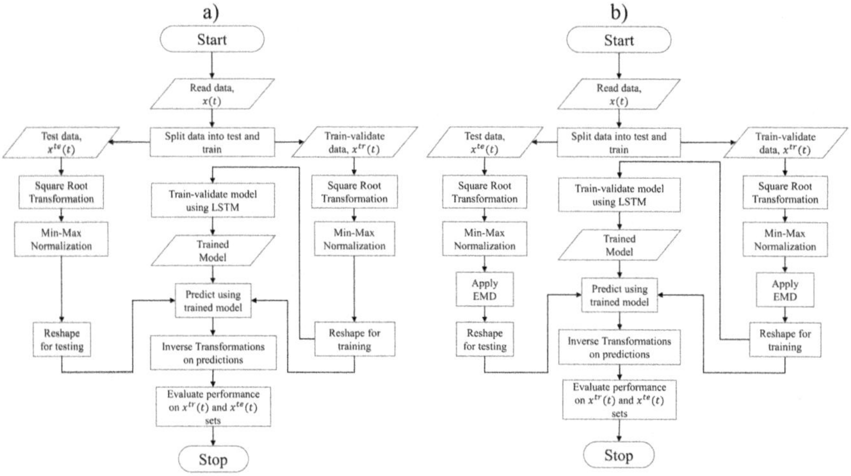

Fig. 6 **a** Flowchart for training stacked LSTM, **b** Flowchart for training EMD+ many-to-one stacked LSTM

Finally, for determining the goodness of the model, the following performance metrics were utilized,

i. *Coefficient of determination*,

$$R^2 = 1 - \frac{\sum_{j=1}^{n}\left(y_j - \hat{y}_j\right)^2}{\sum_{j=1}^{n}\left(y_j - \overline{y}_j\right)^2} \tag{19}$$

where, y_j is the actual value, $\hat{y}_j$ is the model predicted value, $\overline{y}_j$ is the mean value

ii. *Root Mean Square Error*,

$$RMSE = \sqrt{\frac{1}{n}\sum_{j=1}^{n}\left(y_j - \hat{y}_j\right)^2} \tag{20}$$

where, y_j is the actual value, $\hat{y}_j$ is the model predicted value

iii. *Mean Absolute Error*,

$$MAE = \frac{1}{n}\sum_{j=1}^{n}\left|y_j - \hat{y}_j\right| \tag{21}$$

where, y_j is the actual value, $\hat{y}_j$ is the model predicted value.

Additionally, the Day-1 predictions for the year 2021 were evaluated using the following metrics which are used by IMD to measure the quality of forecasts as per the SOP for NWP 2021,

iv. *Pearson's Correlation Coefficient, r*
v. *RMSE*

3 Results and Discussions

The sub-sections below show the results and the insights which were obtained by applying the methods on the available data. The first section shows the results of the trend analysis followed by the results of the rainfall forecasting.

3.1 Trend Analysis

From Fig. 7, we can see that out of the 13 grids covering the Kamrup Rural district, 9 are showing a negative trend, and 4 grids show a positive trend, i.e. 69.23% of the locations show a negative trend. The magnitude for the negative trend lies between −1.172 and −3.448 in 6 locations, of which one is found to be significant by both MK-test and EMD method; one is significant by MK-test and non-significant by EMD method and another where MK-test shows non-significant, and EMD method shows significant, the rest of the three locations were found non-significant by both MK-test and EMD method. The remaining 3 locations which show a negative trend, have magnitudes in the range of −1.172–0.000; however, all are found non-significant by both MK-test and EMD methods. Similarly, for trends showing a positive trend, 3 locations have trend magnitude in the range of 0.000–0.336, of which one is found to be significant using the EMD method and non-significant by the MK-test. The remaining two locations are found to be non-significant by both MK-test and EMD methods. Lastly, there is one location that has a trend magnitude in the range of 0.336–3.190. However, it is found to be non-significant by both MK-test and EMD methods.

3.2 Rainfall Forecasting

From Figs. 8 and 9, we can see that for both the EMD+LSTM model and the LSTM, the validation loss follows the training loss for a certain number of epochs, after which the training loss keeps on decreasing, but no further improvement is seen in the validation loss, or in some cases, the validation loss instead starts increasing, and the model stops the training process as per the configuration.

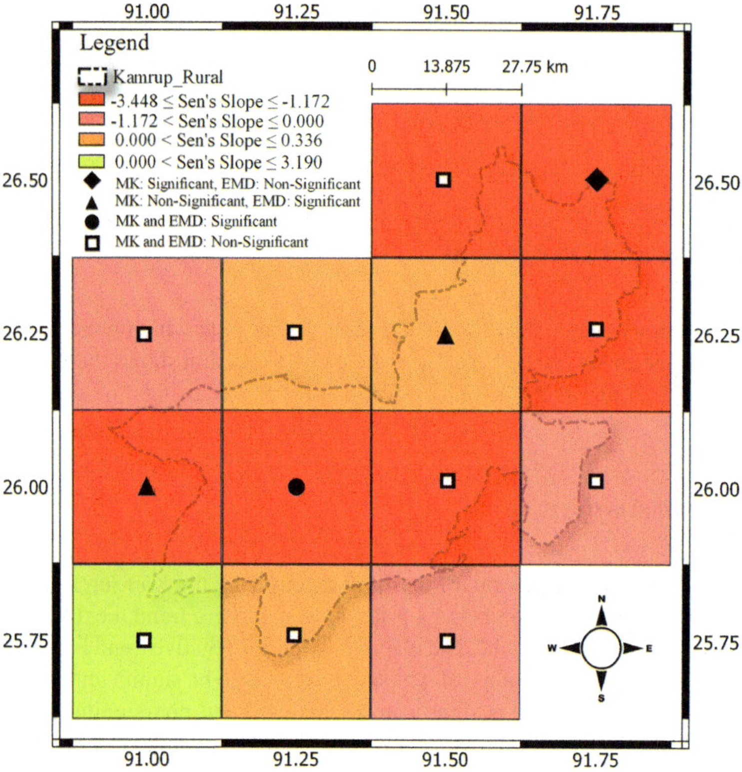

Fig. 7 Trend analysis of rainfall for Kamrup rural district

The results in Fig. 10. show that the R^2 values for testing and training data for all the locations are consistently higher for the EMD+LSTM method as compared to plain LSTM. Also, the testing and training R^2 have similar values in almost all locations for EMD+LSTM, showing that the model has been appropriately trained. Similarly, the RMSE and MAE values are consistently lesser for the EMD+LSTM method as compared to the plain LSTM method.

Once the model is trained using the EMD+LSTM model, Day-1 predictions are made for the year 2021 for all the 13 locations covering the districts. The predicted values were compared with the gridded data, and the compiled results are shown in Fig. 11. The figure shows that only two locations have correlation coefficient values greater than 0.4, which is considered to be good for precipitation forecast as per IMD. However, among the two, one has an RMSE value greater than 20, indicating high error for prediction. 6 locations have correlations between 0.2 and 0.4 with RMSE values in the range 8.306–13.126, and 5 locations have correlation values in the range of 0.1–0.2 with RMSE values in range 9.824–16.811.

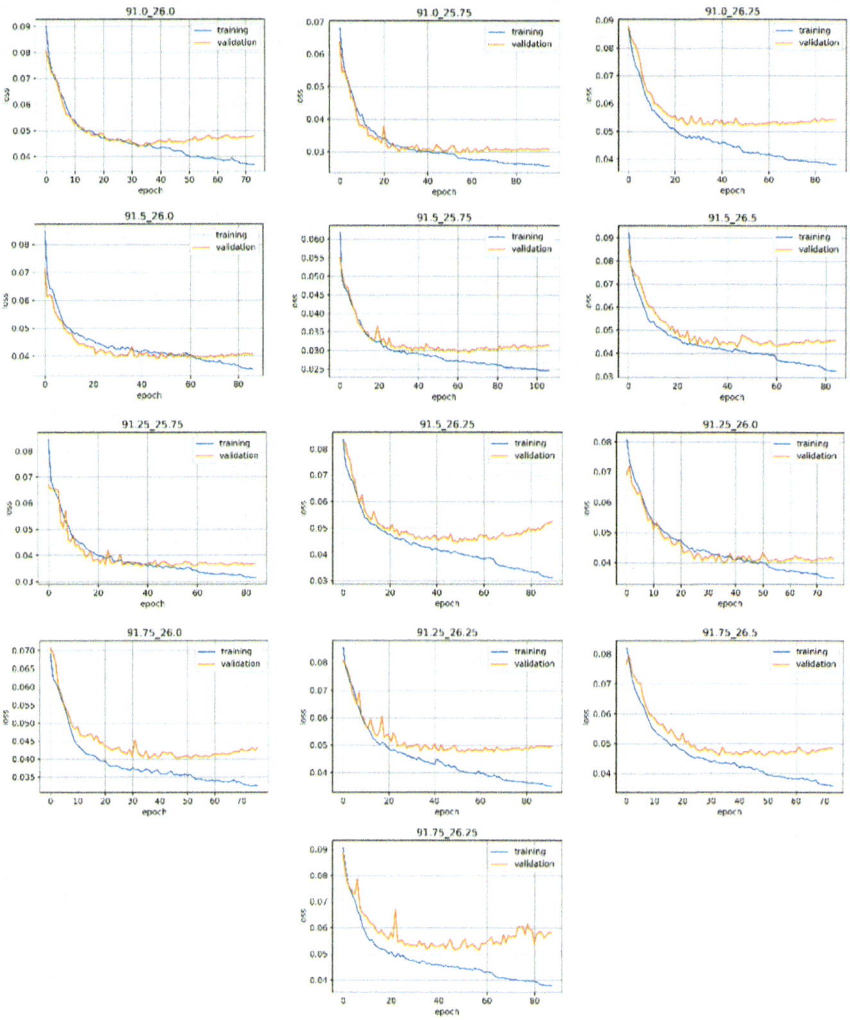

Fig. 8 Loss versus epoch graph while training the model using EMD+LSTM

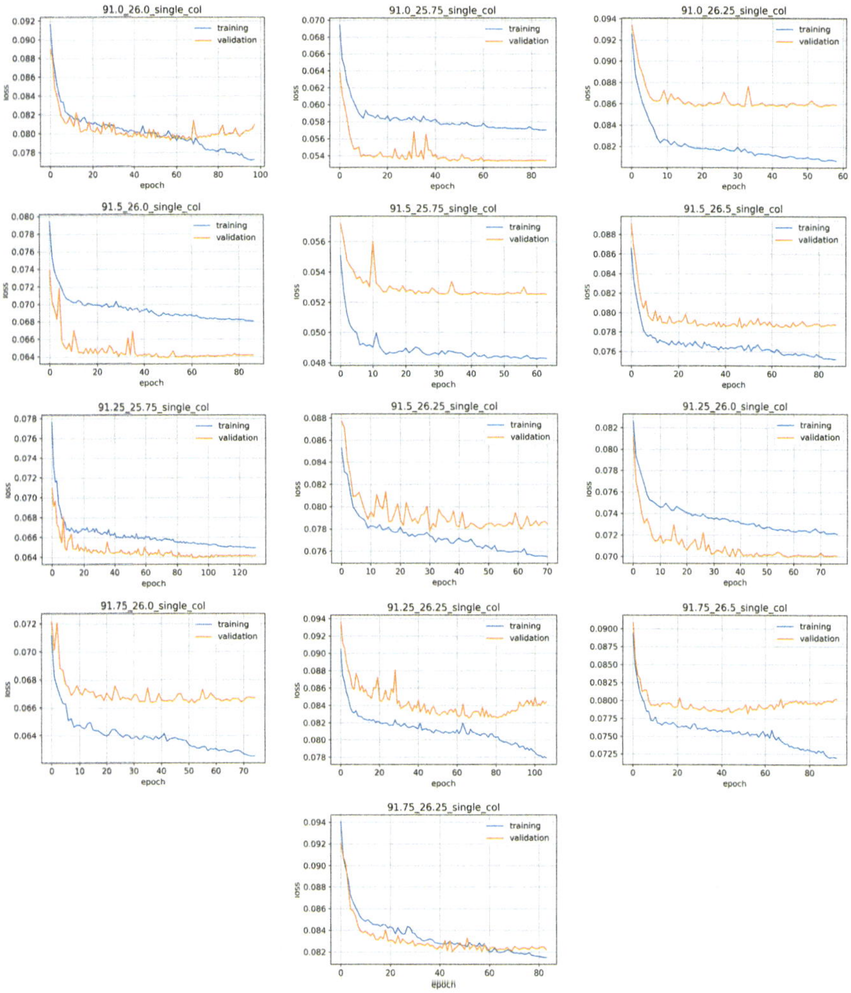

Fig. 9 Loss versus epoch graph while training the model using only LSTM

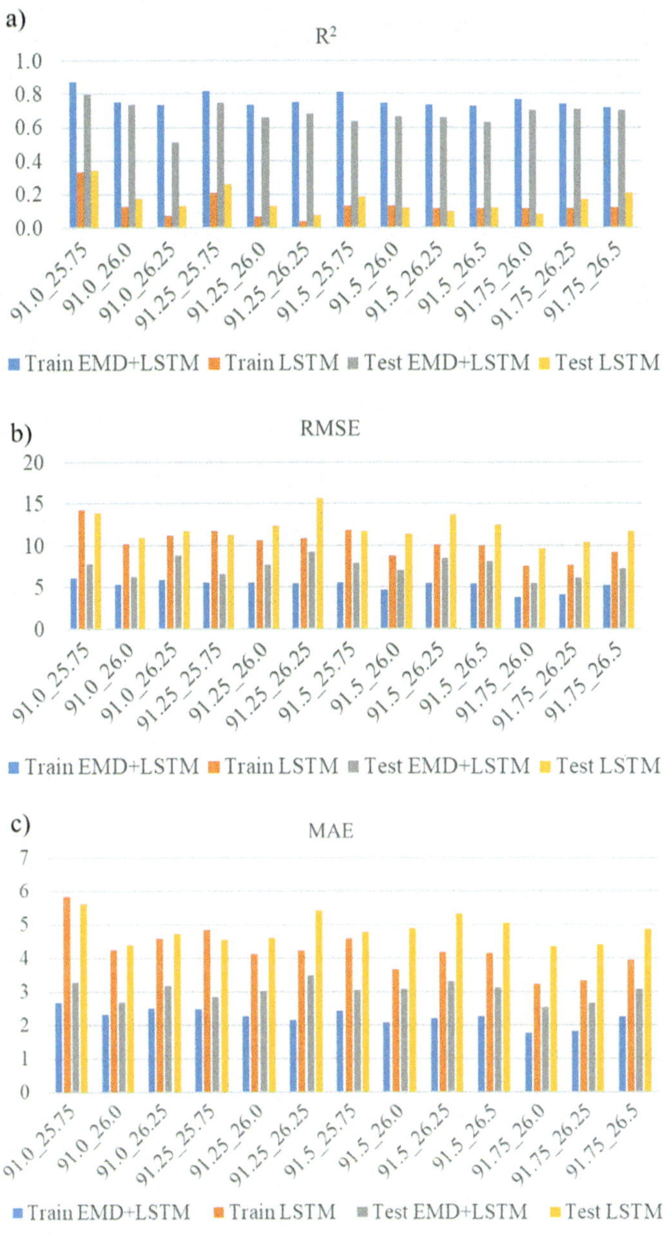

Fig. 10 The figures show the various metric values for all 13 locations **a** R^2 values for test and train dataset for EMD and EMD+LSTM model, **b** RMSE for test and train dataset for EMD and EMD+LSTM model, **c** MAE for test and train dataset for EMD and EMD+LSTM model

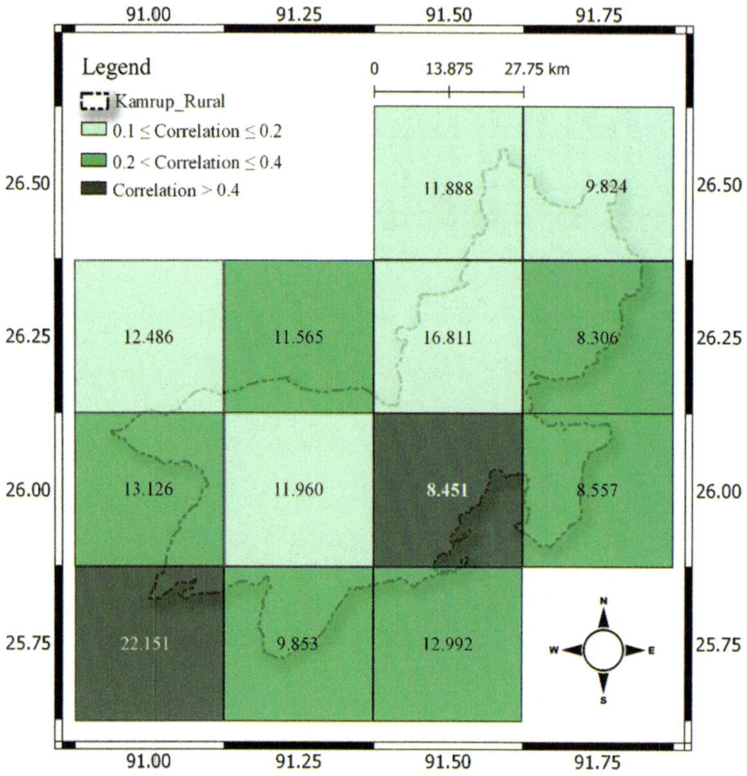

Fig. 11 Day-1 predictions for the year 2021 using trained models. The values in each grid show the RMSE in the respective grid, and the colours represent the Pearson's correlation coefficient

4 Conclusion

Overall, the trend analysis results for the Kamrup Rural district show that there is a decreasing trend in the majority of the locations, with the highest negative trend magnitude being 3.448. However, of the 13 locations, only 3 locations show a significant decreasing trend using either MK-test and EMD method or both, along with one location with a positive trend.

The results of models generated for rainfall forecasting indicate that the EMD+LSTM method performs better in comparison to plain LSTM over metrics such as R^2, RMSE, and MAE. The results of the Day-1 forecasts using the trained models for the year 2021 show that 2 of the 13 locations, i.e. 15.38% of the locations, performed good over Pearson's Correlation Coefficient and has a correlation value over 0.4, one among which also has an RMSE value of 8.451. However, one among them has an RMSE value of over 20. Additionally, 46.15% of the locations have correlation values in the range of 0.2–0.4, and the remaining 38.46% have correlation values in the range of 0.1–0.2. The RMSE values for all locations with a correlation between 0.1 and 0.4 lie below 15 except for one where the RMSE value is 16.811.

The improvement in model training for EMD+LSTM may be due to the fact that the feature vectors used in this method are IMF components of the original time series, which reflects series characteristics under different frequencies or temporal scales [8], thereby allowing a better fit for the seasonal cycles, periodicities, trend component and the inherent noise in hydrological data as opposed to the plain LSTM which uses the original time-series as input and might have struggled to fit the inherent periodicities, noise and nonlinearities in the data.

Acknowledgements The authors would like to thank the School of Agro and Rural Technology, Indian Institute of Technology Guwahati, and Centre for Disaster Management and Research, Indian Institute of Technology Guwahati, for all kinds of logistical support to carry out the work. The authors are thankful to the Indian Meteorological Department (IMD) for the precipitation data. Arnab Paul Choudhury and Debaditya Gupta would like to thank the Ministry of Education (MoE), Government of India, for providing them with the fellowship.

References

1. Chhetri M, Kumar S, Pratim Roy P, Kim B-G (2020) Deep BLSTM-GRU model for monthly rainfall prediction: a case study of Simtokha, Bhutan. . Remote Sens 12(19):3174. https://doi.org/10.3390/rs12193174
2. Diop L, Samadianfard S, Bodian A, Yaseen ZM, Ghorbani MA, Salimi H (2020) Annual rainfall forecasting using hybrid artificial intelligence model: integration of multilayer perceptron with whale optimization algorithm. Water Resour Manag 34(2):733–746. https://doi.org/10.1007/s11269-019-02473-8
3. Elman JL (1990) Finding structure in time. Cogn Sci 14(2):179–211. https://doi.org/10.1207/s15516709cog1402_1
4. Feng R, Fan G, Lin J, Yao B, Guo Q (2021) Enhanced long short-term memory model for runoff prediction. J Hydrol Eng 26(2):04020063. https://doi.org/10.1061/(asce)he.1943-5584.0002035
5. Hochreiter S, Schmidhuber J (1997) Long short-term memory. Neural Comput 9(8):1735–1780. https://doi.org/10.1162/neco.1997.9.8.1735
6. Huang NE, Shen Z, Long SR, Wu MC, Shih HH, Zheng Q, Yen N-C, Tung CC, Liu HH (1998) The empirical mode decomposition and the Hilbert spectrum for nonlinear and non-stationary time series analysis. Proc Royal Soc London. Series A: Math, Phys Eng Sci **454**(1971):903–995. https://doi.org/10.1098/rspa.1998.0193
7. Kendall MG (1975) Rank correlation methods, Charles Griffin, London
8. Kim D, Kim KO, Oh H-S (2012) Extending the scope of empirical mode decomposition by smoothing. EURASIP J Adv Signal Process 2012(1). https://doi.org/10.1186/1687-6180-2012-168
9. Kumar D, Singh A, Samui P, Jha RK (2019) Forecasting monthly precipitation using sequential modelling. Hydrol Sci J 64(6):690–700. https://doi.org/10.1080/02626667.2019.1595624
10. Lei J, Quan Q, Li P, Yan D (2021) Research on monthly precipitation prediction based on the least square support vector machine with multi-factor integration. Atmosphere 12(8):1076. https://doi.org/10.3390/atmos12081076

11. Li W, Gao X, Hao Z, Sun R (2021) Using deep learning for precipitation forecasting based on spatio-temporal information: a case study. Clim Dyn 58(1–2):443–457. https://doi.org/10.1007/s00382-021-05916-4
12. Mann HB (1945) Nonparametric tests against trend. Econometrica 13(3):245. https://doi.org/10.2307/1907187
13. Ouyang Q, Lu W, Xin X, Zhang Y, Cheng W, Yu T (2016) Monthly rainfall forecasting using EEMD-SVR based on phase-space reconstruction. Water Resour Manag 30(7):2311–2325. https://doi.org/10.1007/s11269-016-1288-8
14. Pai DS, Rajeevan M, Sreejith OP, Mukhopadhyay B, Satbha NS (2021) Development of a new high spatial resolution (0.25° × 0.25°) long period (1901–2010) daily gridded rainfall data set over India and its comparison with existing data sets over the region. MAUSAM 65(1):1–18. https://doi.org/10.54302/mausam.v65i1.851
15. Peng Y, Gong D, Deng C, Li H, Cai H, Zhang H (2021) An automatic hyperparameter optimization DNN model for precipitation prediction. Appl Intell 52(3):2703–2719. https://doi.org/10.1007/s10489-021-02507-y
16. Pham BT, Le LM, Le T-T, Bui K-TT, Le VM, Ly H-B, Prakash I (2020) Development of advanced artificial intelligence models for daily rainfall prediction. Atmos Res 237:104845. https://doi.org/10.1016/j.atmosres.2020.104845
17. Ponnoprat D (2021) Short-term daily precipitation forecasting with seasonally-integrated autoencoder. Appl Soft Comput 102:107083. https://doi.org/10.1016/j.asoc.2021.107083
18. Raval M, Sivashanmugam P, Pham V, Gohel H, Kaushik A, Wan Y (2021) Automated predictive analytics tool for rainfall forecasting. Sci Reports 11(1):17704. https://doi.org/10.1038/s41598-021-95735-8
19. Sang Y-F, Wang Z, Liu C (2014) Comparison of the MK test and EMD method for trend identification in hydrological time series. J Hydrol 510:293–298. https://doi.org/10.1016/j.jhydrol.2013.12.039
20. Sen PK (1968) Estimates of the regression coefficient based on Kendall's Tau. J Am Stat Assoc 63(324):1379–1389. https://doi.org/10.1080/01621459.1968.10480934
21. Song C, Chen X (2021) Performance comparison of machine learning models for annual precipitation prediction using different decomposition methods. Remote Sens 13(5):1018. https://doi.org/10.3390/rs13051018
22. Stephenson DB, Kumar KR, Doblas-Reyes FJ, Royer J-F, Chauvin F, Pezzulli S (1999) Extreme daily rainfall events and their impact on ensemble forecasts of the indian monsoon. Mon Weather Rev 127(9):1954–1966. https://doi.org/10.1175/1520-0493(1999)127%3c1954:edreat%3e2.0.co;2
23. Tao L, He X, Li J, Yang D (2021) A multiscale long short-term memory model with attention mechanism for improving monthly precipitation prediction. J Hydrol 602:126815. https://doi.org/10.1016/j.jhydrol.2021.126815
24. Venkatesh R, Balasubramanian C, Kaliappan M (2021) Rainfall prediction using generative adversarial networks with convolution neural network. Soft Comput. https://doi.org/10.1007/s00500-020-05480-9
25. Wu Z, Huang NE (2004) A study of the characteristics of white noise using the empirical mode decomposition method. Proc Royal Soc London. Series A: Math, Phys Eng Sci 460(2046):1597–1611. https://doi.org/10.1098/rspa.2003.1221
26. Zhang X, Wu X, He S, Zhao D (2021) Precipitation forecast based on CEEMD–LSTM coupled model. Water Supply. https://doi.org/10.2166/ws.2021.237

Role of Climatic Variables on Forest Fire in the State of Mizoram

Dhruval Bhavsar, Kasturi Chakraborty, Jakesh Mohapatra, K. K. Sarma, and S. P. Aggarwal

Abstract In a view of increasing fire incidences and risk of climate change on the ecosystem, current study has been carried out in Mizoram state of north east India which is one of the fire-prone states in the country. The study is carried out to analyse the effect of climatic variables (precipitation, temperature and humidity) on forest fire occurrence using trend analysis (MK test) and spatial relation between the two. Active fire points (MODIS) from 2003 to 2020 and gridded climate data (ERA 5 & FLDAS) were used for spatial analysis. Significant (at 95% confidence level) trend analysis showed an increase in temperature (1.97–3.03) and humidity (1.98–3.18) in monsoon and post-monsoon seasons. Precipitation was observed to decrease in September for Lawngtlai (−2.05) and Mamit (−1.97). Overall forest fire occurrence in the state was found to be decreasing while spatial analysis over the state showed an increasing trend in parts of Mamit, Lunglei and Lawngtlai districts. The same region was found to have negative relation for temperature in winter season while rest of the state had positive relation with fire occurrence. Precipitation and humidity both were having negative relation to forest fire occurrence with exception of areas mentioned earlier, with positive relation in winter, premonsoon (precipitation) and monsoon seasons (humidity). Despite the major role of anthropogenic activities in forest fire in the state, the variability in results suggests that forest fire occurrence may be due to the response of climatic conditions, but it is restricted to certain areas.

Keywords Forest fire · Temperature · Precipitation · Humidity · Climate trend

D. Bhavsar (✉) · K. Chakraborty · J. Mohapatra · K. K. Sarma · S. P. Aggarwal
North Eastern Space Applications Centre, Government of India, Department of Space, Umiam, Meghalaya, India
e-mail: bhavsardhruval@gmail.com

S. Mitra et al. (eds.), *Disaster Management and Risk Reduction: Multidisciplinary Perspectives and Approaches in the Indian Context*,
https://doi.org/10.1007/978-981-99-6395-9_20

1 Introduction

Forest fire is a natural component of forest ecosystems but becomes dangerous when uncontrolled. Recently, the frequency and intensity of forest fires have increased throughout the world and hence, have become a threat to the biodiversity and lives of people living in the vicinity of forests. While fire can be important for maintaining ecosystems processes and vegetation in some ecosystems, e.g. savannahs and grasslands [45], it is now occurring more frequently in ecosystems that rarely experience large fires (e.g. tropical forests). An increase in the frequency and intensity of wildfires [48] especially in ecosystems that are not adapted to fire, could lead to lasting changes in vegetation structure and composition [10, 35], wildlife populations [5, 31], soil erosion [39] soil nutrient status [7], atmospheric chemistry and biogeochemical cycles [27] and the benefits humans derive from these landscapes [47]. Precious forest resources including carbon locked in the biomass are lost due to forest fires every year, which adversely impact the flow of goods and services from forests.

Globally, only around 4% of all forest fires have natural causes such as lightning. In the remaining cases, human activities are responsible for the fires directly or indirectly [19]. However, the changes in patterns and increasing intensity of fires are attributed to two factors: climate change-driven changes in extremes [13, 26] and widespread land use change [3, 4, 12], including and especially increasing anthropogenic ignitions [6, 9]. The interaction between climate change and forest fire is like a feedback system. As climate change has an impact on forest fire, the forest fire may also have an impact on climate change. There is mounting evidence that forest fires will increase in number and size due to different impacts of climate change such as increased frequency and intensity of El Niño due to an increase in temperature [29, 43] and climate stress [49]. Effect of trace gas emissions in forest fire have also been well documented on long-term climatic response [42]. High summer temperature, high wind velocity, steep terrain and availability of inflammable material in the forest account for major damage and extensive spread of forest fire [17]. The combination of these factors can cause rapid and intense fire. Such wildfires have become increasingly common in some parts of the world in recent past—California (2017 and 2018), Canada (2014, 2017 and 2018), the Mediterranean (2017 and 2018), Siberia (2003, 2012 and 2019), Australia (2009, 2013 and 2019) and India (2016, 2018 and 2020). The observations from 1987 to 2007 show that the increasing intensity and spread of forest fires in Asia were largely related to rise in temperature and decline in precipitation in combination with increasing intensity of land uses [21].

In India, forest fire is a major cause of forest degradation and particularly in tropical forests as it is a recurrent phenomenon [2]. Based on the forest inventory records by Forest Survey of India (FSI), 54.40% of forests in India are exposed to occasional fires, 7.49% to moderately frequent fires and 2.40% to high incidence levels, while 35.71% of India's forests have not yet been exposed to fires of any real significance [14]. Overall, 22.27% of the total forest cover of India is prone to forest fire from high to extreme level [22]. The Himalayan region has also reported increased fire frequency due to long dry spells of winter and increased temperature as a result of climate change [1]. An analysis on loss from forest fire in India was estimated to be Rs. 440 crore annually [8]. Incidents of fires in Indian forests are increasingly threatening the valuable biodiversity and genetic resources [37].

North East India, one of the biodiversity hotspots of the world, contributes significantly (about 16%) to the total fire occurrence of the country [22]. The tribes of northeast India practicing jhum (shifting cultivation) have unique and varied community traditions. In practice forest is cut and then burned for agriculture which is also called as slash and burn agriculture. Jhum is the major factor for forest fire in the region. Within the north eastern states of India, Mizoram is highly prone to forest fire. Around 76% of the forest cover of Mizoram is under very high to extreme fire-prone category which is highest in the country [22]. Climate change study on Mizoram observed an annual rise in temperature by 0.01 °C since 1951 to 2010. Rainfall in Mizoram during May–August was found to be increasing while during the other months in a year showed a decreasing rainfall departure trend [20]. Earlier studies on climate change and forest fire are either based on non-spatial data or limited to only trend analysis. This gap in the information guided to carry out the present study in the state of Mizoram.

The present study is aimed to analyse the effect of climatic factors (precipitation, temperature and humidity) variation on forest fire occurrence. Spatial trend of climate variable was carried out to understand the distribution pattern. The relation between forest fire occurrence and climate variable was also estimated to understand role of climate in forest fire.

The chapter encompasses different sections where location and description of the study is presented in study area. Material and methodology section describes the source of data and procedure followed to develop relation between forest fire and climate data. In the next section results of trend analysis, spatial relation was presented. The inference of the result is explained in discussion and conclusion section.

2 Study Area

Mizoram has been considered as a study area which is a part of north eastern states of India. It is southernmost landlocked state of north east, sharing borders with Tripura, Assam and Manipur. The state also has 722 km border with neighbouring countries of Bangladesh and Myanmar. The geographical location of Mizoram lies between East Longitude 92°15′ and 93°29′ and North Latitude 21°58′ to 24°35′ and its total geographical area is 21,087 km^2 (Fig. 1). Geography of Mizoram consists hills, valleys, rivers and lakes. Mizoram has a mild climate, comfortable in summer 20 °C to 29 °C and never freezing during winter, with temperature from 7 °C to 21 °C. The region is influenced by monsoons, raining heavily from May to September with little rain in the dry or cold season. The average annual rainfall is 2540 mm.

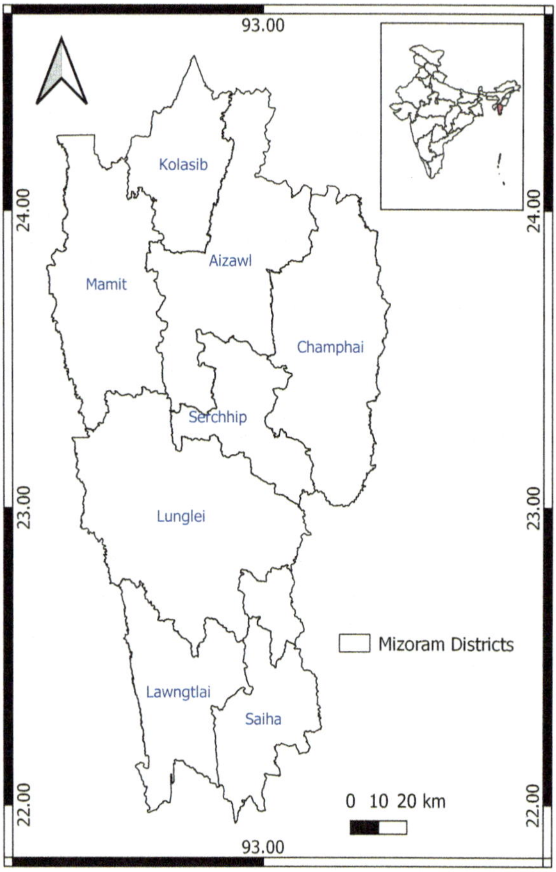

Fig. 1 Geographic location of study area

Forest is one of the most important landuse in Mizoram occupying over 86.27% geographical area of the state. Major forest type in the state include: Tropical semi evergreen, Tropical moist deciduous, Subtropical broadleaved hill, and subtropical pine forest. The Mizos, living in Mizoram, are highly dependent on the forest resources. The main pattern of agriculture followed is Jhum in the state. Jhum is an integral part of socio-cultural life of Mizos. Mizoram is the 2nd least populous state in the country. However, the state is highly literate agrarian economy, where about 80% of the people of Mizoram are engaged in agricultural pursuits. With increase in population, considerable pressure has increased on forest due to agricultural need. This has created a major challenge for the Mizo society to manage fire activities [11]. Mizoram recorded highest number of fire occurrence (12,846) within the north eastern states from the November, 2020 to June, 2021 [22]. The state is also having 76% of its geographical area under high to very high risk of getting forest fire [22] which is highest in the country. Majority of forest area in Mizoram is being converted to the bamboo due to jhum which has become more susceptible of getting fire due to hot temperature and low humidity. With recent climate change and global warming scenario the risk of forest fire has increased. Hence, study on fire and climate relation becomes very important in Mizoram.

3 Material and Methodology

Climate variables and fire occurrence data are the requirement for the present study. Acquisition and processing of these dataset to fulfil the objective of the study is given in subsequent section.

3.1 Material

The study is focused on to identify the role of climate on the forest fire occurrence in the study area at spatial level. Three important parameters important for forest fire occurrence and intensity are temperature, precipitation and humidity. Spatial gridded data of ERA 5 (ECMWF reanalysis v5) data was used for the temperature and precipitation. Specific humidity of FLDAS (FEWS NET Land Data Assimilation System) data was used in the study. Both the datasets have spatial resolution of 0.1° which is the highest available for the study area. Since Mizoram has undulating terrain and geography which impact the changes in the climatic condition, use of such dataset was helpful to analyse the significant spatial variation among the climate variables.

The present study utilizes climate data in addition to the forest fire occurrence. The forest fire occurrence was acquired from Forest Survey of India which is an organization under MoEFCC, GoI, which distributes forest fire alerts all over the India. The historical data for forest fire is available from year 2003 generated from MODIS satellite imagery. This caused to limit the present study for 18 years (2003–2020). The information about the forest fire points and its availability is also presented in Sect. 3.1 Material.

Fire occurrence is one of the most essential data for the present study. Forest Survey of India (FSI) generates daily fire alerts, which was downloaded from its website [15]. Fire alerts are generated based on two satellite data viz. Moderate Resolution Imaging Spectroradiometer (MODIS) and Suomi National Polar-orbiting Partnership (SNPP) Visible Infrared Imaging Radiometer Suite (VIIRS). SNPP-VIIRS has higher spatial resolution (750 m) and provides more number of fire alerts compared to MODIS (1000 m). However, VIIRS data is available from year 2018 which cannot be utilised in the study which is dependent on the information of preceding years for better understanding of the trend and relation. MODIS data is available from year 2003 and was used in the present study. The climate data and forest fire occurrence data was used from year 2003 to 2020 to understand forest fire relation with climate.

3.2 Methodology

Methodology involves the processing of climatic data and fire points which is to make the data homogeneous with space and time. Various steps involved in the study are shown in Fig. 2. The details of each step are as follows:

Climate Data Preparation. The source of data is mentioned in the material Sect 3.1 "Spatial gridded data of ERA 5 (ECMWF reanalysis v5) data was used for the temperature and precipitation. Specific humidity of FLDAS (FEWS NET Land Data Assimilation System) data was used in the study.""Fire occurrence isone of the most essential data for the present study. Forest Survey of India (FSI) generates daily fire alerts, which was downloaded from its website (http://117.239.115.41/smsalerts/index.php)." The data acquired for different climatic variable was in the netCDF (Network Common Data Form) format and to process the data, open source Python platform (Anaconda) was used. Climate data was acquired in monthly accumulated product. For temperature and humidity monthly average product and for precipitation, total rainfall during the month was used. These entire dataset was arranged and grouped for each year in different categories viz. district wise and state as a whole in spatial domain. For temporal domain, monthly and seasonal category of each year was used. The season used in the study was followed as per Indian Meteorological Department (IMD) nomenclature i.e. winter (January and February), premonsoon (March, April and May), monsoon (June, July, August and September) and post-monsoon (October, November and December). Trend analysis was carried out in each group for all the climate variables.

Fire Occurrence Data Preparation. Fire occurrence data was available in the tabular format with information of time, day and location of fire occurrence. This information was converted into the point data (shapefile) using QGIS software. Further to analyse

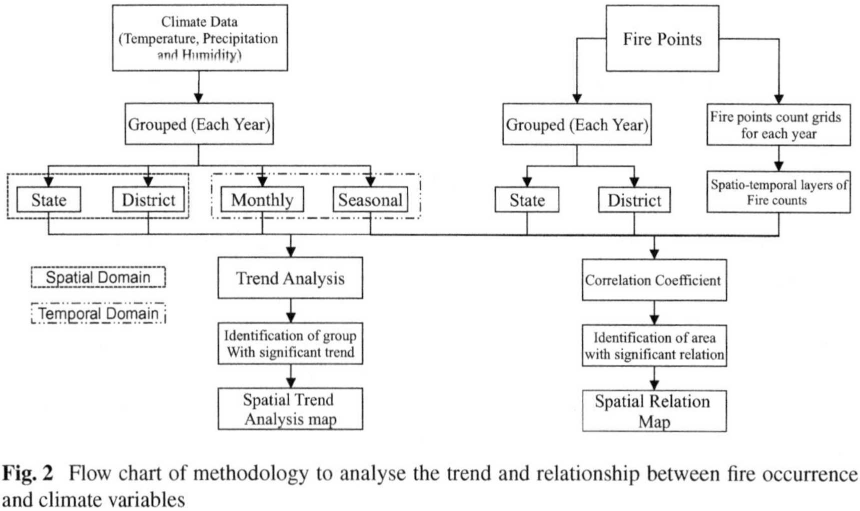

Fig. 2 Flow chart of methodology to analyse the trend and relationship between fire occurrence and climate variables

and make it compatible with climate data, point data was converted to gridded data. The grid size was chosen to be 0.1° which is same as of the climate data spatial resolution. The grids were created using the extent of Mizoram. The value of each grid was computed as total number of fire occurrence in each year. The process was done using spatial analysis tool in QGIS. The outcome of the process generates sptio-temporal data of fire occurrence in Mizoram from year 2003 to 2020.

Trend Analysis. Trends in the data can be identified using parametric and non-parametric methods. Temperature, rainfall and humidity are the non-parametric climatic variables. The non-parametric method does not require normality of time series and less sensitive to outliers [25]. In the present study, Mann–Kendall (MK) test [30, 36] was used for identifying trends in the climatic variable which is one of the most commonly used methods [18], is based on the test statistics, S, is defined as follows:

$$S = \sum_{k=1}^{n-1} \sum_{j=k+1}^{n} sgn\left(X_j - X_k\right) \quad (1)$$

where, n is the number of observations and X_j is the jth observation and $sgn(.)$ is the sign function which can be defined as

$$sgn(\theta) = \begin{matrix} 1 & if\ \theta > 0 \\ 0 & if\ \theta = 0 \\ -1 & if\ \theta < 0 \end{matrix} \quad (2)$$

The variance of S can be computed as follows:

$$VAR(S) = \frac{1}{18}\left[n(-1)(2n+5) - \sum_{p=1}^{g} t_p\left(t_p - 1\right)\left(2t_p + 5\right)\right] \quad (3)$$

where, g is the number of tied groups and t_p is the number of observations in the pth group. MK test statistic, Z can be computed as

$$Z = \begin{matrix} \frac{S-1}{\sqrt{VAR(S)}}\ if\ S > 0 \\ 0\ if\ S = 0 \\ \frac{S+1}{\sqrt{VAR(S)}}\ if\ S < 0 \end{matrix} \quad (4)$$

A positive or negative value of Z indicates the data tends to increase or decrease with time. At 5% level of significance, if the value of MK test statistics lies within the limits of -1.96 to $+1.96$, then the null hypothesis is accepted and there is no trend in the data. If the value lies beyond the above-mentioned range, the null hypothesis is rejected. If $Z > 1.96$, there is an increasing trend and if $Z < -1.96$, there is decreasing trend. To analyse the magnitude of the increasing and decreasing trend, Sen's slope was also calculated as follows

$$Q = \frac{Y_{i'} - Y_i}{i' - i} \tag{5}$$

where, Q is the slope estimate. $Y_{i'}$ and Y_i are the values at times i' and i, where i' is greater that i. Sen's slope is then calculated as the median from all slopes.

The trend analysis was also carried out using linear regression test for non-spatial data. For the non-spatial data, mean of the values of climatic variable as well as active fire points were extracted at state as well as district level. In the spatial data, trend analysis was performed on each pixel of the data. For the fire occurrence, spatial trend analysis was performed on the spatio-temporal gridded data.

Relation between Climate and Fire data. The relation between forest fire and climate data is important to relate with frequency and intensity. Correlation analysis was carried out between fire occurrence and each climatic variable. Linear correlation coefficient was used to determine the magnitude of relationship. Spatial products of both fire occurrence and climate variables were correlated with each other. All the layers were matched for the spatial extent and after that the correlation was performed. Fire occurrence data of entire year was correlated with different season's climatic data. Winter and premonsoon seasons are the period of forest fire in the state hence the relation between fire occurrence and climate data in these two seasons was carried for the same year. While climate condition of monsoon and post-monsoon period can affect the forest fire in upcoming year. Hence the correlation analysis in monsoon and post-monsoon season of fire occurrence in current year was related with climatic variable of previous year. For e.g. monsoon season precipitation of year 2004 was correlated with fire occurrence data of year 2005. As precipitation in monsoon season of year 2005 cannot impact the forest fie occurrence of year 2005 which has already been passed. To analyse the cumulative effect of climate pattern, all the datasets were compared in the interval of two year as well. The comparison of analysis of one year and two year interval can provide more information on their relationship.

4 Result

The result of trend analysis and relation between climatic variable and fire occurrence is presented in the following subsection.

4.1 Trend Analysis

Trend analysis of all the parameters (temperature, precipitation, humidity and fire occurrence) in the study analysed in non-spatially as well as spatially. Non-spatial trend analysis at the district level in each month and season is given in Fig. 3. It has been found that there is no significant trend from February to June for all the three climate variables i.e. temperature, precipitation and humidity. The annual trend analysis also did not show any significant trend result. Detailed result with trend value is given in Appendix 1.

Temperature. Non spatial MK test (Fig. 3) shows that there is positive trend for temperature only in the month of September (1.97–3.03) for all the districts. The remaining months showed no trend in the outcome. However, non-significant trend results were positive as well as negative. The annual temperature variability over the years also showed no significant trend. In the seasonal pattern, premonsoon season showed positive trend for the Mizoram, which indicates temperature is increasing premonsoon season over a period of time. However, for individual districts significant trend results were not obtained. Winter and postmonsoon seasons showed no significant trend for any district or state. There is positive significant trend was found in the monsoon season for the districts Lawngtlai, Lunglei, Mamit, Saiha and Serchhip. Overall, significant temperature trend showed positive value in the outcome which describes that in Mizoram there is increase in the temperature over the period from 2003 to 2020, limited to certain time period and area. Linear trend of mean annual temperature presented in Fig. 4 also shows increase in temperature in Mizoram from 2003 to 2020.

To identify the area with significant trends spatial trend analysis was carried out. The results obtained from the spatial analysis showed significant outcome only in monsoon season. Similar findings were observed in the non-spatial trend analysis, where monsoon season had maximum significant outcomes. Figure 5 describes the magnitude of trend distributed over Mizoram with significant values. In the results only significant results of trend analysis were mapped. The outcome of trend analysis in winter, premonsoon and postmonsoon seasons did not show any significant values. It can be seen from the spatial map of trend (Fig. 5) that most of the part of Aizawl, Champhai, Mamit and complete Kolasib district has no trend. Zero value of Sen's slope in Fig. 5 shows no trend in the data. Temperature in the monsoon season was found to be increasing where magnitude of increment is higher in parts of Mamit, Serchhip and Champhai districts. These are the northern districts of the state. While southern districts of the state (Lunglei, Lawngtlai and Saiha) had lower increment rate compared to others.

Precipitation. Rainfall is discrete in nature its accumulation varies over a period of time and space. The non-spatial trend analysis results showed almost no trend in the entire state except for the two districts i.e. Lawngtlai and Mamit. Negative trend was found in the month of September for both the districts, i.e. −2.05 for Lawngtlai and −1.97 for Mamit (Fig 3). This negative trend indicates the decrease in the rainfall in the given period of time. Linear trend over a period of given time in Mizoram has been presented in Fig. 6 where trend line is almost flat. Precipitation after year 2017 has been decreased in the state. Minimum precipitation was observed in year 2014. Spatial trend analysis using MK test also showed no significant outcome. Therefore, spatial map of the MK test result was not generated. The non-significant results are

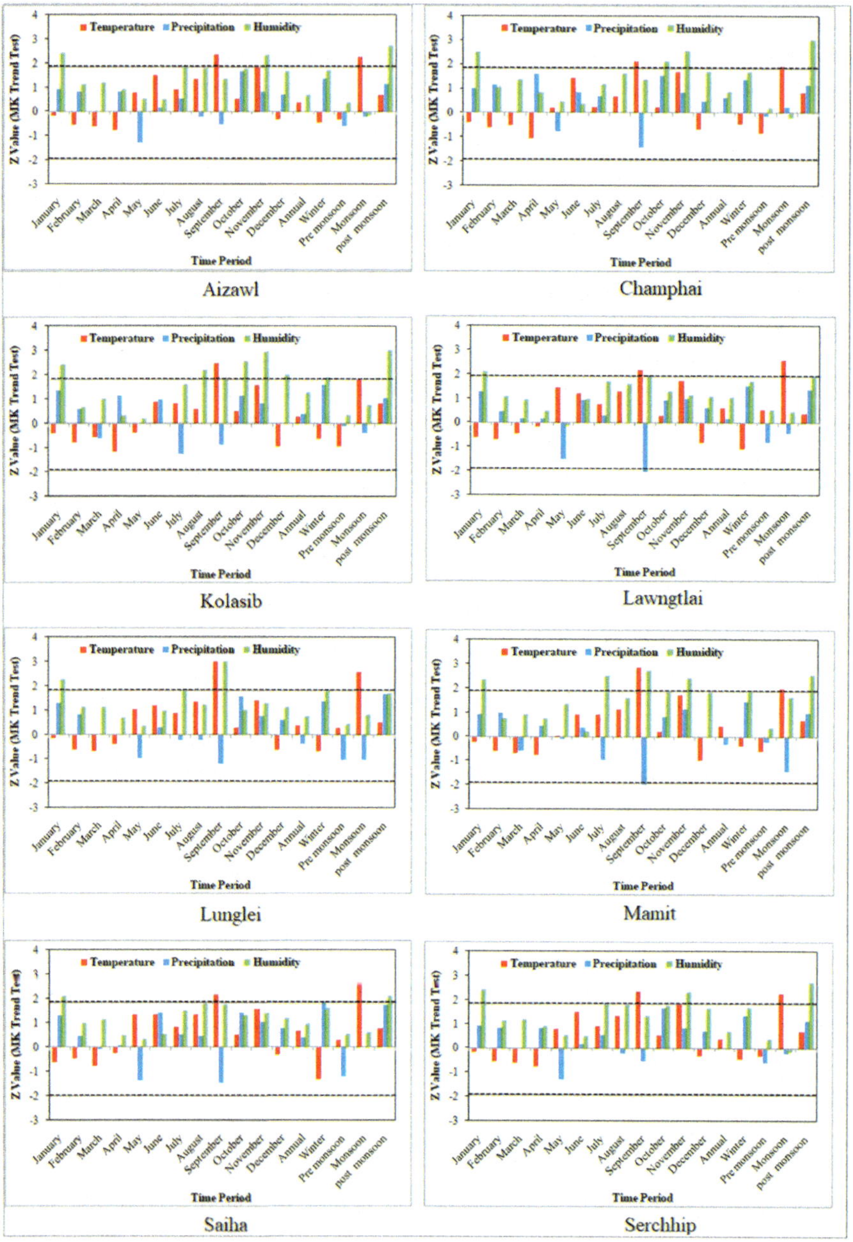

Fig. 3 Plot showing MK trend test (Z) with of various climatic variables over different districts of Mizoram at different time period. The dotted line is the threshold beyond which results are significant

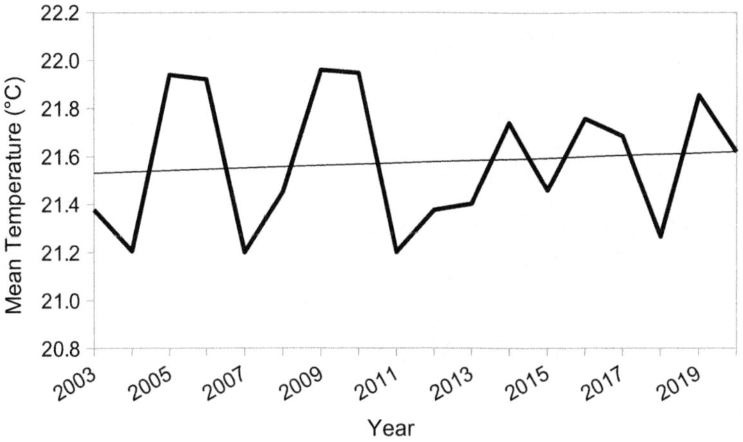

Fig. 4 Linear trend of temperature in Mizoram from 2003 to 2020

mixed of positive and negative trend but most of the districts showed decreasing precipitation in Mizoram.

Humidity. Overall trend for humidity in the study has been found to be positive (1.98 to 3.18) for all the districts and in all the months and seasons. In January month all the district and state has shown positive trend (>1.96 Z score) which is significant at 5% level of significance. Significant positive trend was also observed in July, August, September, October, November and December months (Fig. 3). It has been observed that pre-monsoon period which is very crucial for the forest fire has not shown any significant trend. Aizawl district has shown increase in humidity in winter as well as in post-monsoon seasons. The increase in the humidity in the monsoon season is also observed for entire state only. Linear trend for average annual humidity in Mizoram also shown to be having positive trend. However, humidity was low after 2017 (Fig. 6).

Humidity is the only climatic variable, that showed significant results in all the four climatological seasons using MK test (Fig. 7). However, few of the grids showed no trend at 5% level of significance which is marked as 0 in the map. The result of spatial trend analysis also showed humidity is increasing over a period of time in Mizoram; however the magnitude of the trend is varying which can be seen in Fig. 7.

Fire Points. Trend analysis for fire points was performed only on an annual basis because fire occurs mainly during winter and premonsoon period. During other season or months there is more or less no fire in the state. Fire occurrence in different district of Mizoram is shown in Table 1. Lunglei, Lawngtlai and Mamit are the districts having the maximum fire occurrence in the state from 2003 to 2020. Active fire points did not show any significant trend for the MK test. However, linear regression analysis performed using annual fire count over Mizoram showed decreasing trend as shown in Fig. 9. It can be seen from the data that the total fire occurrence

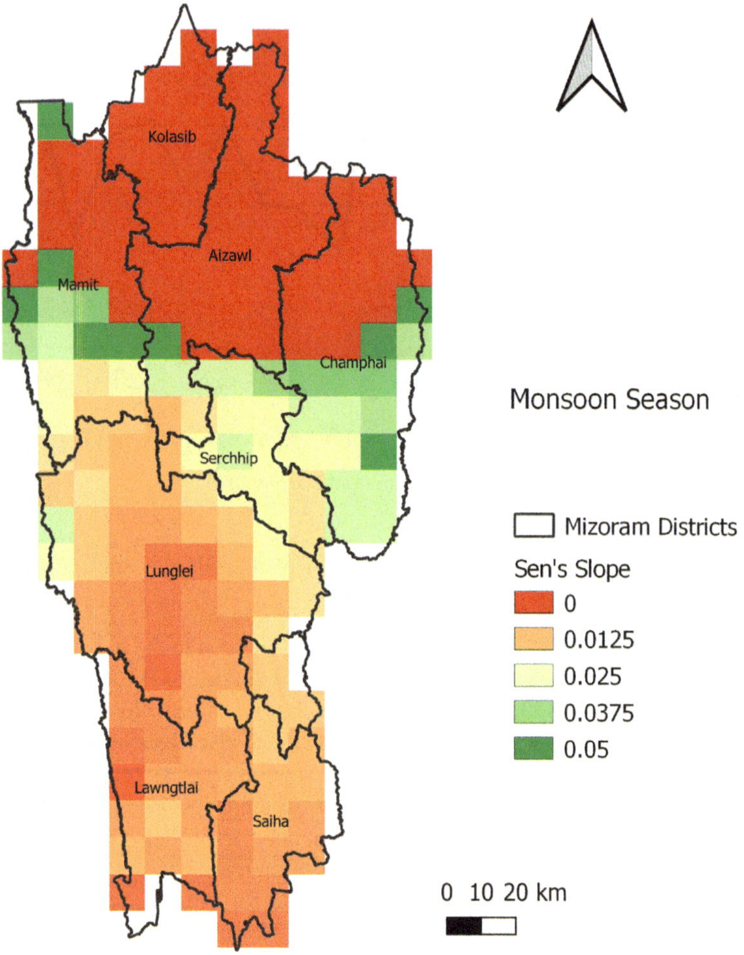

Fig. 5 Magnitude of trend of temperature in monsoon season in Mizoram at 5% level of significance

in the state has been reduced over a period of time. Maximum fire occurrence was found in year 2009 and fire occurrences have been reduced afterwards.

Spatial trend map of fire occurrence prepared using linear regression method is shown in Fig. 10. It can be observed that most of the parts of the state are showing negative trend. However, parts of Lawngtlai, Mamit and Lunglei districts showed increasing trend of fire occurrence. These districts area also having the highest number of fire occurrence, making it highly vulnerable to forest fire.

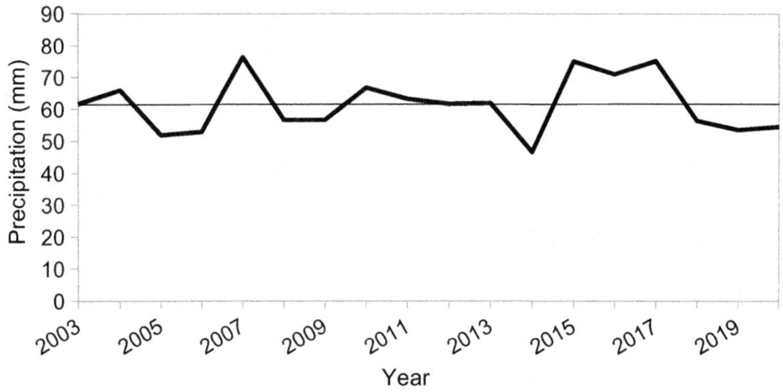

Fig. 6 Linear trend of total precipitation in Mizoram from 2003 to 2020

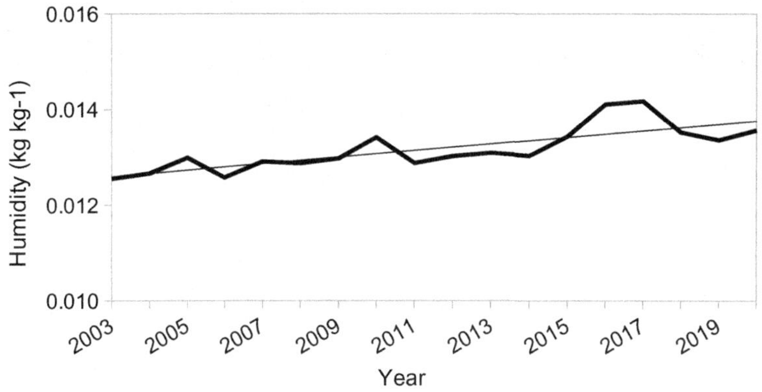

Fig. 7 Linear trend of humidity in Mizoram from 2003 to 2020

4.2 Relation Between Climate Variable and Fire Occurrence

Based on the result of trend analysis it can be inferred that climate variable and fire occurrence had opposite trend result. As temperature and humidity are increasing with decrease in precipitation which is suitable condition for fire occurrence. However, fire occurrence trend in decreasing over a period of given time. Hence, relation analysis between these two parameters will help for further understanding. Correlation analysis was carried out to understand the relation between fire occurrence and climate variable and correlation coefficient value obtained from the analysis is shown in Table 2. The analysis was carried for different seasons as already mentioned in the methodology section.

Temperature. Temperature and fire occurrence relation has been found to be positive for most of the categories except for the monsoon season. During the monsoon season

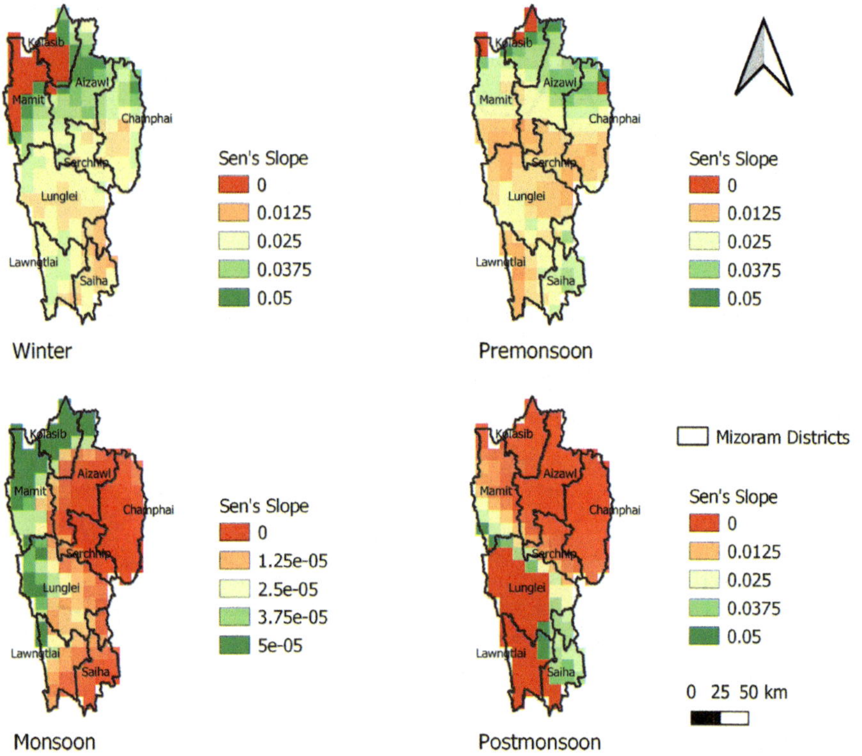

Fig. 8 Magnitude of trend of Humidity in difference season in Mizoram at 5% level of significance

almost all the districts and state have shown negative correlation coefficient value (Table 2). However, these values are not significant. Significant results were obtained in premonsoon (0.50–0.82) and cumulative winter and premonsoon seasons (0.51–0.75). It has also been observed that use of time interval with one year and two year has also shown some differences in the correlation coefficient values. The correlation coefficient value for two year time interval is mostly higher than the one year time interval. The positive relation in winter and premonsoon season shows that increase in temperature in the particular period may increase the chance of forest fire. The significant results were not found to be significant in Lunglei and Lawngtlai districts which are having maximum fire occurrences.

Spatial analysis map was also prepared for each grid in the state based on the correlation coefficient value which is presented in Fig. 11. There mixed kind of result was obtained in the spatial analysis which means that within each season there is positive as well as negative relation was found. During the winter and premonsoon seasons, south-western and eastern part of the state has shown negative relation coming under the Lunglei, Lawngtlai and Champhai districts as can be seen in Fig. 11. Some positive relation has been found in the northern and central districts i.e. Kolasib, Aizawl and part of Mamit. During the monsoon season the result is opposite to what

Table 1 Active fire points in Mizoram and its districts from year 2003 to 2020

Year	District								Mizoram
	Aizawl	Champhai	Kolasib	Lawngtlai	Lunglei	Mamit	Saiha	Serchhip	
2003	649	640	535	768	1112	817	216	294	5031
2004	572	798	460	1109	1418	834	379	363	5933
2005	722	461	403	731	1317	951	273	335	5193
2006	1046	1044	733	1039	1866	1367	416	456	7967
2007	765	566	635	960	1917	1360	503	228	6934
2008	487	464	212	525	885	610	142	179	3504
2009	930	701	573	1421	1992	1623	788	377	8405
2010	972	796	611	746	1395	1265	530	422	6737
2011	307	334	179	589	619	392	112	183	2715
2012	427	573	180	844	904	485	440	202	4055
2013	482	589	210	822	962	616	294	233	4208
2014	424	460	241	675	823	552	192	132	3499
2015	467	616	258	748	852	618	145	240	3944
2016	302	267	215	592	683	493	122	64	2738
2017	294	387	128	394	432	290	91	127	2143
2018	362	401	165	578	690	482	176	192	3046
2019	205	303	113	591	607	415	130	95	2459
2020	251	297	193	611	647	569	124	103	2795
Total	9664	9697	6044	13,743	19,121	13,739	5073	4225	81,306

is found during the winter and premonsoon seasons. Postmonsoon season has also shown negative relation for the entire state except few pockets.

Precipitation. The relation between precipitation (water) and fire is opposite in nature. In the present study also the significant relation between the two has been found negative, however the significant results are very few (Table 2). Annual precipitation relation with fire occurrence was found to be negative (−0.49) for entire Mizoram. The monsoon season which records the maximum precipitation during the year has shown significant negative relation with fire occurrence in only Champhai district (−0.58). The seasonal data showed inconsistent results for e.g. in winter season significant negative relation was found in Serchhip district (−0.55) for monsoon it is in Champhai (−0.58) and for post-monsoon, it is Lunglei (−0.49). In premonsoon season no significant results were found.

The result of spatial analysis in different seasons is shown in Fig. 11. It can be seen from the map that most of the parts of the state is falling under the negative relation. However, the value of the correlation coefficient is towards the lower side. In the winter season, it can be found that the western border of Lunglei and Lawngtlai has positive correlation. However, precipitation is very low during this season. During the monsoon season all parts of the state showed negative correlation where Champhai

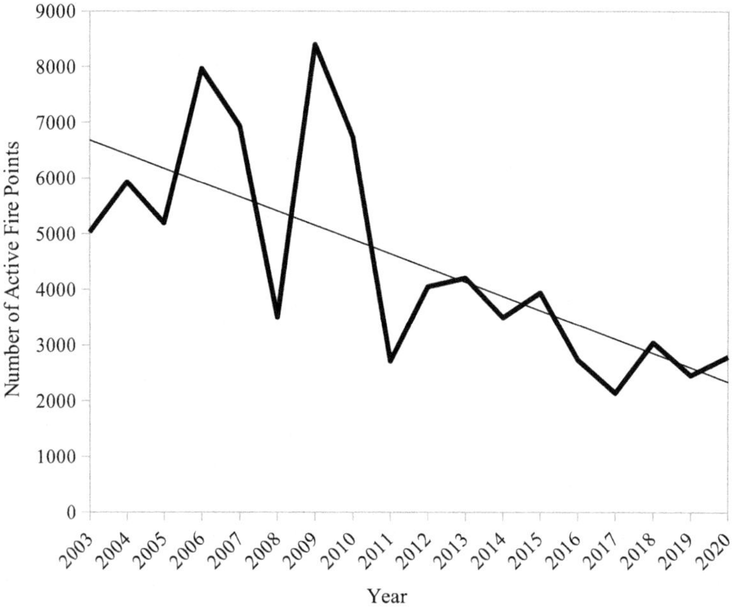

Fig. 9 Linear trend of fire occurrence in Mizoram from year 2003 to 2020

district showed highest negative relationship. While in post-monsoon season, parts of Lunglei showed highest negative correlation coefficient value. Overall it can be inferred from the analysis that the precipitation in monsoon and post-monsoon season might have a role in fire occurrence, however, its role is not that much significant.

Humidity. The relation of humidity with fire occurrence was found to be more consistent than other two climatic variables. The correlation coefficient value for all the season was found to negative and the in most of cases, the value is towards the higher side (Table 2). The monsoon season correlation coefficient values were almost significant except for few districts. It can be inferred that the humidity in monsoon season can have role in fire occurrence in the upcoming year as it can determine the moisture in the environment. Maximum correlation coefficient for the humidity relation was found to be for Kolasib district, i.e. −0.84 with time interval of two years followed by Champhai district which is −0.81. Premonsoon and postmonsoon season found to have almost no role in relation to fire occurrence as the result were almost non-significant.

Spatial analysis results showed that almost all the area is under the negative relation except the few regions in Lunglei and Lawngtlai districts a monsoon season (Fig. 11). The relation is near zero in the winter season for the same region. Premonsoon and postmonsoon seasons maps showed most of the area with negative relationship. During the post-monsoon season, south-eastern part of the state in Lawngtlai and Saiha districts showed a high negative correlation coefficient which describes its role in upcoming fire events.

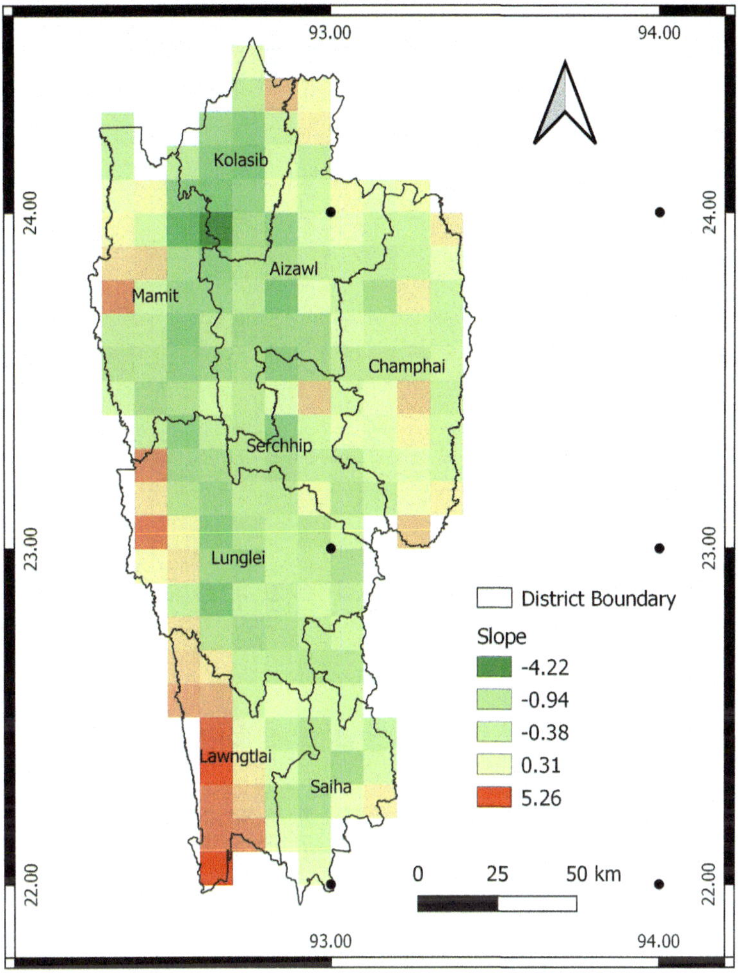

Fig. 10 Annual spatial trend of fire occurrence in Mizoram state

5 Discussion and Conclusion

Climate change being one of the most burning issues in recent times has link with other calamities. Change in the climate condition cause and effect, directly or indirectly to society and its functions. The changes in precipitation patterns causing alternations in timing of runoff into streams can lead to water stress for plants increasing risk of wildfire [44]. Natural phenomenon can also be affected, depending on the climatic conditions. Forest fire is one of such phenomenon associated with the climate or weather. Forest fire throughout the world has demonstrated ecological, social and economic effect in recent times and it has also been affecting the earth's surface and atmosphere for over 350 million years [12]. Temperature rise is one of

Table 2 Correlation coefficient value obviated by the analysis between fire occurrence and climate variables in different season over Mizoram and its districts at time interval of 1 year and 2 year for the period of 2003–2020

District/State	Time interval	Winter			Premonsoon			Monsoon		
		Temperature	Precipitation	Humidity	Temperature	Precipitation	Humidity	Temperature	Precipitation	Humidity
Mizoram	1	0.40	−0.37	**−0.54**	0.45	0.12	−0.48	−0.17	−0.32	**−0.74**
	2	0.48	−0.33	−0.59	**0.71**	0.10	−0.51	−0.16	−0.20	**−0.75**
Mamit	1	0.39	−0.24	−0.39	**0.54**	−0.06	−0.45	−0.12	−0.06	**−0.61**
	2	0.47	−0.30	−0.45	**0.82**	−0.19	−0.48	−0.02	0.07	−0.62
Aizawl	1	0.47	−0.27	**−0.50**	**0.62**	−0.05	−0.43	−0.04	−0.41	**−0.77**
	2	0.53	−0.29	−0.52	**0.79**	−0.19	−0.46	−0.02	−0.31	**−0.73**
Lunglei	1	0.28	−0.17	**−0.50**	0.30	0.21	**−0.52**	−0.29	−0.20	**−0.66**
	2	0.34	−0.04	−0.57	0.56	0.36	−0.60	−0.26	−0.08	**−0.73**
Lawngtlai	1	0.27	−0.38	**−0.50**	0.06	0.26	−0.43	−0.38	−0.23	**−0.61**
	2	0.49	−0.30	**−0.73**	0.45	0.40	−0.50	−0.29	−0.16	**−0.72**
Champhai	1	0.48	−0.47	**−0.64**	0.47	0.08	−0.39	0.14	**−0.58**	**−0.67**
	2	0.53	−0.56	−0.65	0.62	0.05	−0.45	−0.20	−0.28	**−0.81**
Saiha	1	0.36	−0.38	**−0.52**	0.19	0.30	−0.29	−0.45	−0.23	**−0.56**
	2	0.50	−0.24	−0.56	0.47	0.28	−0.35	−0.21	−0.36	−0.48
Kolasib	1	0.38	−0.26	−0.43	**0.50**	−0.11	−0.42	0.08	−0.30	**−0.67**
	2	0.46	−0.42	−0.50	**0.70**	−0.28	−0.49	−0.15	−0.09	**−0.84**
Serchhip	1	**0.52**	**−0.55**	**−0.56**	0.40	0.17	−0.38	−0.16	−0.26	**−0.76**

(continued)

Table 2 (continued)

District/State	Time interval	Winter			Premonsoon			Monsoon		
		Temperature	Precipitation	Humidity	Temperature	Precipitation	Humidity	Temperature	Precipitation	Humidity
	2	0.63	−0.58	−0.58	0.57	0.30	−0.41	−0.13	−0.17	**−0.74**
District/State	Time interval	Postmonsoon			Winter + Pre Monsoon			Total precipitation	Temperature (av.)	Humidity (av.)
		Temperature	Precipitation	Humidity	Temperature	Precipitation	Humidity			
Mizoram	1	0.12	−0.45	−0.47	**0.51**	0.03	**−0.59**	**−0.49**	0.25	**−0.57**
	2	0.19	−0.07	−0.29	0.66	−0.05	−0.57	−0.09	0.50	−0.59
Mamit	1	0.15	−0.37	−0.42	**0.53**	−0.12	**−0.50**	−0.42	0.42	−0.47
	2	0.43	0.00	−0.14	**0.71**	−0.30	−0.48	−0.10	0.60	−0.47
Aizawl	1	0.13	−0.26	0.04	**0.64**	−0.12	**−0.53**	**−0.49**	0.47	0.11
	2	0.16	−0.02	0.11	**0.75**	−0.32	−0.50	−0.20	0.60	0.11
Lunglei	1	0.10	**−0.49**	−0.41	0.35	0.18	**−0.59**	−0.34	0.17	**−0.52**
	2	0.23	−0.06	−0.03	0.50	0.37	−0.61	0.01	0.35	−0.57
Lawngtlai	1	−0.01	−0.41	−0.47	0.23	0.19	**−0.54**	−0.37	0.04	**−0.50**
	2	0.08	−0.18	−0.11	0.58	0.28	−0.65	−0.05	0.37	−0.61
Champhai	1	0.16	−0.29	−0.40	**0.57**	−0.07	**−0.58**	**−0.60**	0.27	**−0.61**
	2	−0.02	−0.28	−0.58	0.66	−0.26	−0.56	−0.15	0.45	**−0.67**
Saiha	1	−0.04	−0.39	−0.31	0.36	0.24	−0.47	−0.34	0.18	−0.34
	2	0.01	0.23	0.07	0.59	0.20	−0.48	−0.02	0.38	−0.38
Kolasib	1	0.19	−0.34	**−0.53**	**0.53**	−0.16	**−0.50**	**−0.58**	0.40	**−0.58**
	2	0.29	−0.08	−0.55	0.66	−0.42	−0.51	−0.24	0.53	−0.64
Serchhip	1	0.19	−0.26	−0.36	**0.57**	0.04	**−0.53**	−0.33	0.29	**−0.61**
	2	0.02	−0.18	−0.48	**0.69**	0.07	−0.51	0.19	0.48	−0.60

Bold values represent 5% level of significance

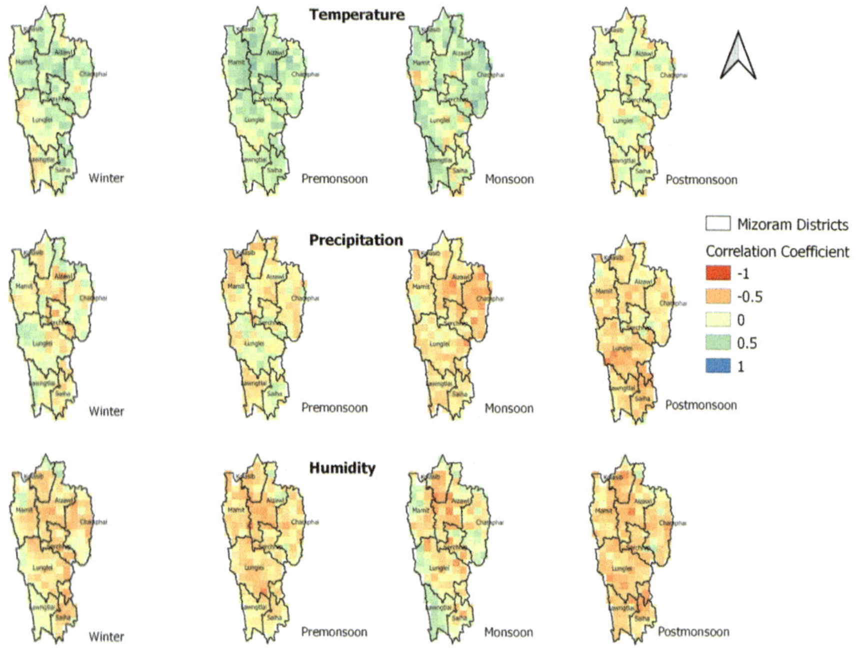

Fig. 11 Spatial correlation analysis between fire occurrence and climatic variables (temperature, precipitation and humidity) from year 2003 to 2020

the important factors in regulating climate conditions. Global annual temperature has been found to increase at an average rate of 0.08 °C per decade since 1880. The rate has been found to increase over twice the rate since 1981 (+0.18 °C) compared to previous one. The year 2020 was found to have recorded highest northern hemisphere land temperature in 141 year, i.e. +1.28 °C above average [38]. There are studies show that changes in climate increases the chance of forest fire occurrence [16]. The present study also reveals the trend in the change of climate condition over Mizoram. Temperature, precipitation and specific humidity were considered for the trend analysis, since these are important factors affecting forest fire the most.

The first objective of the study was to analyse the trend of climatic variables in the study area which showed very less significant results. Temperature trend from year 2003 to 2020 showed significant positive result of MK test. However, it was found only for September month (1.97–3.03) and few districts (Lawngtlai, Lunglei, Mamit, Saiha and Serchhip) in monsoon season. The spatial trend analysis using MK test also found to be significant only in monsoon season. In the other seasons trend results were not significant. Earlier study by Jain et al. [23] also showed positive or no trend from year 1901 to 2003 for entire north east India. They used monthly surface air temperature data provided by IITM, Pune for their analysis. Global trend analysis results show increase in temperature from 1990 to 2020 for the north eastern states of India [38]. The non-significant trend of temperature in winter season was completely negative which means there is decrease in temperature over a given period of time. Similar results were also obtained by Jain et al. [23] on trend analysis of temperature from 1951 to 2003 over north east India where winter and premonsoon season showed negative trend while monsoon and post-monsoon results were positive. However, statistical significance trend results were found to be increasing. The correlation analysis between temperature and fire occurrence showed positive relation for annual mean temperature for all the districts and state as well. Kale et al. [29] also investigated role of climate extremities on forest fire refines in India and they found forest fire was significantly correlated with dry days and maximum average temperature from 2003 to 2013. In the present study, temperature relation with fire occurrence was positive in winter and premonsoon season for entire state except south eastern part of state. The region is under Lunglei and Lawngtlai districts. The region was also found to have increasing fire occurrence trend while rest of the state had negative trend for fire occurrence.

Precipitation trend analysis using MK test was found to have almost no significance, only Lawngtlai and Mamit districts showed significant negative trend. The spatial trend analysis at grid level also showed no significant results. The analysis of trend of rainfall from year 1913–2012 in 9 stations of north east India also showed negative or no trend [34]. The rainfall trend for India from 1901 to 2015 was also mostly negative [40]. Kumar et al. [32] also analysed trend for long-term rain in India for 135 year (1871–2005), for which, India was divided into subdivisions. The results showed that half of the subdivisions were having increasing trend and others including north east India having negative trend. The relation of precipitation with fire occurrence was found to be positive as well as negative, however, it was mostly negative. The effect of precipitation in monsoon season is found to affect mostly Champhai district, where highest negative correlation coefficient value is obtained.

Humidity trend was found to have more number of statistical significant results than temperature and precipitation. Humidity was found to be increasing over a period of time in the present study. Singh et al. [46] in their study on changes in relative humidity over northwest and central India found positive in 100 years of time interval. Spatial trend analysis in the present study was also found to be positive for entire state however, the magnitude of slope varies. Recent study by Putatunda [41] also showed positive trend using ERA-interim data over Indian land from year 1980 to 2017. The study also showed that humidity is increasing at all the pressure level. Premonsoon season which is critical for forest fire occurrence did not have any significant trend for humidity. Postmonsoon season and January month was found to have high significant results for humidity except Lunglei and Lawngtlai districts. The spatial trend analysis using MK test was also found to be insignificant in parts of these districts, which highlights the difference in climatic condition in these districts. Jaswal and Koppar [24] also studied trend in humidity over a time period from 1969 to 2007 and showed a positive trend in the Mizoram and nearby area. The correlation analysis also showed negative relation which indicates that lower the humidity and higher the forest fire occurrence. This relation is almost negative for all the districts except the parts of western border districts, which has inconsistent results.

The prime cause of forest fire in north eastern region is mainly because of traditional practice of jhum cultivation [11, 28, 33]. Jhum cultivation and human interference in forest with addition to climate change scenario can increase the chances of spread of forest fire. Overall climate variables trend favours forest fire however, annual trend of forest fire in the study showed decrease fire occurrence from year 2003 to 2020. Only south eastern and eastern region (part of Lunglei, Lawngtlai and Mamit districts) of the state showed po'sitive trend for fire occurrence which are also having favouring trend for climate variables. However, the correlation analysis results over this region had low correlation coefficient values compared to other parts of the state at for different climate variable and in different season.

Overall it can be concluded from the study there is significant inclusion of human interference and climate condition in the forest fire occurrence in the state. However, the sole reason cannot be attributed to cause of forest fire and its danger on environment and society. There are few pockets in the state which is in response to the climate variable while few are not following the trend of natural phenomenon.). It can be inferred from the present study that in addition to the anthropogenic factors, climatic conditions also crucial for forest fire and alteration in these conditions may affect the forest fire incidences. It may also be suggested from that local level study in this region can be carried out to find out the ambiguities in the study.

Acknowledgements Authors are thankful to Director, North Eastern Space Applications Centre, Umiam, Meghalaya to facilitate and carry out the present work.

Appendix 1

Results of the MK trend test (Z) of various climatic variables over different districts of Mizoram and state as well on monthly, seasonal and annual basis

District	Climate Variable	Month												Annual	Season			
		January	February	March	April	May	June	July	August	September	October	November	December		Winter	Pre monsoon	Monsoon	Post monsoon
Aizawl	Temperature	0	−0.61	−0.61	−1.21	0.37	1.14	0.76	0.68	**2.20**	0.68	1.89	−0.76	0.61	−0.45	−0.76	1.74	0.91
	Precipitation	1.83	−0.11	−0.23	1.18	−0.11	0.72	−0.87	−0.04	−1.77	0.99	0.84	0.92	0.34	1.06	−0.19	0	1.18
	Humidity	**2.50**	0.98	1.06	0.76	0.87	0.37	1.85	1.77	1.44	**2.35**	**2.95**	**2.20**	0.87	**2.12**	0.21	0.29	**3.18**
Champhai	Temperature	-0.38	−0.61	−0.53	−1.06	0.21	1.44	0.23	0.68	**2.12**	0.23	1.67	−0.68	0	−0.45	−0.83	1.89	0.83
	Precipitation	0.98	1.14	0	1.59	−0.76	0.83	0.68	0	−1.44	1.52	0.83	0.45	0.61	1.36	−0.15	0.23	1.14
	Humidity	**2.50**	1.06	1.36	0.84	0.45	0.37	1.19	1.61	1.36	**2.13**	**2.55**	1.67	0.87	1.69	0.21	−0.21	**3.03**
Kolasib	Temperature	-0.38	−0.76	−0.53	−1.14	−0.37	0.91	0.83	0.61	**2.50**	0.53	1.59	−0.91	0.3	−0.61	−0.91	1.82	0.83
	Precipitation	1.36	0.61	−0.61	1.14	0	0.98	−1.22	0	−0.87	1.14	0.83	0	0.38	1.59	−0.11	−0.38	1.06
	Humidity	**2.42**	0.68	1.03	0.34	0.21	0	1.61	**2.21**	1.85	**2.55**	**2.96**	**2.01**	1.28	1.89	0.37	0.78	**3.03**
Lawngtlai	Temperature	-0.61	−0.68	−0.45	−0.15	1.44	1.21	0.76	1.28	**2.18**	0.3	1.74	−0.83	0.61	−1.06	0.53	**2.58**	0.38
	Precipitation	1.25	0.45	0.15	0.15	−1.52	0.91	0.3	0	**−2.05**	0.91	0.98	0.61	0.15	1.52	−0.83	−0.45	1.36
	Humidity	**2.12**	1.06	0.95	0.46	−0.12	0.99	1.69	1.61	1.94	1.29	1.14	1.06	1.03	1.69	0.54	0.45	1.89
Lunglei	Temperature	-0.15	−0.61	−0.68	−0.38	1.06	1.21	0.91	1.36	**3.03**	0.3	1.44	−0.61	0.38	−0.68	0.3	**2.58**	0.53
	Precipitation	1.29	0.83	0	0	−0.98	0.3	−0.23	−0.23	−1.21	1.59	0.76	0.61	−0.38	1.36	−1.06	−1.06	1.67
	Humidity	**2.28**	1.14	1.14	0.72	0.37	0.99	1.85	1.24	**3.01**	1.03	1.29	1.14	0.78	1.85	0.45	0.82	1.71
Mamit	Temperature	-0.23	−0.61	−0.68	−0.76	0.04	0.91	0.91	1.14	**2.88**	0.23	1.74	−0.98	0.45	−0.38	-0.61	**1.97**	0.68
	Precipitation	0.91	0.98	−0.61	0.45	−0.08	0.38	−0.98	0	**−1.97**	0.83	1.14	0	−0.3	1.44	−0.23	−1.44	0.98
	Humidity	**2.35**	0.76	0.91	0.76	1.36	0.21	**2.52**	1.61	**2.73**	1.87	**2.41**	1.82	1.03	1.93	0.37	1.65	**2.54**
Saiha	Temperature	-0.61	−0.45	−0.76	−0.23	1.36	1.36	0.83	1.36	**2.18**	0.53	1.59	−0.3	0.68	−1.29	0.3	**2.58**	0.76
	Precipitation	1.29	0.45	-0.08	0.08	−1.36	1.44	0.53	0.45	−1.44	1.44	1.06	0.76	0.38	1.82	−1.21	0	1.74
	Humidity	**2.12**	0.98	1.14	0.49	0.33	0.54	1.53	1.81	1.77	1.33	1.44	1.21	0.95	1.61	0.54	0.62	**2.12**
Serchhip	Temperature	-0.15	−0.53	−0.61	−0.76	0.78	1.52	0.91	1.36	**2.35**	0.53	1.82	−0.3	0.38	−0.45	−0.3	**2.27**	0.68

(continued)

(continued)

District	Climate Variable	Month												Annual	Season			
		January	February	March	April	May	June	July	August	September	October	November	December		Winter	Pre monsoon	Monsoon	Post monsoon
	Precipitation	0.91	0.83	0	0.83	−1.29	0.15	0.53	−0.23	−0.53	1.67	0.83	0.68	0	1.36	−0.61	−0.23	1.14
	Humidity	**2.42**	1.14	1.21	0.91	0.54	0.5	1.85	1.81	1.36	1.75	**2.32**	1.67	0.7	1.69	0.37	−0.16	**2.73**
Mizoram	Temperature	-0.23	−0.61	−0.68	−0.6	1.03	1.21	0.61	1.29	**2.80**	0.53	1.67	−0.68	−0.57	−0.23	**2.08**	0.72	0.46
	Precipitation	1.14	0.61	−0.15	0.61	−0.98	0.98	−0.23	0.23	−1.29	1.36	1.14	0.61	1.52	−0.53	−0.61	1.36	0.08
	Humidity	**2.50**	1.06	1.14	0.69	0.62	0.45	1.81	1.9	1.69	1.9	**1.98**	1.71	1.69	0.37	0	**2.58**	0.78

Bold values represent 5% level of significance

References

1. Ahmed M, Suphachalasai S (2014) Assessing the costs of climate change and adaptation in South Asia. Asian Development Bank
2. Ankila H (2007) Forest fires in India: Extent, justification and policy. In Schmerbeck J, Hiremath A, Ravichandran C (eds) Fires in Indian forests, Workshop proceedings ATREE, Bangalore, India. Freiburg, Germany, Institute of Silviculture
3. Aragao LEO, Malhi Y, Barbier N, Lima A, Shimabukuro Y, Anderson L, Saatchi S (2008) Interactions between rainfall, deforestation and fires during recent years in the Brazilian Amazonia. Philos Trans Royal Soc B: Biol Sci 363(1498):1779–1785
4. Archibald S, Roy DP, van Wilgen BW, Scholes RJ (2009) What limits fire? An examination of drivers of burnt area in Southern Africa. Glob Change Biol 15(3):613–630
5. Barlow J, Peres CA (2004) Avifaunal responses to single and recurrent wildfires in Amazonian forests. Ecol Appl 14(5):1358–1373
6. Bistinas I, Harrison SP, Prentice IC, Pereira JMC (2014) Causal relationships versus emergent patterns in the global controls of fire frequency. Biogeosciences 11:5087–5101
7. Certini G (2005) Effects of fire on properties of forest soils: a review. Oecologia 143(1):1–10
8. Chandra KK, Bhardwaj AK (2015) Incidence of forest fire in India and its effect on terrestrial ecosystem dynamics, nutrient and microbial status of soil. Int J Agric Forest 5(2):69–78
9. Clarke H, Gibson R, Cirulis B, Bradstock RA, Penman TD (2019) Developing and testing models of the drivers of anthropogenic and lightning-caused wildfire ignitions in south-eastern Australia. J Environ Manag 235:34–41
10. Cochrane MA, Schulze MD (1999) Fire as a recurrent event in tropical forests of the eastern amazon: effects on forest structure, biomass, and species composition 1. Biotropica 31(1):2–16
11. Darlong VT (2002) Traditional community-based fire management among the Mizo shifting cultivators of Mizoram in northeast India. In: Communities in flames: proceedings of an international conference on community involvement in fire management. FAO, Bangkok, Thailand, pp 119–124
12. Doerr SH, Santín C (2016) Global trends in wildfire and its impacts: perceptions versus realities in a changing world. Philos Trans Royal Soc B: Biol Sci 371(1696):20150345
13. Flannigan M, Cantin AS, De Groot WJ, Wotton M, Newbery A, Gowman LM (2013) Global wildland fire season severity in the 21st century. For Ecol Manag 294:54–61
14. Forest Survey of India, Forest fire Activities. https://fsi.nic.in/forest-fire-activities?pgID=forest-fire-activities. Last Accessed 08 Dec 2021
15. Forest Fire Alerts System 3.0, Forest Survey of India from http://117.239.115.41/smsalerts/index.php. Last Accessed 05 Sep 2021
16. Goss M, Swain DL, Abatzoglou JT, Sarhadi A, Kolden CA, Williams AP, Diffenbaugh NS (2020) Climate change is increasing the likelihood of extreme autumn wildfire conditions across California. Environ Res Lett 15(9):094016
17. Gupta S, Roy A, Bhavsar D, Kala R, Singh S, Kumar AS (2018) Forest fire burnt area assessment in the biodiversity rich regions using geospatial technology: Uttarakhand forest fire event 2016. J Indian Soc Remote Sens 46(6):945–955
18. Hess A, Iyer H, Malm W (2001) Linear trend analysis: a comparison of methods. Atmos Environ 35(30):5211–5222
19. Hirschberger P (2016) Forests ablaze: causes and effects of global forest fires. Berlin, Germany, WWF
20. IMD (2013) State level climate change in India. In: Rathore LS, Attri SD, Jaswal AK (eds) Indian Meteorological Department (IMD), Government of India
21. IPCC (2007) Fourth assessment report of the intergovernmental panel on climate change. Cambridge University Press, Cambridge
22. ISFR (2021) Indian state of forest report 2021. Forest Survey of India, MoEFCC, Government of India, Delhi, India 15
23. Jain SK, Kumar V, Saharia M (2013) Analysis of rainfall and temperature trends in northeast India. Int J Climatol 33(4):968–978

24. Jaswal AK, Koppar AL (2011) Recent climatology and trends in surface humidity over India for 1969–2007. Mausam 62(2):145–162
25. Jhajharia D, Yadav BK, Maske S, Chattopadhyay S, Kar AK (2012) Identification of trends in rainfall, rainy days and 24 h maximum rainfall over subtropical Assam in Northeast India. CR Geosci 344(1):1–13
26. Jolly WM, Cochrane MA, Freeborn PH, Holden ZA, Brown TJ, Williamson GJ, Bowman DM (2015) Climate-induced variations in global wildfire danger from 1979 to 2013. Nat Commun 6(1):1–11
27. Joseph S, Anitha K, Murthy MSR (2009) Forest fire in India: a review of the knowledge base. J For Res 14(3):127–134
28. Joshi PK, Chakraborty A, Shukla R (2018): Building a resilient community against forest fire disasters in the northeast India. In: Development and disaster management. Palgrave Macmillan, Singapore, pp 187–200
29. Kale MP, Ramachandran RM, Pardeshi SN, Chavan M, Joshi PK, Pai DS, Bhavani P, Ashok K, Roy PS (2017) Are climate extremities changing forest fire regimes in India? An analysis using MODIS fire locations during 2003–2013 and gridded climate data of India meteorological department. Proc Natl Acad Sci, India, Sect A 87(4):827–843
30. Kendall MG (1975) Rank correlation methods, 4th edn. Charles Griffin, London
31. Kennedy PL, Fontaine JB (2009) Synthesis of knowledge on the effects of fire and fire surrogates on wildlife in US dry forests. Oregon State University Agricultural Extension Station
32. Kumar V, Jain SK, Singh Y (2010) Analysis of long-term rainfall trends in India. Hydrol Sci J 55(4):484–496
33. Lamat R, Kumar M, Kundu A, Lal D (2021) Forest fire risk mapping using analytical hierarchy process (AHP) and earth observation datasets: a case study in the mountainous terrain of Northeast India. SN Appl Sci 3(4):1–15
34. Laskar SI, Kotal SD, Roy Bhowmik SK (2014) Analysis of rainfall and temperature trends of selected stations over North East India during last century. Mausam 65(4):497–508
35. Lovejoy TE, Nobre C (2018) Amazon tipping point. Sci Adv 4(2):2340
36. Mann HB (1945) Nonparametric tests against trend. Econometrica: J Econometric Soc 245–259
37. Mukhopadhyay D (2009) Impact of climate change on forest ecosystem and forest fire in India. In: IOP Conference Series. Earth and Environmental Science 6(38). IOP Publishing
38. NOAA (2021) National Centers for Environmental Information, The state of Global Climate 2020. https://www.ncdc.noaa.gov/sotc/global/202013. Last Accessed 10 Oct 2022
39. Pierce JL, Meyer GA, Jull AT (2004) Fire-induced erosion and millennial-scale climate change in northern ponderosa pine forests. Nature 432(7013):87–90
40. Praveen B, Talukdar S, Mahato S, Mondal J, Sharma P, Islam ARMT, Rahman A (2020) Analyzing trend and forecasting of rainfall changes in India using non-parametrical and machine learning approaches. Sci Rep 10(1):1–21
41. Putatunda I (2021) Multiscale temporal analysis and trends of relative humidity over India and Indian Ocean. J Atmos Solar Terr Phys 218:105551
42. Randerson JT, Liu H, Flanner MG, Chambers SD, Jin Y, Hess PG, Pfister G, Mack MC, Treseder KK, Welp LR, Chapin FS (2006) The impact of boreal forest fire on climate warming. Science 314(5802):1130–1132
43. Rowell A, Moore PF (2000) Global review of forest fires. Forests for Life Programme Unit, WWF International, pp 66–66
44. Sahoo UK, Singh SL, Lalnundanga L, Devi AS, Zothanzama J (2018) Climate change impacts on forest and its adaption study in Mizoram
45. Scholes RJ, Archer SR (1997) Tree-grass interactions in savannas. Annu Rev Ecol Syst 28(1):517–544
46. Singh P, Kumar V, Thomas T, Arora M (2008) Changes in rainfall and relative humidity in river basins in northwest and central India. Hydrol Process 22(16):2982–2992
47. Turner MG, Donato DC, Romme WH (2013) Consequences of spatial heterogeneity for ecosystem services in changing forest landscapes: priorities for future research. Landsc Ecol 28(6):1081–1097

48. Westerling AL, Hidalgo HG, Cayan DR, Swetnam TW (2006) Warming and earlier spring increase western US forest wildfire activity. Science 313(5789):940–943
49. Williams AP, Allen CD, Macalady AK, Griffin D, Woodhouse CA, Meko DM, Swetnam TW, Rauscher SA, Seager R, Grissino-Mayer HD, Dean JS (2013) Temperature as a potent driver of regional forest drought stress and tree mortality. Nat Climate Change 3(3):292–297

Spatiotemporal Analysis of Groundwater Status Using RS-GIS Technique in Assam

Debaditya Gupta, Satyam Raj, and Sudip Mitra

Abstract Groundwater is essential for energy and food security, human health, and ecosystems. The precise prediction of groundwater depth is essential for the adequate utilization and management of groundwater resources. The present study aims toward the temporal trend analysis of groundwater depth for pre and post-monsoons by performing Mann–Kendall's (MK) test and linear regression method for Greater Guwahati Area (GGA). The study also aims to prepare spatial groundwater table fluctuation using a geographic information system (GIS) by using spatial interpolation techniques for different years. The current study shows that the average annual groundwater level (GWL) during all the monsoon seasons in the southeastern parts of the study area has fallen below 8 m below ground level (bgl), highlighting low rainfall infiltration and excessive exploitation of groundwater. The average water level ranges from 1.24 m to 20.07 m during the pre-monsoon and from 0.33 m to 10.96 m during the post-monsoon season. The study revealed that the water level during pre-and post-monsoon seasons lies between 4 and 8 m bgl in 23% of the wells. Spatial variation maps describe that the GWL in the southeastern parts of the basin falls by more than 8 m in all monsoon seasons. Pettitt's change point tests also depict that most change points occurred during the 2012–2014 period in the GGA. Therefore, this study suggests implementing a large-scale rainwater harvesting system in the GGA to augment groundwater resources.

D. Gupta · S. Mitra (✉)
School of Agro and Rural Technology, Indian Institute of Technology Guwahati, Assam, India
e-mail: sudipmitra@iitg.ac.in

D. Gupta
e-mail: dgupta@iitg.ac.in

S. Raj
Ministry of Skill Development and Entrepreneurship, Government of India, New Delhi, India
e-mail: satyam.raj21-05@iimv.ac.in

S. Mitra
Centre for Disaster Management and Research, Indian Institute of Technology Guwahati, Assam, India

S. Mitra et al. (eds.), *Disaster Management and Risk Reduction: Multidisciplinary Perspectives and Approaches in the Indian Context*,
https://doi.org/10.1007/978-981-99-6395-9_21

Keywords Groundwater · Trend analysis · Change point analysis · Spatial interpolation · GIS

1 Introduction

Groundwater is the primary source of freshwater, which is used for the subsistence of living creatures and also to fulfill the requirement of domestic, industrial, and agricultural urgency [31, 36]. The priority of groundwater is constrained in those regions where the adoption of surface water is uncertain either due to insufficiency of quality and quantity or for the absence of a framework for arranging supply from surface sources [22]. Groundwater withdrawal can produce significant surface deformation, leading to extensive infrastructure damage [38]. Groundwater depletion is an increasingly severe issue in aquifers around the planet [14]. Fluctuation of GWL occurs in majorly variation in rainfall, soil moisture, evapotranspiration, recharge and discharge from surface water bodies, and anthropogenic withdrawal of groundwater.

The international debate has been conducted towards environmental degradation due to the threat it poses to living individuals and the standard of living [11]. India ranks 13 among the 17 countries where water stress is extremely high. More than 80% of the available groundwater has been withdrawn in India, out of which approx. 70% is used for agriculture [19]. Water insufficiency would also account for a 6% fall in India's gross domestic product (GDP) [7]. Groundwater elevation or depletion exhibits cyclical and random aspects under the impact of several climatic factors and anthropogenic activities. Hence, it is crucial to accurately predict GWLs for the potential utilization of groundwater resources and the sustainable development of the social economy in arid areas [20]. It is prudent that a system is devised in the form of an early warning system, such that farmers can be made aware of any irregularities in rainfall in advance, thereby avoiding massive losses to crops.

It has been estimated that a sum of almost 23 million cubic km of groundwater has adhered in rock and soil on the planet [3]. However, groundwater provides nearly 50% of all drinking water worldwide, 66% of baseflow to surface water, and 43% of all water for irrigation in agriculture. For this reason, estimating groundwater recharge and its temporal variability with emphasis on droughts is an essential scientific endeavor with socioeconomic importance [25]. In the quantification and forecasting of subsurface processes, the complex scaling behavior remains a challenge, significantly when it changes with the temporal and spatial scales [35]. The GWL is affected by some significant factors, viz. precipitation, evaporation, surface water, and anthropogenic activities. These factors are periodical and random [21]. On comparing public water demands, it is estimated that nearly one-quarter of global river basins are under severe water stress conditions. Also, it has been found that there is a shortage of 36% and 50% in projected global blue (renewable surface and groundwater) and green water (precipitation that naturally infiltrates), respectively [34]. The collaborative development and agricultural production of Assam and other northeastern Indian states empower valuable groundwater resources. Despite the highly developed agricultural areas of Brahmaputra and Barak Valley of Assam, active water wells are not currently enforced to monitor or report withdrawn water volumes to any public agencies. Due to the nonappearance of such information, it becomes very difficult to implement sustainable management policies for over-ratified groundwater basins.

Out of the different temporal trend analysis techniques, Mann-Kendall's test and linear regression methods have wide applications for GWL, groundwater quality trend analysis as well as other hydroclimatic variables [4, 23, 30]. The most common method for change point detection is Pettitt's test is applied mostly in climate change studies on hydroclimatic variables [16, 17], but it has very limited applications in the field of groundwater hydrology. Several researchers have performed different geostatistical interpolation techniques for spatial mapping of GWL such as Ordinary Kriging (OK), and Inverse Distance Weight (IDW) interpolations [39]. An extensive literature review revealed that minimal studies were conducted to understand the GWL behavior throughout the year, particularly in the Northeastern part of India. The above status, therefore, necessitates conducting GIS-based studies on groundwater depth data and exploring their potential for improving and maintaining the depth and quality of GWLs. In the present study, an analysis of GWL is carried out using RS-GIS techniques in the state of Assam along the Brahmaputra River. Considering the knowledge gaps above, the current research effort is to investigate the spatiotemporal GWL to improve the status and management of water resources. (1) To study the GWL behavior by using GIS with the help of the spatial interpolation technique. (2) To carry out temporal trend analysis of groundwater table depth of GGA and (3) Change-point analysis of GWL in GGA.

2 Materials and Methods

2.1 Study Area Description

The study was conducted in the GGA of the state of Assam, which includes some parts of Kamrup Rural district and most parts of Kamrup Metropolitan district, located in the north eastern (NE) part of India. The study area is surrounded by latitudes 25°43′ and 26°51′ north and 90°36′ and 92°12′ east longitudes. Guwahati is the headquarters of the Kamrup Metropolitan district, which is located at latitudes 26°7′ N to 26°30′ N and from 91°38′ E to 91°51′ E longitude. It is the largest commercial, industrial and educational center in the NE region, located on the banks of the Brahmaputra river. The city is situated on an undulating plain with elevations from 49.5 m to 55.5 m above mean sea level (MSL). The study area is shown in Fig. 1. The city is surrounded by hillocks on the southern and eastern sides. There are small hills in the central part of the city, such as Saranya hill (193 m), Nabagrah hill (217 m), Neelachal hill (193 m), and Chunasali hill (293 m). Apart from hilly areas, water bodies like Swamp, Deepor Beel, Silpukhuri, Dighali Pukhuri, Borsolol, and Silasukil also cover the city [8].

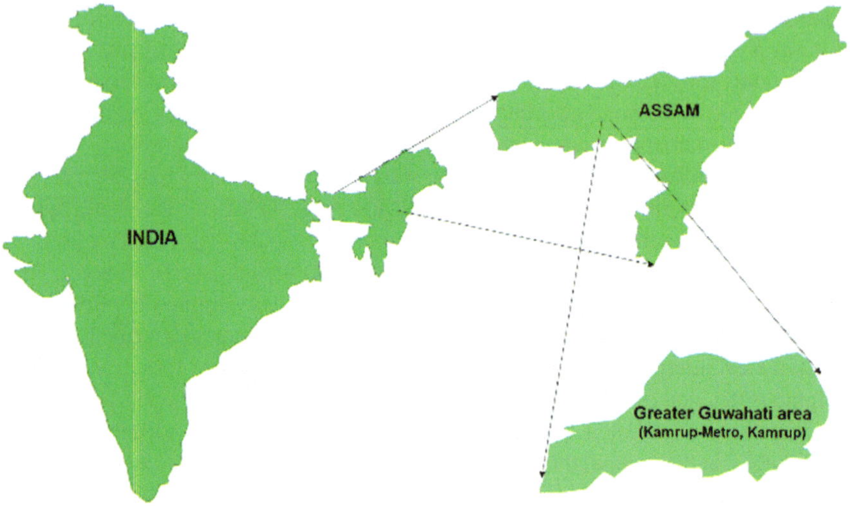

Fig. 1 Study area

The climate of the region is subtropical, with semi-arid summers and cool winters. The maximum and minimum temperatures in the district range from 37–39 °C and 6–7 °C, respectively. The annual rainfall in the district ranges from 1500 mm to 2600 mm, with an average rainfall of about 1752 mm. The rivers Bharalu, Digaru, Kulsi, Umaru, Jagaliya, and Kolong usually join the Brahmaputra river in the north, which flows along the northern boundary of the district. In contrast, the Ghorjan River flows in the south, and the river connects the Brahmaputra along its northern coast. The recent and older Alluvium and Shillong pre-Cambrian age group of rocks constitute the major geological formations of the district. The alluvial sediment is the underlying major water-bearing formation in the district.

2.2 Methodology

In the current study, pre-monsoon and post-monsoon seasonal groundwater depth data (for bore wells and dug wells) related to the study field were gathered from the Central Ground Water Board (CGWB), North Eastern Region, Betkuchi, Guwahati, Assam for the study period 2008–2019.

Water table variations occur in response to changes in the recharge and discharge factors of the groundwater network. Contour maps of pre-monsoon and post-monsoon period water level (m bgl) were prepared for the years 2008-2019 to examine water level variations in the study area. The water level trend in the research field during the pre-monsoon and post-monsoon weather conditions over a period of

12 years (2008–2019) was investigated using the Mann-Kendall test, linear regression, and Sen's slope estimator. The maps of average annual fluctuations during 2008–2019 were drawn. The major land use land cover (LULC) map has been prepared for the years 2008 and 2019 by using LANDSAT images. The LULC map has been classified into five different classes namely "Water body", "Dense vegetation", "Habitation", "Barren land" and "Agricultural land". The class "Water body" consists of the Natural and artificial water bodies such as rivers, lakes, streams, reservoirs, canals and tanks. "Dense vegetation" includes the areas in which forests, shrubs, climbers exist, majorly in the hillocks of the GGA. "Habitation" encompasses the impervious areas where all the major and minor construction works have been done like residential, industrial, commercial areas and transportation networks. "Barren land" comprises of areas that are devoid of vegetation cover such as bare vegetation areas and sand areas. "Agricultural land" consists of those areas that have vegetation of anthropogenic origin such as cultivable lands, orchards, fallow lands which require human activities to maintain it.

Spatial fluctuation assessment of groundwater depth

The spatial and temporal variation in groundwater depth accounts for the sustainable development of groundwater reserves. An essential method for geospatial technology is groundwater management analysis in the contemporary world. Spatial interpolation tools were adopted in GIS to analyze the spatial changes of groundwater depth over a period of 12 years from 2008–2019. From the seasonal groundwater depth, information of 35 wells, seasonal and annual groundwater depth variation was spatially estimated using the IDW method because the depth details of the collected water are very limited.

Seasonal data of GWLs were examined for trends. Given that the non-parametric Mann-Kendall's test and Sen's slope estimator are widely adopted in climate and hydrological time series analysis, these two methods were used in the current study to analyze seasonal trends in GWLs. The functioning adopted in the current study is given in the flowchart (Fig. 2).

2.3 *Statistical Tests for Trend Analysis*

The use of statistical methods such as Man-Kendall's Test, Sen's slope estimator, and Linear regression method was done to estimate the trend and significance of variation and importance in the depth of groundwater by using the slope of linear regression. These techniques are described as follows.

Mann–Kendall test

The Mann–Kendall (MK) test [18, 24] is a rank-based distribution-free test that is widely adopted in trend analysis [4]. The MK tests statistics for a series $y_1, y_2, \ldots y_n$ is given as

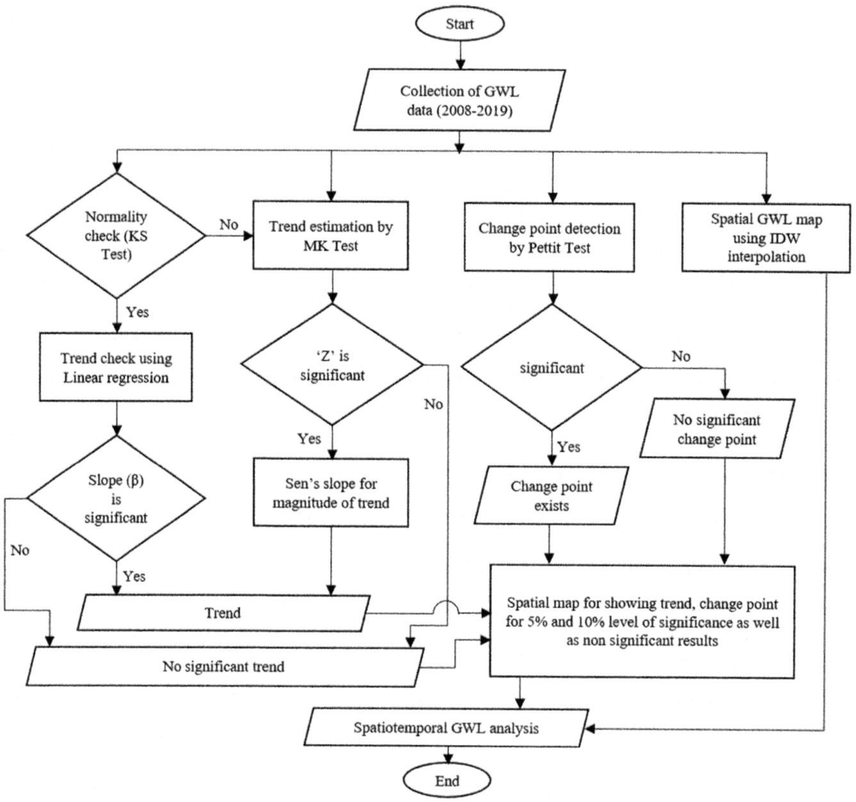

Fig. 2 Flow chart of the methodology

$$S = \sum_{i=1}^{n-1} \sum_{j=i+1}^{n} sgn(y_j - y_i) \tag{1}$$

$$sgn(y_j - y_i) = \begin{cases} 1 \ if \ y_j > y_i \\ 0 \ if \ y_j = y_i \\ -1 \ if \ y_j < y_i \end{cases} \tag{2}$$

$$Var(S) = \frac{1}{18}[n(n-1)(2n+5)] \tag{3}$$

where, y is the yearly groundwater depths In case of the presence of tied events in y variables, the variance is obtained by the formula given as:

$$Var(S) = \frac{1}{18}\left[n(n-1)(2n+s) - \sum_{i=1}^{m} t_i i(i-1)(2i+5)\right] \tag{4}$$

$$Z = \begin{cases} \frac{S-1}{\sqrt{Var(S)}} & if\ S > 0 \\ 0 & if\ S = 0 \\ \frac{S+1}{\sqrt{Var(S)}} & if\ S < 0 \end{cases} \tag{5}$$

Positive and negative values of 'Z' indicate increasing and decreasing trends, respectively.

The trend is significant if,

$$|Z| > Z_{1-\alpha/2} \tag{6}$$

where, α is the level of significance and $\alpha/2$ is for two-tailed tests.

The non-parametric MK test is very suitable for detecting trends in hydrological fluctuations due to various reasons and this test follows the following assumptions:

(a) It does not depend on the data being distributed normally,
(b) It helps with multiple observations per time period,
(c) It allows for missing values and restricted comments in the time series and,
(d) It is less susceptible to sudden breakdown due to homogeneous time series.

Sen's slope estimator

Sen's slope estimator is a focused method used to deal with the unbiased estimator of trends where the data are highly heterogeneous [15]. If the trend is straightforward, the calculated slope change per time assessment was predicted by a moderately non-parametric procedure [32]. Mathematically, it is represented as:

$$Q = median\left(\frac{y_j - y_i}{j - i}\right) \tag{7}$$

where Q indicates the slope of the data points between y_j and y_i. Where, y_j and y_i are the values of growth rate at concentrations j and i respectively.

Linear regression method

The linear regression method is also practiced to find and evaluate trends in time series. The advantage of parametric linear regression analysis is its integrity [33]. The main statistical parameter extracted from the regression test is the slope that reflects the average temporal change of the variables studied. Positive values of the slope indicate an increasing trend, while negative values of the slope indicate a decreasing trend. The total change during the period of inspection is obtained by multiplying the slope by the number of years [33].

$$\widehat{y_i} = a + b.x_i \tag{8}$$

The above equation specifies a line with slope b, y-intercept a. This relationship between the real basic components a and b and the data points is called a linear regression model.

$$b = \frac{\sum_{i=1}^{n} x_i y_i - \frac{\sum_{i=1}^{n} x_i \sum_{i=1}^{n} y_i}{n}}{\sum_{i=1}^{n} x_i^2 - \frac{\left(\sum_{i=1}^{n} x_i\right)^2}{n}} \tag{9}$$

$$a = \overline{y} - b.\overline{x} \tag{10}$$

where y_i and x_i are the groundwater depth and year respectively, n = total number of years and $\overline{y}$ and $\overline{x}$ are mean of y and x respectively

Trend for linear regression i.e., the slope (b) is significant if

$$|t| > t_{\alpha/2,n-2} \tag{11}$$

where, the test statistics' t' is calculated by

$$t = \frac{b}{SE_b} \tag{12}$$

$$SE_b = \sqrt{\frac{\sum_{i=1}^{n} \left(y_i - \widehat{y_i}\right)^2 / (n-2)}{\sum_{i=1}^{n} (x_i - \bar{x})^2}} \tag{13}$$

Pettitt's Change-Point Test

The path followed by Pettitt (1979) is usually tested to find single change points in a hydrological series or climate series with continuous data. It tests the null hypothesis, H_0: The T variables against options that hold the same location criteria (no change) after one or more sharing: one change point survives.

The change point equation is given as

$$U_{t,n} = \sum_{i-1}^{t} \sum_{j-i+1}^{n} sgn\left(y_j - y_i\right) \tag{14}$$

where,

$$sgn\left(y_j - y_i\right) = \begin{cases} 1 \ if \ y_j > y_i \\ 0 \ if \ y_j = y_i \\ -1 \ if \ y_j < y_i \end{cases} \tag{15}$$

$$K_n = Max\left|U_{t,n}\right|, 1 \leq t < n \tag{16}$$

$\left|U_{t,n}\right|$ will continue to rise when change point does not exist in a number series. However, when change points do exist in a series, $\left|U_{t,n}\right|$ will decrease. Therefore, to determine the existence of change points the P value should be calculated as

$$P = 1 - exp\left(\frac{-6K_n^2}{n^2 + n^3}\right) \tag{17}$$

Change point will be significant if,

$$P > 1 - \alpha/2 \tag{18}$$

The M-K test, Sen's slope estimator, linear regression test for trend analysis and Pettitt's change point test were performed codes written in MATLAB 2013a software. Before performing linear regression, the normality of the data was checked by using the Kolmogorov-Smirnov (KS) test in IBM-SPSS 23 software. After performing all the statistical test, the special map for trend, change points were prepared by using ArcGIS 10.3 software. Simultaneously the spatial maps for maximum minimum and average groundwater table depth maps were prepared by using the Inverse Distance Weight (IDW) interpolation technique in GIS environment.

Inverse distance weight (IDW) interpolation

IDW is an extensively used geostatistical interpolation technique that is applied in many GIS packages as it is a straightforward and lesser computationally intensive method [5]. It is regarded as one of the standard interpolation techniques in GIS and is widely applied in GWL depth mapping [2, 27]. It estimates the value of a point by using a linear combination of the value at a sampled data point, weighted by an inverse function of the distance between these two points. The method assumes that observation points closer to the point of prediction are more similar to it as more distant points [28]. The standard IDW is expressed as follows

$$Z_0 = \frac{\sum_{i=1}^{N} w_i Z_i}{\sum_{i=1}^{N} w_i} \tag{19}$$

where the weighing factor w_i is calculated by

$$w_i = \frac{1}{d_i^p} \tag{20}$$

where, Z_0 is the measured samples (GWL) value of the variable at an unknown location, Z_i is the GWL value at location i, d_i is the distance of unknown GWL locations to the measured GWL location. N is the number of samples in the sum, and p is the non-negative exponent of the distance. In this present study, the value of the exponent p is taken as 2 as 2nd power is most commonly used and default in ArcGIS [10, 37].

3 Results and Discussion

In Guwahati, the Assam Urban Water Supply and Sewerage Board (AUWSSB), the Guwahati Municipal Corporation (GMC) and the Public Health Engineering Department (PHED) are primarily engaged in the allocation of drinking water supply from the Brahmaputra River. Nevertheless, through the prevailing system of these agencies, only 30% of the city's population meets the water demand. Indian Oil, railways, and defense agencies also have their own separate supply systems for their regions. The fully installed capacity for drinking water under the GMC sector is around 98 MLD (million liters per day or megaliter per day), while demand is as high as 132 MLD. Water requirement in the current Guwahati municipal area is expected to increase to 425 MLD by the year 2025 [12].

The water stratum of the Brahmaputra River fluctuates about 10 m during the winter and monsoon periods. Even though the Brahmaputra River can meet the water demands of all the residents of this city, it seems that due to unorganized mapping and inadequate development of treatment plants over the years, a large part of the city's population relies on groundwater to meet their water needs. Also, the arrangement and development of aquifers are not the same in the study area, and often during dry climate, many parts of the city dry up hand pumps and deep tube wells, leading to a huge depletion of groundwater. The temporal trend analysis and change-point test results are given in Table 2.

3.1 *Temporal Groundwater-Depth Study*

With the help of GIS techniques, an in-depth mean GWL map was prepared for different periods for the study area. The spatial changes of mean water depth during different monsoon conditions, such as pre-monsoon and post-monsoon seasons, were drawn from the data of 35 observation wells. The average water level ranges from 1.24 m to 20.07 m below ground level (bgl) during the pre-monsoon and from 0.33 m to 10.96 m bgl during the post-monsoon season. The study indicated that in pre and post-monsoon seasons, the water level in 23% of the wells is 4–8 m bgl. The temporal GWL behavior is mentioned in Table 1.

The data for twelve years (2008–2019) of GWL were analyzed under observation. This study showed that the water level depth in the pre-monsoon season in the study area ranges between 1.24 m and 20.07 m bgl, while in the post-monsoon season, it ranges between 0.33 m and 10.96 m bgl. Contour maps of water level depth were drawn for the period before and after the monsoon. These maps in Fig. 3a, b depict mean GWL for pre-monsoon and post-monsoon seasons, Fig. 3c, d depicts maximum GWL for pre-monsoon and post-monsoon seasons and Fig. 3e, f depicts minimum GWL for pre-monsoon and post-monsoon seasons respectively for the GGA.

Table 1 Groundwater level fluctuation in well located at GGA

Well No.	Location	Pre-monsoon			Post-monsoon		
		Max (m bgl)	Min	Mean	Max	Min	Mean
W1	Azara PHC	5.980	4.570	5.397	3.460	1.530	2.240
W2	Boragaon	8.920	2.140	6.466	5.500	0.830	1.677
W3	Narengi Forest gate	8.090	4.420	7.312	9.280	5.470	7.483
W4	Patharquery	9.830	1.070	3.214	5.340	0.530	2.088
W5	Viswakarma Temple	4.150	1.280	2.381	2.190	0.910	1.438
W6	Wireless	2.300	0.400	1.372	1.420	0.700	1.012
W7	Basistha FG	11.870	8.590	10.599	11.460	9.270	10.157
W8	Choonsali, Madhabpur	5.130	2.820	4.096	3.980	2.080	3.319
W9	Khanapara Sc. Museum	10.90	1.560	7.576	8.800	1.210	2.929
W10	Lakhra Chariali	6.340	3.350	5.182	5.170	2.200	3.605
W11	Maligaon	6.880	0.480	3.313	3.730	0.330	1.705
W12	Patgaon	2.430	0.960	1.652	1.870	0.600	1.088
W13	Bakarapara	7.100	2.090	4.330	11.810	0.750	2.636
W14	Survey Odalbakra	8.170	4.030	6.268	5.390	1.840	3.430
W15	Ashwaklanta Temple	5.020	2.390	3.485	3.550	2.020	2.754
W16	Lakshmi Mandir	11.150	1.060	6.991	9.230	0.460	5.617
W17	Zoo Narengi Road HS	8.930	2.110	6.19	9.950	1.790	5.316
W18	AAU, Kahikutchi	7.600	3.630	5.753	3.300	1.770	2.571
W19	Adagudam	29.500	12.100	20.071	21.47	9.15	16.14
W20	GMC	6.680	0.830	2.498	4.86	0.49	2.023
W21	Kacharibasti Christian Basti	11.280	2.240	8.477	8.97	2.15	5.602
W22	Kahilipara L.P. School	8.370	2.580	4.734	2.57	1.4	2.091
W23	Mairapatti	7.730	5.090	6.610	5.58	1.59	3.299
W24	Paltan bazar	3.190	0.540	1.245	1.15	0.33	0.859
W25	Panjabari	12.250	3.690	9.542	11.98	0.74	6.933
W26	Hengrabari FG	6.120	1.590	2.529	3.6	1.1	1.921
W27	North Guwahati	5.090	2.540	4.054	3.6	1.97	2.778

(continued)

Table 1 (continued)

Well No.	Location	Pre-monsoon			Post-monsoon		
		Max (m bgl)	Min	Mean	Max	Min	Mean
W28	Panbazar Circuit House	17.40	13.180	14.44	11.96	10.96	11.492
W29	Amingaon	9.080	3.700	6.322	8.31	1.09	4.443
W30	Ganesh Mandir, Narengi	6.070	4.300	5.171	5.78	0.74	3.648
W31	Avayapuri	4.490	1.530	2.908	6.83	1.7	2.951
W32	Dte of Agri	10.540	2.300	7.397	6.05	0.68	3.71
W33	Fatasil-Ambari	6.040	2.400	3.108	5.9	1.22	2.166
W34	Sijubari	8.870	4.500	7.093	8.31	3.82	5.401
W35	AAU, Khanapara	19.640	7.040	11.86	10.82	3.5	6.143

The spatially varying maps actually show that the GWL in the southeastern (SE) parts of the study area go beyond 8 m during all monsoon seasons. Higher runoff and less infiltration of rainwater as well as widespread pumping of groundwater for drinking and construction purposes, are some of the identified causes of greater groundwater variation [12]. However, the depth of water in the northwestern (NW) part of the study area was always less than 4 m bgl in all seasons. This is basically due to the replenishment of aquifers by the Brahmaputra River flowing in the northern part of the study area. The GWL depth instability in the central parts of the study area was moderate, where the average water level was between 4 m and 8 m bgl.

3.2 *Annual Groundwater-Level Changes*

According to an analysis of annual GWL data for the pre-monsoon season, the year 2018 showed a minimum depth of GWL of 0.4 m bgl and a maximum level of 29.5 m bgl in the year 2008. The post-monsoon data stated that the lowest water level was seen in the year 2018 (0.33 m bgl), and the peak water level was seen in the year 2008 (21.47 m bgl). The spatial fluctuations of the mean annual GWL of the study area during the pre-monsoon season recorded for 12 years (2008–2019) are depicted in Fig. 3a. Similarly, spatially varying maps of the mean annual GWL during the post-monsoon season were depicted in Fig. 3b from 12 years of data using ArcGIS. Spatial maps declare that the southeastern part of the study area has a downfall in GWL by more than 10 m bgl, while the north eastern part of the study area has a change of less than 5 m bgl.

Table 2 Results for trend analysis and change point test of the groundwater level at GGA

Well No.	Location	Pre-monsoon				Post-monsoon			
		MK Test Z	Sens slope Q	Pettitt Change Point	Linear regression (β)	MK Test Z	Sens slope Q	Pettitt Change Point	Linear regression (β)
W1	Azara PHC	–	–	–	–	2.469	0.155	2014	–
W2	Boragaon	1.714*	0.198	–	–	2.331	0.090	–	–
W3	Narengi Forest gate	–	–	2014*	–	1.851*	0.202	–	0.207
W4	Patharquery	−2.400	−0.303	2013	–	–	–	–	–
W5	Viswakarma Temple	–	–	–	–	1.920*	0.050	–	0.061
W6	Wireless	−1.714*	−0.065	–	−0.066*	–	–	–	–
W7	Basistha FG	1.783*	0.155	2014*	0.139*	2.811	0.170	2015	0.165
W8	Choonsali, Madhabpur	–	–	–	–	2.537	0.080	2011	0.085
W9	Khanapara Sc. Museum	–	–	–	–	–	–	–	–
W10	Lakhra Chariali	–	–	–	–	–	–	–	–
W11	Maligaon	−3.291	−0.680	2013	–	−2.606	−0.300	2012	–
W12	Patgaon	–	–	–	–	3.497	0.077	2014	0.085
W13	Bakarapara	3.703	0.404	2012	–	2.674	0.266	2011*	–
W14	Survey Odalbakra	1.646	0.292	2013	0.274	–	–	–	–
W15	Ashwaklanta Temple	–	–	–	–	–	–	–	–

(continued)

Table 2 (continued)

Well No.	Location	Pre-monsoon				Post-monsoon			
		MK Test Z	Sens slope Q	Pettitt Change Point	Linear regression (β)	MK Test Z	Sens slope Q	Pettitt Change Point	Linear regression (β)
W16	Lakshmi Mandir	–	–	–	–	–	–	–	–
W17	Zoo Narengi Road HS	–	–	–	–	2.537	0.448	2014*	0.485
W18	AAU, Kahikutchi	–	–	–	–	–	–	–	–
W19	Adagudam	−1.646*	−0.536	–	−0.644*	−1.714*	−0.602	–	–
W20	GMC	−2.880	−0.367	2012		−3.086	−0.326	2012	–
W21	Kacharibasti Christian Basti	2.331	0.535	2012	0.597	2.263	0.515	2015*	0.346
W22	Kahilipara L.P. School	–	–	–	–	–	–	–	–
W23	Mairapatti	–	–	–	–	2.126	0.210	2014*	–
W24	Paltan bazar	−3.771	−0.[illegible]28	2012	–	–	–	–	–
W25	Panjabari	3.154	0.563	2012	0.560	2.263	0.856	2011	0.770
W26	Hengrabari FG	–	–	–	–	–	–	–	–
W27	North Guwahati	–	–	–	–	–	–	–	–
W28	Panbazar Circuit House	2.674	0.254	2013	–	–	–	–	–
W29	Amingaon	−3.497	−0.466	2013	−0.485	−2.537	−0.218	2012	−0.321

(continued)

Table 2 (continued)

Well No.	Location	Pre-monsoon				Post-monsoon			
		MK Test Z	Sens slope Q	Pettitt Change Point	Linear regression (β)	MK Test Z	Sens slope Q	Pettitt Change Point	Linear regression (β)
W30	Ganesh Mandir, Narengi	–	–	–	–	–	–	–	–
W31	Avayapuri	–	–	–	–	–	–	–	–
W32	Dte of Agri	–	–	–	–	–	–	–	–
W33	Fatasil-Ambari	–	–	–	–	–	–	–	–
W34	Sijubari	–	–	–	–	–	–	–	–
W35	AAU, Khanapara	–	–	–	–	–	–	–	–

"*" denotes significant up to 10% level of significance and "–" denotes non-significant

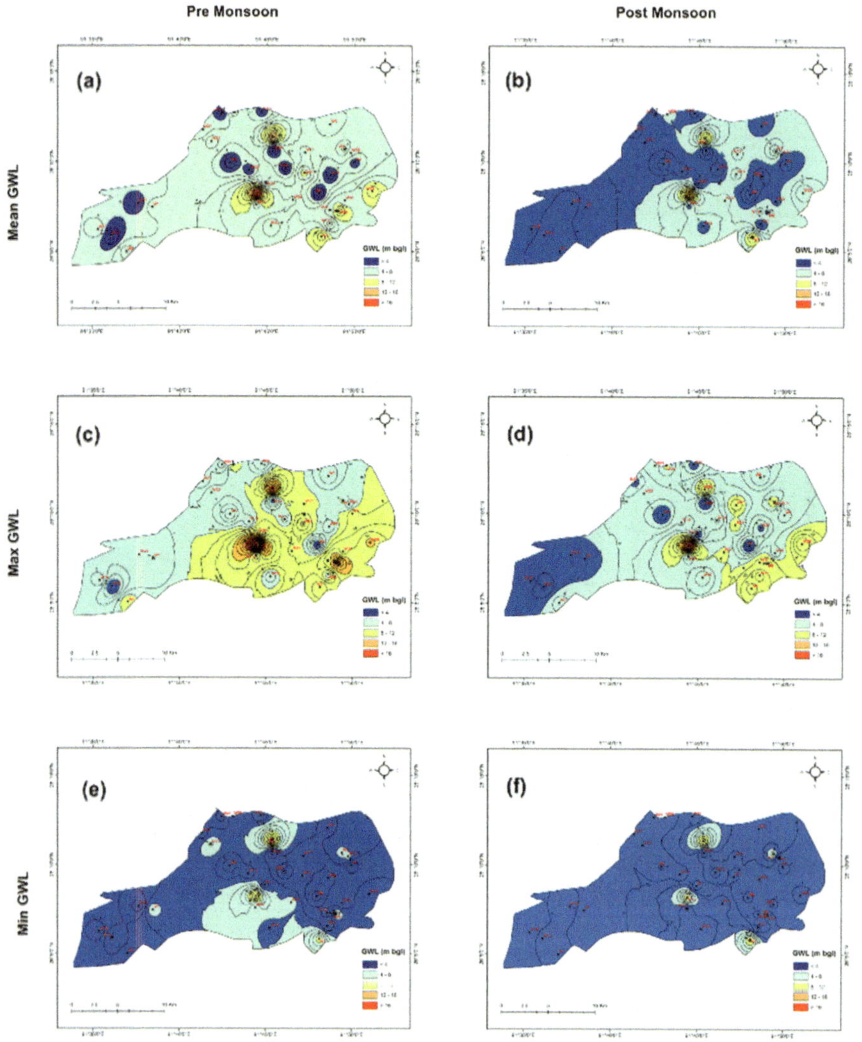

Fig. 3 GWL fluctuations of pre and post-monsoon seasons of GGA

3.3 Groundwater-Level Trend Analysis

In the case of MK trend test, depressions of water table were observed around Bakarapara, Basistha FG, Survey Odalbakra, Kacharibasti Christianbasti, Punjabari, and Panbazar Circuit House during pre-monsoon seasons while at Azara PHC, Bakrapara, Boragaon, Nagengi Forest Gate, Vishwakarma Temple, Basistha FG, Choonsali Madhabpur, Patgaon, Zoo Narengi Road HS, Kacharibasti Christianbasti, Mairapatti, and Panjabari during post-monsoon seasons as shown in Fig. 4a, b respectively.

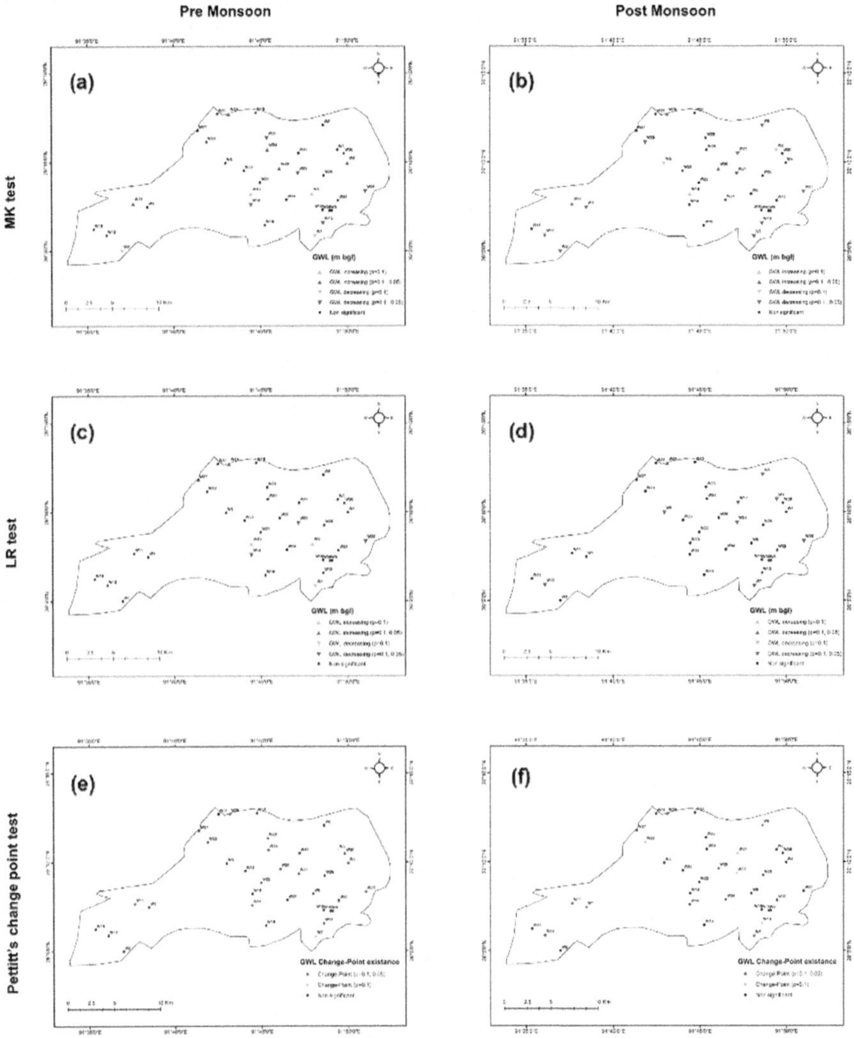

Fig. 4 Trend analysis of GWL for pre and post-monsoon of GGA

The test statistic results have been used by the findings of the water level trend in the GGA. The Z-statistic results of the GWL study were obtained by employing the MK test. The analysis of the trend was done using the 5% and 10% significance level. The trend line slope results for each monsoon season and GWL was calculated using the slope estimator method. The slope values obtained from the test for different monsoon seasons are: pre-monsoon (−0.680 to 0.563 myr^{-1}) and pre-monsoon (−0.602 to 0.856 myr^{-1}).

In the case of the Linear Regression, depressions of the water table were observed around Basistha FG, Survey Odalbakra, Kacharibasti Christianbasti, and Panjabari during the pre-monsoon season while around Narengi Forest gate, Vishwakarma Temple, Basiostha FG, Choonsali Madhabpur, Patgaon, Zoo Narengi Road HS, Kacharibasti, Christianbasti, and Panjabari during post-monsoon season as shown in Fig. 4c, d, respectively.

The statistical trend analysis results for the MK test and linear regression are shown in Table 2. In 35 observation wells, a diminishing trend in changeable water level (2008–2019) was noticed in wells 4, 6, 11, 19, 20, 24 and 29 during the pre-monsoon period and in wells 11, 19, 20, and 29 during the post-monsoon period. However, during the pre-monsoon season, an increasing trend was observed in wells 2, 7, 13, 14, 21, 25, and 28 and in wells 1, 2, 3, 5, 7, 8, 12, 13, 17. 21, 23 and 25 during the post-monsoon season. Analysis of 12-year water level data showed no specific trend in wells 9, 10, 15, 16, 18, 22, 26, 27, 30, 31, 32, 33, 34, and 35, which might be due to stable groundwater table in those fields.

The effects of the trend test revealed that a diminishing trend was detected in 20% of the wells, an increasing trend was detected in 20% of the wells, and no trend was detected in 60% of the wells during the pre-monsoon climatic condition. Similarly, it was found that a decreasing trend was observed in 12% of the observation wells, while an increasing trend was observed in 34% of the observation wells, and no trend was observed in 54% of the wells during the post-monsoon climatic conditions. Increasing trends here indicate greater groundwater depletion, while decreasing trends mean a decrease in groundwater depletion.

The spatial trend changes study of water level showed that most places like Bakarapara, Panbazar Circuit House, Survey Odalbakra, Boragaon, Kacharibasti Christianbasti, Panjabari, and Basistha FG had indicated an increasing trend and the decreasing trend is observed in the places, namely Patharquery, Wireless, Maligaon, Adagudam, GMC, Paltan Bazar, and Amingaon during the pre-monsoon season. Similarly, during the post-monsoon season, the spatial trend change study of water level indicated that an increasing trend is frequently observed in the places, namely Azara PHC, Boragaon, Narengi Forest Gate, Vishwakarma Temple, Basistha FG, Choonsali Madhabpur, Patgaon, Bakarapara, Zoo Narengi Road HS, Kacharibasti Christian Basti, Mairapatti and Panjabari. No trend in water level fluctuations has been found during the 12-year data (2008–2019) of the rest of the study area. According to the reports of Guwahati Metropolitan Development Authority (2009), increasing population and urbanization in the study area could be one of the major influencing factors for the declining trend coupled with increasing anthropogenic activities.

3.4 Pettitt's Change-Point Test Results

The Pettit's change-point test results have been analyzed for different monsoon conditions. The pre-monsoon and post-monsoon results for the study area are presented in

Fig. 4e, f. In the case of Pettitt's change point test, change points were observed around Nagengi Forest Gate, Patharquery, Basistha FG, Maligaon, Bakarapara, Survey Odalbakra, GMC, Kacharibasti Christianbasti, Paltan bazar, Panzabari, Panbazar Circuit house, and Amingaon in different years for the pre-monsoon season as shown in Fig. 4e. It was reported that the change points were mostly shown over 2012 to 2014 for this season (Table 2).

The results showed that nearly 14% change points were observed for both the years 2012 and 2013. Similarly, for the post-monsoon season, the change points were observed around Azara PHC, Basistha FG, Choonsali Madhabpur, Patgaon, Bakarapara, Zoo Narengi Road HS, GMC, Kacharibasti Christianbasti, Mairapatti, and Amingaon, as shown in Fig. 4f. The change points were observed mostly in 2012 and 2014 (Table 2). The results depicted that all four years had shown a nearly equal percentage of change points, whereas the remaining locations had non-significant test results.

3.5 *Significance of the Consequences of Water Resources Management*

The outcomes of the spatial GWL analysis were associated with the existing records of the Central Ground Water Board [13]. It showed that around 52% of the wells fluctuate in 4–8 m GWL during pre-monsoon. Similarly, about 68% of wells have fluctuating GWLs of less than 4 meters in the post-monsoon season. The results correspond to the findings of the CGWB yearbook report (2019–20). The approaches adopted in the present study are similar to the statistical trend analysis implemented by Anand et al. [4] in the Lower Bhavani River Basin, Tamil Nadu. Furthermore, the results of the present study were compared with those already done by Singh et al. [33] in the Jalandhar district of Punjab.

The collective findings of the present research proposed that the problem of water scarcity is a serious threat to agriculture and associated activities in the study area. The study highlights the water-stressed areas during various monsoon seasons. It will be helpful for the policymakers and executor to go for the working procedures which are essentially required to raise the GWL in the area. Prediction of trends through statistical methods will have a great impact on groundwater resource development in this region. The results of the analysis will be useful to administrators and planners for better planning of water resources management. It has been observed from the statistical tests that,

1. Boragaon well (W2) is located in the southwestern zone of the study area. In place of the Borgaon well, the GWL is decreasing for both the pre-monsoon and post-monsoon seasons, indicating an increasing trend of groundwater at that location. This may be due to developmental activities like the construction of national highways, setting up of industries. It was observed that a considerable amount of construction activities had been carried out in recent years, including

the widening of road and bridge, building bypass of National Highway NH-37 with several overbridges in that area. No change point has been observed in this location for any of the monsoon seasons.

2. Wireless well (W6) is located in the central zone of the study area. It was found that GWL in this well location is increasing for the pre-monsoon season, whereas it was stable for the post-monsoon season. This clearly shows that the GWL trend is decreasing in place of that well. This may be due to the presence of Silsakko Lake and Vashishta-Bahini River near the well.
3. The Basistha FG well (W7) is located in the southeastern zone of the study area. The GWL in this area is decreasing for both the pre and post-monsoon seasons, indicating that the groundwater trend is increasing at that location. This may be due to the presence of Kanyadhara hills and hills of Meghalaya as it provides an impermeable surface hence there will be more runoff and less infiltration. Also, the construction of highways can also be responsible for the increasing trend of groundwater in the region. Change point has also been observed in this location for both the pre and post-monsoon seasons.
4. Chunsali, Madhavpur well (W8), is located in the north eastern part of the study area. At this well located, the GWL for the post-monsoon season is decreasing, indicating that the groundwater trend for the post-monsoon season at this location is increasing. This may be due to the presence of Ramchai hills in this region. It was also observed that the GWL is stable for the pre-monsoon season. This may be due to the replenishment of aquifers by the Brahmaputra River flowing in the northern part of the study area.
5. Azara PHC (W1) is located in the southwestern of the study area. In this region, the GWL for the post-monsoon season is decreasing, indicating an increasing trend of GWL. The point of change is also seen in this location. This may be due to the ongoing development works and construction works in this area. The development from the airport to the international airport, construction of highways, widening of roads could be the main factors for the increasing trend of GW in this region. The increasing population nearby this well location also plays a big role in increasing trends of groundwater
6. Narengi FG (W3) is located in the northeastern region of the study area. The GWL has been found to decrease in the post-monsoon season, indicating an increasing trend in groundwater. This might be due to undulating land and the presence of Muja hills and Kharghuli hills near this location. A railway line is also passing through this place. Industries have also been established at this location which may be the reason for the decrease of GWL in this area. The change point is also observed for the pre-monsoon season.
7. Patharquari well (W4) is located in the northeastern region of the study area. The GWL has been found to increase in the pre-monsoon season, indicating a decreasing trend in groundwater. This may be due to the presence of Lake Silsako near this well. The change point is also seen for this place in the pre-monsoon season.
8. Vishwakarma temple well site (W5) is located in the central area of the study area. The GWL is found to be low in the post-monsoon season, indicating an

increasing trend in groundwater. This may be due to the presence of Nilanchal hills near the well. Dense population, undulating land, construction of over-bridges, widening of the main road of the city, development of railway line can be some of the important factors for the increasing trend of GW depth.

9. Maligaon well site (W11) is located in the central zone of the study area. The GWL is found to be increasing in both pre and post-monsoon seasons, indicating a decreasing trend in GWL. This may be due to the presence of fallow land in this area. Bharalu river is also flowing near this place. The location of this well also sees change points for both the pre and post-monsoon seasons.
10. Patgaon well site (W12) is located in the southwest region of the study area. The GWL is found to decrease in the post-monsoon season, indicating an increasing trend in GWL. This may be due to developmental activities, including widening of roads and bridges, construction of hospitals, schools, army buildings. The growing population in this area is also a major factor in the increasing trend of GWL. The location of this well also sees a change-point for the post-monsoon season.
11. Bakrapara well site (W13) is located in the southeastern part of the study area. The GWL is found to be decreasing in both pre and post-monsoon seasons, indicating an increasing trend in GWL. This may be due to the presence of the Kanyadhara hills and hills of Meghalaya as it provides an impermeable surface; hence there will be more runoff and less infiltration. Also, the construction of highways can also be responsible for the increasing trend of groundwater in the region. Change-point has also been observed in this location for both the pre and post-monsoon seasons.
12. The survey Odalbakra well site (W14) is located in the central zone of the study area. At this location, the GWL for the pre-monsoon season is decreasing, indicating an increasing trend of GWL at this location. This place is highly populated. There are many construction works going on in this area. This area is also developing as an industrial site. It has been observed that a considerable amount of construction activities, including the widening of roads and bridges and other buildings, have been taken up in recent years. This place has seen the change point for the pre-monsoon season.
13. Zoo Narengi Road HS well site (W17) is located in the central zone of the study area. The GWL is found to be decreased in the post-monsoon season, indicating an increasing trend in GWL. This may be due to overexploitation of groundwater in this area as this location is highly populated. Multi-story buildings, widening of roads, and other developmental works are also major factors for reducing GWL in this area.
14. Adagudam well site (W19) is located in the central zone of the study area. At this well location, the GWL for both the pre-and post-monsoon season is increasing, indicating that the groundwater trend for the post-monsoon season at this location is decreasing. This might be due to the presence of the Bharalu river and Basistha-Bahini river nearby the well location, which led to the flow of water from the river to the aquifer system.

15. GMC well site (W20) is located in the central zone of the study area. The GWL is found to increase in both the pre and post-monsoon season, indicating a decreasing trend in GWL. This may be due to the presence of the Bharalu river near the well location, which plays a vital role in the replenishment of aquifers. Change-point has also been observed in this location for both the pre and post-monsoon seasons.
16. Kacharibasti-Christian Basti well site (W21) is located in the central zone of the study area. At this well location, the GWL for both the pre and post-monsoon season is increasing, indicating that the groundwater trend for the post-monsoon season at this location is decreasing. This might be due to the overexploitation of groundwater in this area as this location is highly populated. Multi-story buildings, widening of roads and other developmental works are also major factors for reducing GWL in this area. Change-point has also been observed in this location for both the pre- and post-monsoon seasons.
17. Mairapati well site (W23) is located in the northwestern zone of the study area. The GWL is found to be decreasing in the post-monsoon season, indicating an increasing trend in GWL. It was observed that a considerable amount of construction activities had been carried out in recent years, including widening of road and bridge, the building of Barpeta-Hojo road in that area. Change-point has been observed in this location for the post-monsoon season.
18. Paltan Bazar well site (W24) is located in the central zone of the study area. The GWL is found to increase in the pre-monsoon season, indicating a decreasing trend in GWL. This may be due to the presence of the Bharalu river near the well location, which plays a vital role in the replenishment of aquifers. Change point has also been observed in this location for the pre-monsoon season.
19. Panjabari well site (W25) is located in the southeastern zone of the study area. The GWL is found to be decreased in both the pre and post-monsoon seasons, indicating an increasing trend in GWL. This may be due to overexploitation of groundwater in this area as this location is highly populated. Multi-story buildings, widening of roads, and other developmental works are also major factors for reducing GWL in this area. This area is regarded as both residential as well as and industrial area. Change point has also been observed in this location for the pre-monsoon season.
20. Panbazar Circuit house well site (W28) is located in the central zone of the study area. At this well location, the GWL for the pre-monsoon season is decreasing, indicating that the groundwater trend for the pre-monsoon season at this location is increasing. This may be due to overexploitation of groundwater as the area is highly populated.
21. Amingaon well site (W29) is located in the northwestern zone of the study area. The GWL is found to increase in both the pre-and monsoon season, indicating a decreasing trend in GWL. This may be due to the presence of the Brahmaputra river near the well location, which might play a vital role in the replenishment of aquifers. Change-point has also been observed in this location for both the pre- and post-monsoon seasons.

22. Rest of the well locations have shown stable GWL for both the pre and post-monsoon seasons.
23. The average water level ranges from 1.245 m to 20.071 m during the pre-monsoon and from 0.859 m to 16.140 m in the post-monsoon season. The study indicated that in pre and post-monsoon seasons, the water level in 23% of the wells is 4 m and 8 m below ground level (bgl), respectively. It was reported that the change points were mostly shown over 2012–2014 for this season. The results showed that nearly 14% change points were observed for both the years 2012 and 2013. It showed that around 52% of the wells fluctuate in 4–8 m GWL during pre-monsoon. Similarly, about 68% of wells have fluctuating GWLs of less than 4m in the post-monsoon season. It was observed that Adagudam (W19) had the highest GW depth of 29.5 m bgl in the year 2008, and the lowest GW depth was observed in the Wireless (W6) well site, which is 0.4 m bgl in the year 2018. It was also observed that the mean GW depth ranges from 1.245 m to 20.071 m bgl for the pre-monsoon season. Similarly, for the post-monsoon season, the maximum GW depth was found to be 21.47 m bgl for the well location present in the locality of Adagudam (W19) in the year 2008, and minimum GW depth was observed as 0.33m bgl for Paltan Bazar well location (W24) in the year 2018. The mean GW level varies from 0.859 m to 16.14 m bgl. Paltan Bazar and Panbazar are considered as the oldest business areas of Guwahati city and are considered as the core of the city. From the spatial maps, it has been indicated that these areas are under the highest water shortage condition. In Khanapara well locations (W9, W16 and W35), the GW table seemed to be stable. Still, there was a water stress problem throughout the study period i.e., from 2008–2019. Bhaskar Nagar slum, a notified slum of Guwahati, located in the eastern part of the city in Ward no. 20, is a densely populated area. This slum is located near a busy commercial and industrial area (Ahmad et al., 2021). It was observed that the built-up, as well as cultivated and managed areas towards the periphery of the city, increased over the years while natural and semi-natural vegetation recorded significant losses. These unplanned built-up establishments have emerged along National Highways 31, and 27, which pass through the south and southwest ends of the city, respectively [9]. Rapid population growth has pushed the expansion of urban areas in southern Guwahati into suburban areas, and most of these expansions are unplanned. It is evident that in recent years the built-up area has expanded into areas that were previously either natural vegetated or human-induced vegetation-managed land. From the LULC map of GGA of 2008 and 2019 (Fig. 5a, b respectively) it is observed that between 2008 and 2019 there was an increase in habitation and agricultural land of about 16.74% and 3.70%, respectively, whereas declining of dense vegetation, barren lands and water bodies are of 11.30%, 8.76% and 0.46%, respectively.

The increase in habitat percentage follows a similar pattern of nearly 17% as carried out by Pawe and Saikia [29]. A large part of Guwahati has been developed by filling wetlands, and the process of filling and degradation of wetlands

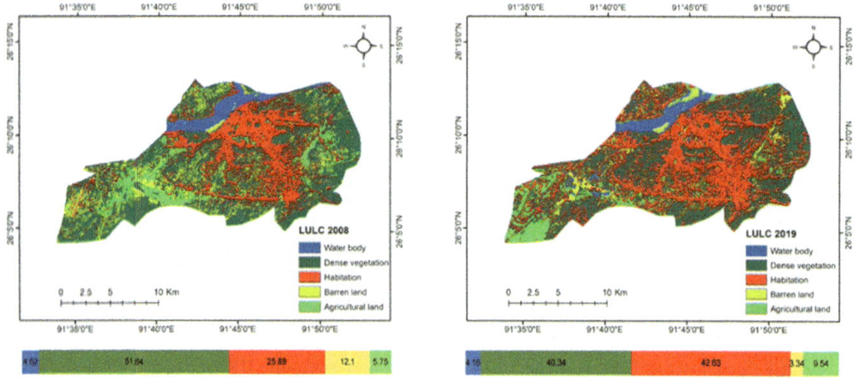

Fig. 5 LULC map of GGA of 2008 and 2019

still continues [6]. The rapid growth of population and anthropogenic activities are increasing rapidly to change the land use land cover (LULC), therefore going to an alarming level [26].

The expected hypothesis was that the developmental works would be more attached to groundwater depletion than the increasing population. In a study done by Singh et al. [33], it was found that the declination of GWL might affect most of the water-dependent activities, especially the agriculture sector, in the near future. In a study done by Abdullahi et al. [1], it was found that anthropogenic activities play a major role for years which are wholly responsible for the instability of groundwater tables. There was a significant trend found in the depth of groundwater for both the pre and post-monsoon seasons. However, there was a non-significant trend found in some parts of the study area. The results from the survey of 35 wells at the GGA, Assam, supported the expected hypothesis. The findings showed a significant difference in the level of GW in the southeastern and southwestern parts of the study area. It was found that the areas having hard rock/soil formation are significantly more affected in terms of GWL than the areas which are having plane lands. There was also a significant difference found in the level of groundwater in the central part of the study area, where the population is dense. Paltan Bazaar, Fancy Bazar were significantly more affected than Azara PHC, Zoo Narengi Tiniali. The results from this study seem consistent with the previous testing done on similar subjects.

The results from this study seem consistent with the previous testing done on a similar subject. In a study done by Anand et al. [4], they found that due to increasing population and industrialization, groundwater resources get over-exploited in different parts of the world, which led to a rapid depletion in GWL, especially in hard rock areas.

4 Conclusion

The present study was conducted to examine the seasonal trends, if any, in groundwater occurrence in the Greater Guwahati region of Kamrup Metropolitan District in Assam. The study was conducted at 35 well locations in the Kamrup metropolitan district over the period 2008 to 2019, with available data on pre-monsoon and post-monsoon seasonal average GWL (m bgl). The MK test statistic, Sen's slope estimator, and Linear Regression method were used to examine the trends. GIS-based maps have been verified as an important asset for the estimation of GWL volatility, although this study was a constraint only for the Greater Guwahati region as GWL data for Kamrup metropolitan and Kamrup rural districts of Assam were not available.

The spatial analysis of the water level for the study area was carried out with the help of spatial interpolation techniques and statistical procedures. The GWL data from 35 wells collected for a period of 12 years from 2008 to 2019 were imputed in the analysis to understand groundwater fluctuations in space and time. Following the spatial interpolation of seasonal GWL data, there was a severe rate of groundwater degradation from 2012 to 2014.

The annual depth analysis of mean GWL shows that between 2013 and 2014, there has been an over-pumping that stresses the aquifer in many parts of the study area. Nevertheless, it is naturally overgrown over the next few years due to substantial rainfall, which is clearly reflected in the GWL instability. This declination was observed in both the pre-monsoon and post-monsoon seasons and witnessed mostly in the southeastern sides of the study area. Further, the study concludes that the GWL in the southeastern and northeastern regions of the study area has decreased by more than 8 m bgl during all the monsoon seasons. High surface runoff, low infiltration, and over-exploitation of groundwater are the main reasons for high groundwater instability. The GWL fluctuations are more in the central parts of the study area, where the mean water level is between 1 and 20 m bgl. Nevertheless, in the northwestern and southwestern parts of the study area, the GWL is always less than 5 m bgl during all pre-monsoon and post-monsoon climatic conditions. This is mainly due to the presence of alluvial formation and recharge by the Brahmaputra, which flows in the northern parts of the study area.

According to trend analysis, 20% of wells during the pre-monsoon season and 12% of wells during post-monsoon show a decreasing trend in GWL over a period of 12 years (2008–2019). The places, viz. Panbazar, Zoo Narengi Road HS, Kacharibasti Christianbasti, Panjabari, and Survey Odalbakra are primarily affected by the declining trend in GWL. The findings also revealed that over-exploitation of groundwater for domestic and industrial purposes in the southeastern and central parts of the study area leads to more disturbance of groundwater during summer, resulting in aquifer stress. These changes in GWL with respect to space, time, and depth mainly bother the common people. Thus, it is necessary to augment the groundwater resources in the respective areas by constructing suitable structures through artificial recharge facilities.

This study has successfully exhibited a GIS-based methodology along with statistical analysis for evaluation of GWL variation of the GGA of Assam. The study concluded that remote sensing and GIS is a potentially effective tool and very useful for hydrogeologist to conduct groundwater exploration mapping faster and more efficiently at the same time.

Acknowledgements The authors are very grateful to Central Ground Water Board (CGWB), Betkuchi, Guwahati, Assam for providing the GWL data to carry out our research works. Debaditya Gupta and Satyam Raj also acknowledge the Ministry of Education (MoE) to provide fellowships to carry out the research work.

References

1. Abdullahi MG, Toriman ME, Gasim MB, Garba I (2015) Trends analysis of groundwater: using non-parametric methods in Terengganu Malaysia. J Earth Sci Clim Chang 6:2–25. https://doi.org/10.4172/2157-7617.1000251
2. Adhikary PP, Dash CJ (2017) Comparison of deterministic and stochastic methods to predict spatial variation of groundwater depth. Appl Water Sci 7:339–348. https://doi.org/10.1007/s13201-014-0249-8
3. Amos J (2015) Earth's underground water quantified. BBC Science Correspondent. https://www.bbc.com/news/science-environment-34837461
4. Anand B, Karunanidhi D, Subramani T, Srinivasamoorthy K, Suresh M (2020) Long-term trend detection and spatiotemporal analysis of GWLs using GIS techniques in Lower Bhavani River basin, Tamil Nadu, India. Environ Dev Sustain 22:2779–2800. https://doi.org/10.1007/s10668-019-00318-3
5. Barbulescu A, Bautu A, Bautu E (2020) Optimizing inverse distance weighting with particle swarm optimization. Appl Sci 10:2054. https://doi.org/10.3390/app10062054
6. Baruah P (2020) Potential of urban wetlands for ecotourism development—A case of Deepor Beel, Guwahati. Nat Environ Pollut Technol 19:611–625. https://doi.org/10.46488/NEPT.2020.v19i02.016.
7. BBC News (2018) India facing the 'worst water crisis in its history'. https://www.bbc.com/news/world-asia-india-44492994.
8. Das N, Goswami DC (2013) A geo-environmental study on groundwater recharge zones and groundwater management in the Guwahati municipal area. Int J Environ Sci 4:66–75. https://doi.org/10.1007/978-3-319-74494-0_29
9. Devi U, Bhattacharyya KG (2018) Mobility and bioavailability of Cd Co, Cr, Cu, Mn and Zn in surface runoff sediments in the urban catchment area of Guwahati, India. Appl Water Sci 8:1–14. https://doi.org/10.1007/s13201-018-0651-8
10. Elumalai V, Brindha K, Sithole B, Lakshmanan E (2017) Spatial interpolation methods and geostatistics for mapping groundwater contamination in a coastal area. Environ Sci Pollut Res 24:11601–11617. https://doi.org/10.1007/s11356-017-8681-6
11. Emenike PC, Tenebe I, Ogarekpe N, Omole D, Nnaji C (2019) Probabilistic risk assessment and spatial distribution of potentially toxic elements in groundwater sources in Southwestern Nigeria. Sci Rep 9:1–15. https://doi.org/10.1038/s41598-019-52325-z
12. GMDA (2009) Master Plan for Guwahati Metropolitan Area—2025, Part-I. https://gmda.assam.gov.in/documents-detail/master-plan-guwahati-2025-maps. Accessed 24 April 2021
13. Goswami M, Rabha D (2020) Trend analysis of ground-water levels and rainfall to assess sustainability of groundwater in Kamrup Metropolitan District of Assam in Northeast India, p 17. Roorkee Water Conclave-2020, February 26–28

14. Gurdak JJ (2017) Climate-induced pumping. Nat Geosci 10:71–71. https://www.nature.com/articles/ngeo2885
15. Hirsh RM, Slack JR, Smith RA (1982) Techniques of trend analysis for monthly water quality data. Water Resour Res 18:107–121. https://doi.org/10.1029/WR018i001p00107
16. Huo Z, Dai X, Feng S, Kang S, Huang G (2013) Effect of climate change on reference evapotranspiration and aridity index in arid region of China. J Hydrol 492:24–34. https://doi.org/10.1016/j.jhydrol.2013.04.011
17. Jaiswal RK, Lohani AK, Tiwari HL (2015) Statistical analysis for change detection and trend assessment in climatological parameters. Environ Process 2:729–749. https://doi.org/10.1007/s40710-015-0105-3
18. Kendall MG (1975) Rank correlation methods. Grifn, London. http://www.sciencedirect.com/science/refhub/S0895-9811(15)00014-0/sref18
19. Khadka NS (2019) India water crisis flagged up in global report. Environment correspondent, BBC World Service. https://www.bbc.com/news/world-asia-india-49232374
20. Li B, Rodell M, Kumar S, Beaudoing HK, Getirana A, Zaitchik BF, de Goncalves LG, Cossetin C, Bhanja S, Mukherjee A, Tian S, Tangdamrongsub N, Long D, Nanteza J, Lee J, Policelli F, Goni IB, Daira D, Bila M, de Lannoy G, Mocko D, Dunne SCS, Tian S (2019) Global GRACE data assimilation for groundwater and drought monitoring: advances and challenges. Water Resour Res 55:7564–7586. https://doi.org/10.1029/2018WR024618
21. Li F, Wei W, Zhao Y, Qiao J (2017) Groundwater depth prediction in a shallow aquifer in north China by a quantile regression model. Hydrogeol J 25:191–202. https://doi.org/10.1007/s10040-016-1473-0
22. Long D, Chen X, Scanlon BR, Wada Y, Hong Y, Singh VP, Chen Y, Wang C, Han Z, Yang W (2016) Have GRACE satellites overestimated groundwater depletion in the Northwest India Aquifer? Sci Rep 6:1–11. https://doi.org/10.1038/srep24398
23. Machiwal D, Islam A, Kamble T (2019) Trends and probabilistic stability index for evaluating groundwater quality: the case of quaternary alluvial and quartzite aquifer system of India. J Environ Manag 237:457–475. https://doi.org/10.1016/j.jenvman.2019.02.071
24. Mann HB (1945) Non-parametric tests against trend. Econometrica 13:245–259. https://doi.org/10.2307/1907187
25. Manna F, Walton KM, Cherry JA, Parker BL (2019) Five-century record of climate and groundwater recharge variability in southern California. Sci Rep 9:1–8. https://doi.org/10.1038/s41598-019-54560-w
26. Nath B, Ni-Meister W, Choudhury R (2021) Impact of urbanization on land use and land cover change in Guwahati city, India and its implication on declining GWL. Groundw Sustain Dev 12:100500. https://doi.org/10.1016/j.gsd.2020.100500
27. Nistor MM, Rahardjo H, Satyanaga A, Hao KZ, Xiaosheng Q, Sham AWL (2020) Investigation of groundwater table distribution using borehole piezometer data interpolation: case study of Singapore. Eng Geol 271:105590. https://doi.org/10.1016/j.enggeo.2020.105590
28. Ohmer M, Liesch T, Goeppert N, Goldscheider N (2017) On the optimal selection of interpolation methods for groundwater contouring: an example of propagation of uncertainty regarding inter-aquifer exchange. Adv Water Resour 109:121–132. https://doi.org/10.1016/j.advwatres.2017.08.016
29. Pawe CK, Saikia A (2018) Unplanned urban growth: land use/land cover change in the Guwahati metropolitan area India. Geografisk Tidsskrift 118:88–100. https://doi.org/10.1080/00167223.2017.1405357
30. Pradhan S, Kumar S, Kumar Y, Sharma HC (2021) Implication of intensive agriculture on groundwater utilization in Ganga-Ramganga interbasin, India. Environ Earth Sci 80:1–21. https://doi.org/10.1007/s12665-021-09602-w
31. Rodell M, Velicogna I, Famiglietti JS (2009) Satellite-based estimates of groundwater depletion in India. Nature 460:999–1002. https://doi.org/10.1038/nature08238
32. Sen PK (1968) Estimates of the regression coefficient based on Kendall's tau. J Am Stat Assoc 63:1379–1389. https://doi.org/10.1080/01621459.1968.10480934

33. Singh A, Sharma CS, Jeyaseelan AT, Chowdary VM (2015) Spatio-temporal analysis of groundwater resources in Jalandhar district of Punjab state, India. Sustain Water Resour Manag 1:293–304. https://doi.org/10.1007/s40899-015-0022-7
34. Solander KC, Reager JT, Wada Y, Famiglietti JS, Middleton RS (2017) GRACE satellite observations reveal the severity of recent water over-consumption in the United States. Sci Rep 7:1–8. https://doi.org/10.1038/s41598-017-07450-y
35. Sun H, Gu X, Zhu J, Yu Z, Zhang Y (2019) Fractal nature of GWL fluctuations affected by riparian zone vegetation water use and river stage variations. Sci Rep 9:1–9. https://doi.org/10.1038/s41598-019-51657-0
36. Thomas BF, Famiglietti JS (2019) Identifying climate-induced groundwater depletion in GRACE observations. Sci Rep 9:1–9. https://doi.org/10.1038/s41598-019-40155-y
37. Varouchakis EA, Hristopulos DT (2017) Comparison of spatiotemporal variogram functions based on a sparse dataset of GWL variations. Spat Stat 34:100245. https://doi.org/10.1016/j.spasta.2017.07.003
38. Vasco DW, Farr TG, Jeanne P, Doughty C, Nico P (2019) Satellite-based monitoring of groundwater depletion in California's central valley. Sci Rep 9:1–14. https://doi.org/10.1038/s41598-019-52371-7
39. Xiao Y, Gu X, Yin S, Shao J, Cui Y, Zhang Q, Niu Y (2016) Geostatistical interpolation model selection based on ArcGIS and spatio-temporal variability analysis of GWL in piedmont plains, northwest China. Springerplus 5:1–15. https://doi.org/10.1186/s40064-016-2073-0

The manufacturer's authorised representative in the EU is Springer Nature Customer Service Centre GmbH, Europaplatz 3, 69115 Heidelberg, Germany. If you have any concerns regarding our products, please contact ProductSafety@springernature.com

Printed and bound by CPI Group (UK) Ltd, Croydon, CR0 4YY
01/10/2026
02225867-0014